# Principles of Engineering Instrumentation

# Principles of Engineering Instrumentation

D. C. Ramsay

OXFORD AUCKLAND BOSTON JOHANNESBURG MELBOURNE NEW DELHI

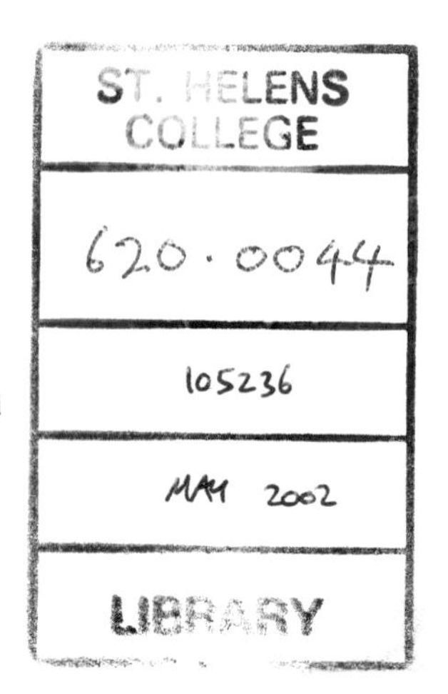


Butterworth-Heinemann
Linacre House, Jordan Hill, Oxford OX2 8DP
225 Wildwood Avenue, Woburn, MA 01801-2041
A division of Reed Educational and Professional Publishing Ltd

A member of the Reed Elsevier plc group

First published in 1996
Reprinted by Butterworth-Heinemann 2001



**British Library Cataloguing in Publication Data**
A catalogue record for this book is available from the British Library

**Library of Congress Cataloguing in Publication Data**
A catalogue record for this book is available from the Library of Congress

ISBN 0 340 645569 5

Typeset in 10/11 pt Times by GreenGate Publishing Services, Tonbridge, Kent
Printed and bound in Great Britain by J W Arrowsmith Ltd, Bristol

# Contents

Preface vii

Acknowledgements viii

**1 THE STRAIN GAUGE 1**

Types of strain gauge. Bonded gauges. Unbonded gauges. Strain calculation. The bridge circuit. Rosettes. Temperature-compensated gauges. The direct-reading bridge. The twin-active-gauge bridge. The all-active-gauge bridge. Semiconductor strain gauges. Possible sources of error in strain gauge signals. Exercises.

**2 TEMPERATURE TRANSDUCERS 16**

What is temperature? Methods of temperature measurement. Time constant. Sinusoidally varying temperatures. Thermocouples. Thermocouple materials. Thermistors. Positive temperature coefficient thermistors. Resistance thermometers. Integrated circuit temperature sensor. Radiation pyrometers. Factors which effect radiation pyrometry. Temperature sensors for radiation pyrometers. Calibration. Sealed-fluid thermometers. The bimetallic strip. Exercises.

**3 DISPLACEMENT TRANSDUCERS 37**

Mechanical devices. The potentiometer. Pentiometer loading. Potentiometer noise. The linear variable differential transformer (LVDT). LVDT limitations. Variable inductance transducers. Inductance bridges. Capacitive transducers. Binary coded disc. Sensors. Exercises.

**4 FORCE, TORQUE AND PRESSURE TRANSDUCERS 52**

Elastic sensors. Hydraulic load cell. Piezoelectricity. A piezoelectric force transducer. The step response of a piezoelectric force transducer. The charge amplifier. The electromagnetic force balance. Combinations of two or more transducers. Torque measurement. Measurement of torque in a rotating shaft. Dynamometers. The Froude hydraulic dynamometer. The electric dynamometer. The eddy current dynamometer. Pressure measurement. U-tube manometers. Vacuum measurement. Exercises.

**5 VELOCITY TRANSDUCERS 69**

Linear velocity measurement. Velocity from distance/time. Doppler effect. The pitot-static tube. Rotational velocity measurement. The stroboscopic lamp. Rotational velocity from toothed wheel and proximity pickup. Hall-effect transducers. The eddy current drag-cup tachometer. DC tachogenerators. AC tachogenerators. Exercises.

6 **ACCELERATION AND VIBRATION TRANSDUCERS** **78**

Seismic pickups. Displacement pickups. Velocity pickups. Acceleration pickups (accelerometers). Piezoelectric accelerometers. Types of low-frequency accelerometer. Servo accelerometers. The calibration of accelerometers. Exercises.

7 **FLOW MEASUREMENT** **86**

The venturi meter. Reynolds number. The orifice plate. Mechanical flowmeters. The turbine flowmeter. The rotameter. The electromagnetic flowmeter. The hot-wire anemometer. The vortex flowmeter. Doppler flowmeters. Ultrasonic Doppler flowmeters. Laser Doppler flowmeters. The calibration of flowmeters. Exercises.

8 **ELECTRONICS FOR INSTRUMENTATION** **102**

The decibel. Bandwidth. Amplifiers. Transformers. Integrated circuits. The operational amplifier. Noise. Screening techniques. Suppressors. Filters. Amplitude modulation. Frequency modulation. Exercises.

9 **DIGITAL INSTRUMENTATION PRINCIPLES** **124**

Why digital? Binary digit voltage levels. Data signals and control signals. Series/parallel data transfer. Integrated circuit families. Logic gates. Schmitt trigger. Oscillators. Binary counters. BCD codes. Decade counters. Decimal digit display. Scalers. Counter-timers. Multiplexing. Sampling. The sampling theorem. Analogue-to-digital conversion. Digital-to-analogue conversion. A complete digital system. Polarity indication. Companding. Pulse code modulation. Exercises.

10 **DISPLAY AND RECORDING OF OUTPUT** **151**

The moving coil meter. Voltmeters. Moving-coil multimeters. Digital display multimeters. The ultraviolet recorder. Servo recorders.The XY plotter. Chart recorders. The cathode-ray oscilloscope. Magnetic tape recording. Data acquisition systems. Exercises.

**APPENDIX: A LIBRARY OF CONTROL SYSTEM ELEMENTS** **179**

Linear hydraulic actuators (hydraulic jacks). Rotary hydraulic actuators. Spool valve. Flapper and nozzle. Solenoid. DC electric motor types: permanent magnet, shunt-wound, series-wound, compound-wound, separately excited. DC servo-motor. Stepper motor. AC electric motor types: 3-phase induction, single phase induction. AC servo-motors. Thyristors and triacs. Thermostats. Process control valves.

Suggested further reading 203

Answers to exercises 204

Symbols and SI units 207

The Greek alphabet 211

Index 213

# Preface

Some textbooks are written by academics for other academics. But not this book. It is unashamedly written for students. It aims to give the broadest possible introduction to engineering instrumentation for students studying this subject on HNC, HND and degree courses.

Because students will have arrived on these courses by various routes, some via mechanical or electrical ONCs or ONDs, others via A levels at school, I have assumed very little in the way of common knowledge: a few simple mechanical and electrical principles and a working knowledge of SI units. Anything more is explained in the text or in the worked examples.

Formulae are stated where necessary, but their proofs have been omitted, so calculus is not needed. In fact all the mathematics required can be done on a scientific calculator.

Because electronics now plays such an important part in instrumentation there are chapters on electronics for instrumentation and digital electronics; mechanical engineering students should find them easy enough to digest since concepts are explained as we come to them.

There is also a chapter on display and recording devices. Students should find this helpful in their labwork; for instance the example on using an oscilloscope, which is rather different from watching somebody else use one.

Finally there is an appendix on control system elements. It covers hydraulic actuators, electric motors, electrical control devices and control valves. These tend to be neglected in control system theory, but they are an essential part of 'real life' control systems. Students should find the information useful in project and design work. Not only students but engineers of all disciplines should also find this book useful as a reference.

I am grateful to everybody who helped or put up with me while I was writing the book, and especially to:

Mr R W Taplin, of Airbus, British Aerospace, Bristol, and Mr R T G Colclough, R & D, Lister-Petter, Dursley, Glos., who gave of their valuable time to answer my questions.

# 1
# The strain gauge

## Types of strain gauge

A strain gauge is, strictly speaking, any device for measuring mechanical strain (i.e. the ratio: elongation/original length) but in instrumentation the term is generally taken to mean an *electrical resistance strain gauge*. The full name gives us a clue to its principle of operation; it is an electrical conductor whose resistance changes as it is strained. It is thus a *transducer*, converting strain into change of electrical resistance.

## Bonded gauges

The first strain gauges were wound with fine copper wire on a thin paper tube, which was then flattened and stuck (bonded) to a metal surface to determine the strain in the metal. Similar strain gauges (called *wrap-around wire strain gauges*) are still available, and are used where the requirement is for high resistance gauges, of several hundred ohms resistance. But most strain gauges are of either the *flat grid wire* type, or of the *etched foil* type, and generally have resistances of not much more than one hundred ohms. The three strain gauge types are shown for comparison in Figures 1.1, 1.2 and 1.3.

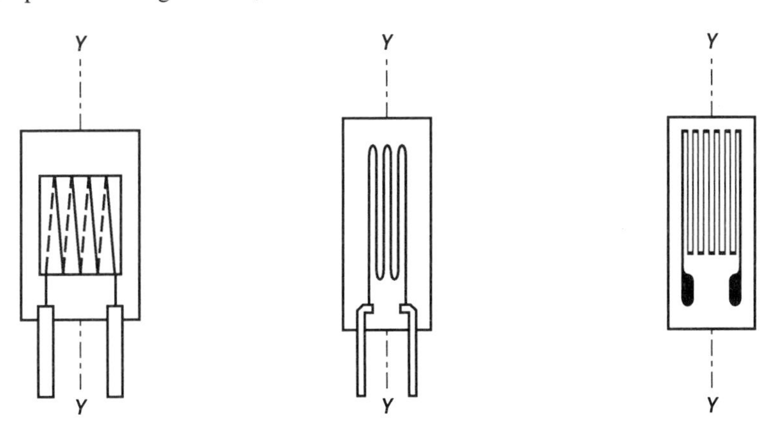


**Figure 1.1** Wrap-around wire strain gauge

**Figure 1.2** Flat-grid strain gauge

**Figure 1.3** Foil strain gauge

These bonded strain gauges must be fixed with a suitable adhesive to the surface whose strain they are to measure. When they are properly bonded, the electrical conductor is embedded in the bonding material, so that not only does tensile strain cause the strain gauge's resistance to increase, but compressive strain causes it to decrease in exactly the same proportion.

## Unbonded gauges

Another type of strain gauge is the *unbonded strain gauge*. This is used to give an electrical output signal proportional to a very small displacement of one body relative to another body. Figure 1.4 shows the principle applied to an accelerometer. The very fine wire forming the strain gauge resistances is stretched between insulating pegs on the two bodies; in this case between the seismic mass of the accelerometer (see p.78) and the casing around it. The tension in the wires must be sufficient to ensure that some tension still remains at maximum displacement of the seismic mass in either direction.

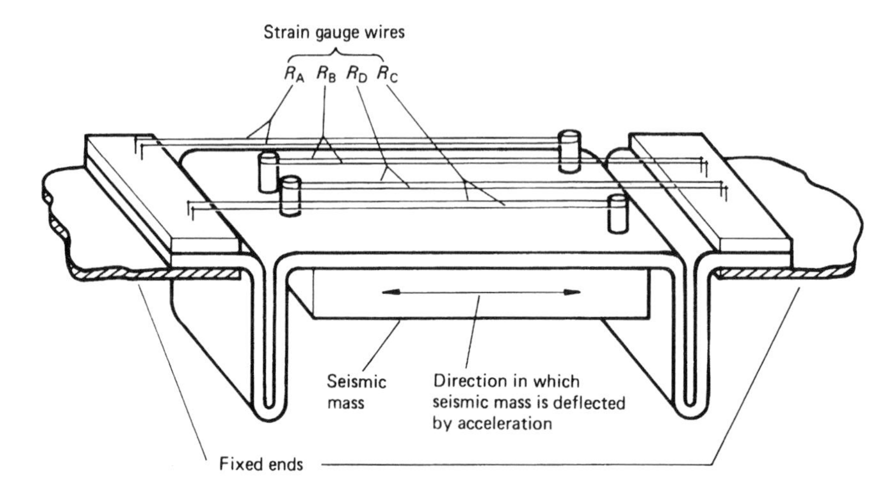


**Figure 1.4** A strain gauge accelerometer. The lettering of the strain gauge wires relates to the Wheatstone bridge circuit in Figure 1.5

This type of strain gauge is extremely delicate and must be protected by the casing, which may also contain an amplifier to boost the output voltage to a reasonable value. Thus unbonded strain gauges can only be seen by taking the transducer apart – that is never, as far as most of us are concerned. They are always part of some other transducer – usually an accelerometer or an extensometer – which to us will be just a 'black box' with a known output/input ratio.

## Strain calculation

The vast majority of strain gauges are of the bonded type previously described. The relationship between strain and electrical resistance is given by the equation:

$$\frac{\delta R}{R} = k\varepsilon$$

where:

$\varepsilon$ is strain. That is, $\varepsilon = \dfrac{\text{elongation}}{\text{original length}}$

$R$ is the original resistance of the strain gauge.

$\delta R$ is the change in resistance due to $\varepsilon$.

$k$ is a constant of proportionality, called the *gauge factor*.

If we think of ε as

$$\frac{\delta l}{l}$$

then

$$\frac{\delta R}{R}$$

is a similar concept which we could call 'electrical strain', and the equation is really saying that the electrical strain of a strain gauge is proportional to the mechanical strain. The constant of proportionality, the gauge factor, usually has a value between 2.0 and 2.2. The manufacturer determines the gauge factor for a particular batch of gauges by testing specimens taken from the batch.

The actual change of resistance is very small, as the following example shows.

*Example 1.1*
A strain gauge has an unstrained resistance of 120.2 Ω and a gauge factor of 2.15. It is bonded to a tie-bar which is then loaded to a stress of 260 N/mm$^2$. Determine the percentage change in resistance of the gauge. Young's modulus for the tie-bar material is 200 GN/m$^2$

*Solution*
First we must determine the strain in the tie-bar from:

$$\frac{\text{stress}}{\text{strain}} = E$$

$$\therefore \frac{260 \times 10^6}{\text{strain}} = 200 \times 10^9$$

$$\therefore \text{strain} = \frac{260 \times 10^6}{200 \times 10^9} = 1.3 \times 10^{-3}$$

(This may also be written as 1300 microstrain. One microstrain is a strain of 0.000 001)

$$\therefore \frac{\delta R}{120.2} = 2.15 \times 1.3 \times 10^{-3}$$

$$\therefore \delta R = 120.2 \times 2.15 \times 1.3 \times 10^{-3} = 0.336\ \Omega$$

$$\therefore \text{percentage change} = \frac{0.336}{120.2} \times 100\% = 0.280\%$$

## The bridge circuit

A change of resistance amounting to a fraction of one percent would be quite difficult to measure accurately by direct measurement. We get over the difficulty by using a *Wheatstone bridge* circuit. This compares the small difference in voltages between two potential dividers, each composed of two resistances in series. This principle, the *bridge circuit*, is often employed in electrical measurements, in both DC and AC circuits. Figure 1.5 shows the conventional representation of the circuit, and Figure 1.6 shows its application to a strain gauge as in the previous example. $R_A$ is the *active gauge*, bonded to the material whose strain is to be measured. $R_B$ is a variable resistance – a decade resistance box, which indicates the resistance value it is set to, could be used here. $R_C$ is a fixed resistance, approximately equal to the strain gauge's resistance. $R_D$ is the *dummy gauge*, another strain

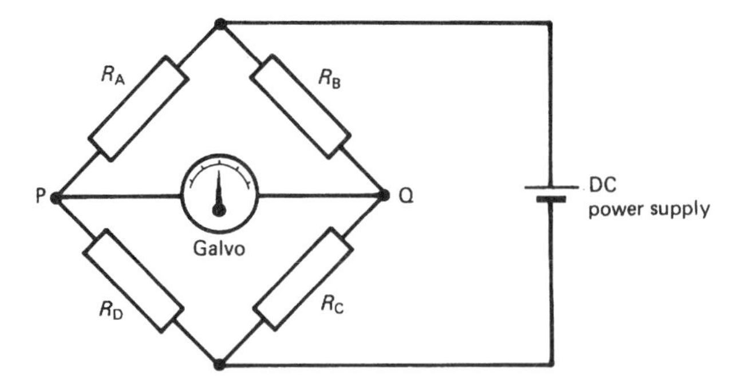


**Figure 1.5** The Wheatstone bridge

gauge, identical to $R_A$ but bonded to an unstrained piece of the same material as $R_A$ is bonded to, and placed as close as possible to $R_A$ so that it shares the same temperature. Its purpose is to cancel out any resistance change in $R_A$ caused by change of temperature, so that the output of the potential divider $R_A$ $R_D$ will remain unaffected by any temperature variations.

To measure the resistance of $R_A$ the bridge is balanced by adjusting $R_B$ so that the voltages at P and Q are equal. This is indicated by a reading of zero on a galvanometer connected between P and Q – a centre-zero microammeter could be used for this purpose. Then:

$$R_A = R_B \times \frac{R_D}{R_C}$$

from which $R_A$ may be found. If, instead of being equal to $R_D$, $R_C$ is made say, ten times as great, then $R_B$ will be ten times as great as $R_A$, which will increase the resolution of the measurement by a factor of ten.

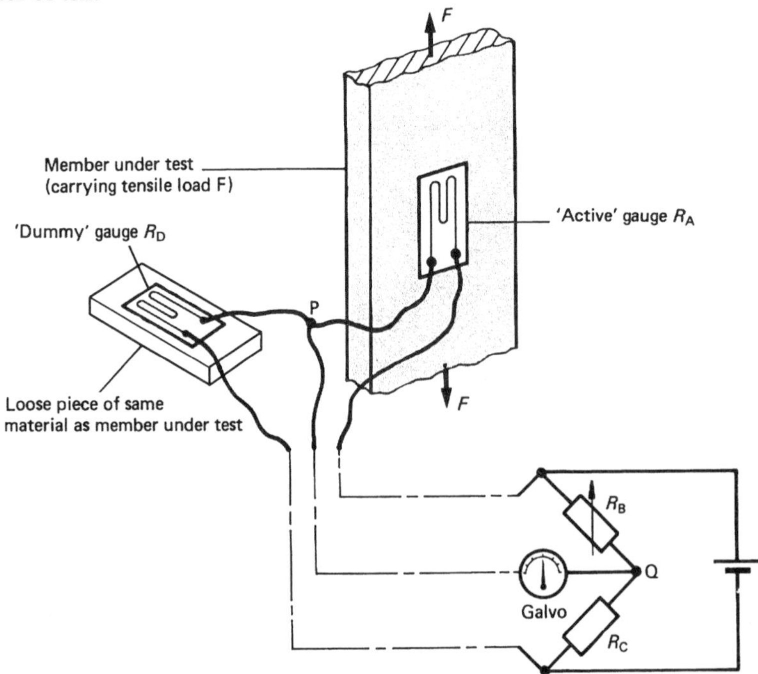


**Figure1.6** Wheatstone bridge as used for strain-gauging

*Example 1.2*

In a test using the circuit of Figure 1.6 on a rectangular block of light alloy loaded in compression, the following results were obtained: $R_C = 1003\ \Omega$, $R_D = 119.8\ \Omega$, $R_B = 1005.1\ \Omega$ at balance with the block under no load, and $1002.4\ \Omega$ when it was loaded. Young's modulus for the material of the block was 70 GN/m$^2$, and the gauge factor of the strain gauges was 2.07. Determine the stress in the block.

*Solution*

$$R_A = R_B \times \frac{R_D}{R_C}$$

$$\therefore \text{ with load applied, } R_A = 1002.4 \times \frac{119.8}{1003} = 119.73\ \Omega$$

$$\text{and with no load, } R_A = 1005.1 \times \frac{119.8}{1003} = 120.05\ \Omega$$

$$\frac{\delta R}{R} = k\varepsilon$$

$$\therefore \quad \frac{119.73 - 120.05}{120.05} = 2.07\varepsilon; \qquad \therefore \quad \varepsilon = \frac{-0.32}{120.05 \times 2.07} = -0.001288$$

$$\frac{\text{stress}}{\text{strain}} = E$$

$$\therefore \text{stress} = -0.001288 \times 70 \times 10^9 = -90\ 200\ 000\ \frac{\text{N}}{\text{m}^2}\ or - 90.2\ \frac{\text{N}}{\text{mm}^2}$$

The negative answer indicates compressive stress. Since the difference in values of $R_B$, 2.7 Ω, is only accurate to two significant figures, we are only justified in stating our result to two significant figures. Thus we can say that the stress was 90 N/mm$^2$

## Rosettes

When the stress distribution is complex, *strain rosettes* may be used to determine the principal stresses. These are arrangements of two or three identical strain gauges, usually of the foil type, formed on a single plastic sheet base, with their axes at 90° in the case of the two-gauge rosette, and at either 45° or 60° to each other in three-gauge rosettes. The three types are shown in Figures 1.7, 1.8 and 1.9.

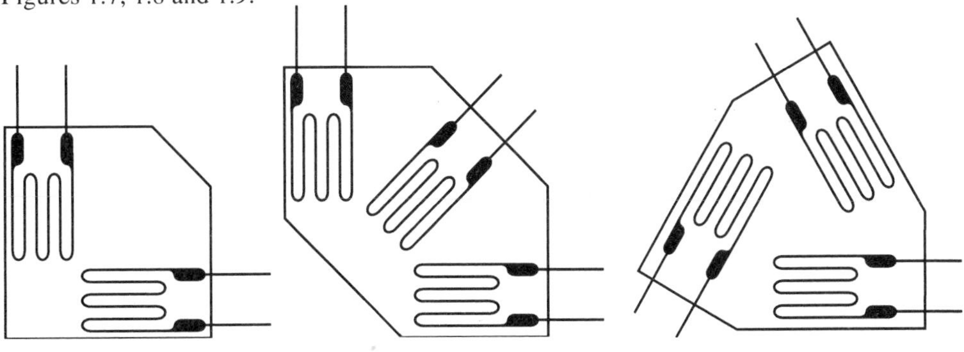

**Figure 1.7** Two-gauge rosette

**Figure 1.8** 45° three-gauge rosette

**Figure 1.9** 60° three-gauge rosette

The two-gauge rosette is used where the directions of the two principal stresses are known, the axes of the gauges being lined up with the directions of the principal stresses. Three-gauge rosettes are necessary where the directions of the principal stresses are not known. Either the 45° or the 60° rosettes may be used in this case, and the principal strains in the material, and their directions, may be determined by constructing a Mohr strain circle from the strain values given by the three strain gauges. The principal stresses can then be calculated from the principal strains.

## Temperature-compensated gauges

Sometimes, due to limitations of space or other constraints, it is difficult or impossible to accommodate a 'dummy' gauge on its loose piece of metal, close enough to the active gauge to ensure that it will remain under all conditions at the same temperature as the active gauge. Temperature-compensated strain gauges have therefore been introduced which do not need a dummy gauge *provided that they are used only on the metal for which they have been compensated.* The 'dummy' gauge, $R_D$, is replaced by an ordinary fixed resistance, as shown in Figure 1.10.

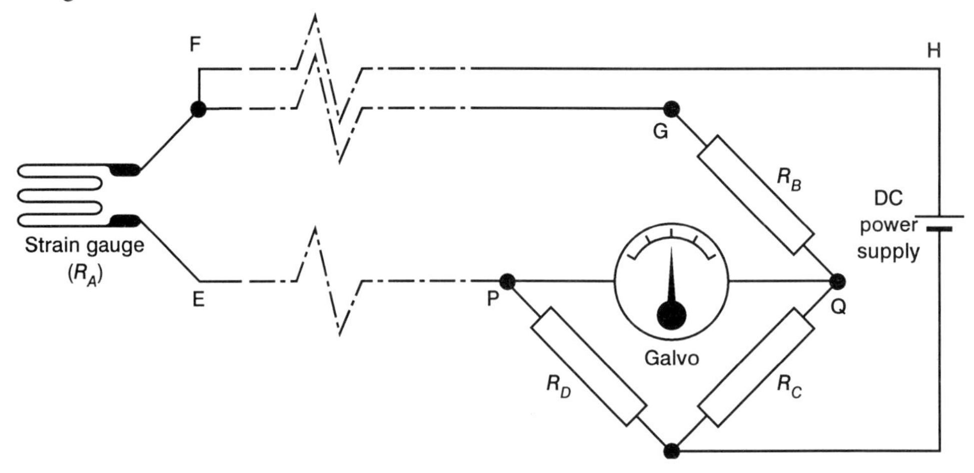


**Figure 1.10** Bridge circuit using a single temperature-compensated gauge

To understand how temperature compensation works, consider the effects of a temperature increase as it acts on a strain gauge alone, and on a steel surface to which it is to be bonded. These effects are:

1 an increase in the resistance of the strain gauge. This is a function of the temperature coefficient of resistance of the alloy from which it is made
2 a linear expansion of the metal of the strain gauge
3 a linear expansion of the steel surface.

As expansion 2) is usually greater than expansion 3), (because of the difference in the coefficients of linear expansion of the two metals) the effect of the bond between strain gauge and steel is to put the strain gauge in compression, causing a decrease in its resistance. By a choice of suitable constituents of the strain gauge alloy, and suitable heat treatment during its manufacture, the decrease in resistance can be made to cancel out the increase in resistance from effect 1). This temperature compensation is not quite perfect, but over a temperature range of about –20° to +120°C the error is small enough to be corrected from a graph of strain error versus temperature, supplied by the manufacturers of the gauges.

Temperature-compensated gauges are available in both single gauge and rosette form, with compensation characteristics to suit steel, aluminium and various other structural materials.

In the circuit of Figure 1.6, any change of resistance due to temperature gradient in the three leads connecting the active and 'dummy' gauges to the rest of the Wheatstone bridge cancels out, but when temperature-compensated gauges are used without a 'dummy' gauge close to them, resistance change in their leads could introduce an error. This is the reason for the arrangement shown in Figure 1.10. Because the power supply lead goes to the strain gauge itself, instead of to $R_B$, the resistance changes in leads EP and FG cancel out, because they are in opposite arms of the bridge, while a resistance change in lead FH does not affect the balance of the circuit.

## The direct-reading bridge

The method we have considered so far – measuring strain by balancing a bridge circuit so that its output voltage, $V_{PQ}$, is zero and then reading the change of resistance, is called the *null method*. It has one great advantage; the results are unaffected by variations in the voltage of the power supply because we only take a reading of resistance when the voltages at *P* and *Q* are equal, and in that case it doesn't matter what that voltage actually is.

The trouble with balancing a bridge circuit in this way is that it takes the time and attention of an operator, and while the balancing process is going on, transient changes of strain may pass unnoticed. In industry, strain gauges are more often used to measure a load causing strain (e.g. in weighing machines) than to measure strain itself, and for this purpose the *direct-reading bridge* is much more suitable. Its great advantage is that it gives a continuous output signal.

In the direct-reading bridge, $R_B$ is a fixed resistance, not a variable one, and so the bridge is left unbalanced. The output is the voltage between P and Q. By applying known loads to the structure and noting the corresponding output voltage, $V_{PQ}$, a calibration curve may be plotted (it should be a straight line), from which the load corresponding to any value of $V_{PQ}$ may be found.

However, the calibration will only be correct for one particular voltage of the power supply – *the direct-reading bridge MUST have a constant-voltage power supply.*

If desired, the output voltage may be zeroed at zero load by adjusting a low-resistance potentiometer, $R_Z$, as shown in Figure 1.11.

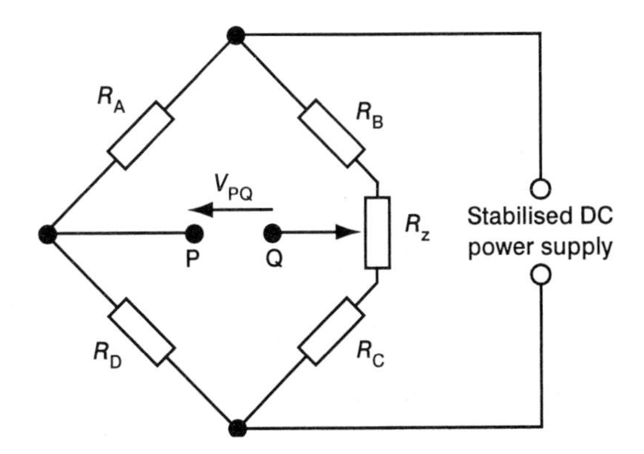


**Figure 1.11** Potentiometer $R_z$ enables bridge output voltage $V_{PQ}$ to be zeroed

The direct-reading bridge may still be used for the direct measurement of strain if it is calibrated by switching a range of high resistances in parallel with the active gauge while it is unstrained, and plotting output voltage against change of resistance in the active gauge. Figure 1.12 shows the circuit and the following example shows how it works.

*Example 1.3*
In Figure 1.12, the active gauge, $R_A$, has a resistance of 100.72 Ω while unstrained, and the output voltage of the bridge, unstrained, was –1.3 mV. The shunt resistors, of 10 kΩ, 15 kΩ and 33 kΩ gave bridge output voltages of +13.7 mV, +8.7 mV and +3.3 mV respectively, when each in turn was temporarily connected across the active gauge.

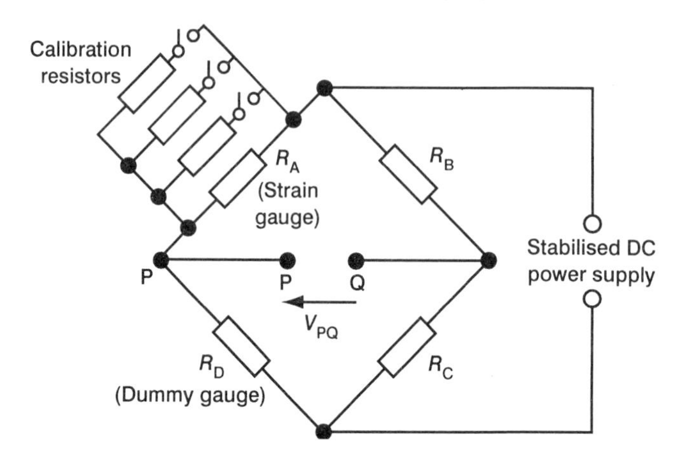


**Figure 1.12** Direct-reading bridge with calibrating resistors

A load was now applied to the structure to which the active gauge was bonded, and a bridge output voltage of –6.6 mV was recorded. Determine the strain at the active gauge. The gauge factor of the strain gauge is 2.12.

*Solution*
We must first calculate the equivalent resistance when each of the shunt resistances is connected in parallel with $R_A$:

$$10\ \text{k}\Omega \text{ gives } \frac{100.72 \times 10\,000}{100.72 + 10\,000} = 99.72\ \Omega$$

$$15\ \text{k}\Omega \text{ gives } \frac{100.72 \times 15\,000}{100.72 + 15\,000} = 100.05\ \Omega$$

$$33\ \text{k}\Omega \text{ gives } \frac{100.72 \times 33\,000}{100.72 + 33\,000} = 100.41\ \Omega$$

These equivalent resistances, being less than the resistance of the unstrained strain gauge alone, simulate compressive strain. In order to be able also to read tensile strains from the calibration curve, we must extend the axes sufficiently to accommodate a similar range of resistance and output voltage in the opposite direction. Therefore put output voltage on the horizontal axis, with a scale going from –20 mV to +16 mV, and resistance on the vertical axis with a scale from 99.6 Ω to 102.0 Ω. Then plot each of the equivalent resistances against its corresponding output voltage. Plot also the point representing the initial state of the bridge (the point –1.3 mV, 100.72 Ω). The calibration curve is the straight line through these four points.

Reading up to the line from –6.6 mV and then across from the line to the resistance axis we get 101.06 Ω. Then, using

$$\frac{\delta R}{R} = k\varepsilon$$

$$\frac{101.06 - 100.72}{100.72} = 2.12\varepsilon$$

$$\varepsilon = \frac{0.34}{100.72 \times 2.12} = 0.001592$$

## The twin-active-gauge bridge

If we are going to use the direct-reading bridge to measure loads instead of strains, we might as well make the 'dummy' gauge do something useful. If it is bonded to the same piece of metal as $R_A$, but in such a way that it is strained in the opposite sense to $R_A$, it will still cancel out temperature effects in $R_A$, but it will also increase the output voltage of the bridge, and get rid of the loose piece of metal to which it would otherwise be bonded.

Figure 1.13 shows how this may be done. Where the gauges are required to measure bending moment, the 'dummy' gauge is positioned on the opposite side of the neutral axis to the active gauge, as shown in (a). For a cross-section symmetrical about the neutral axis, this

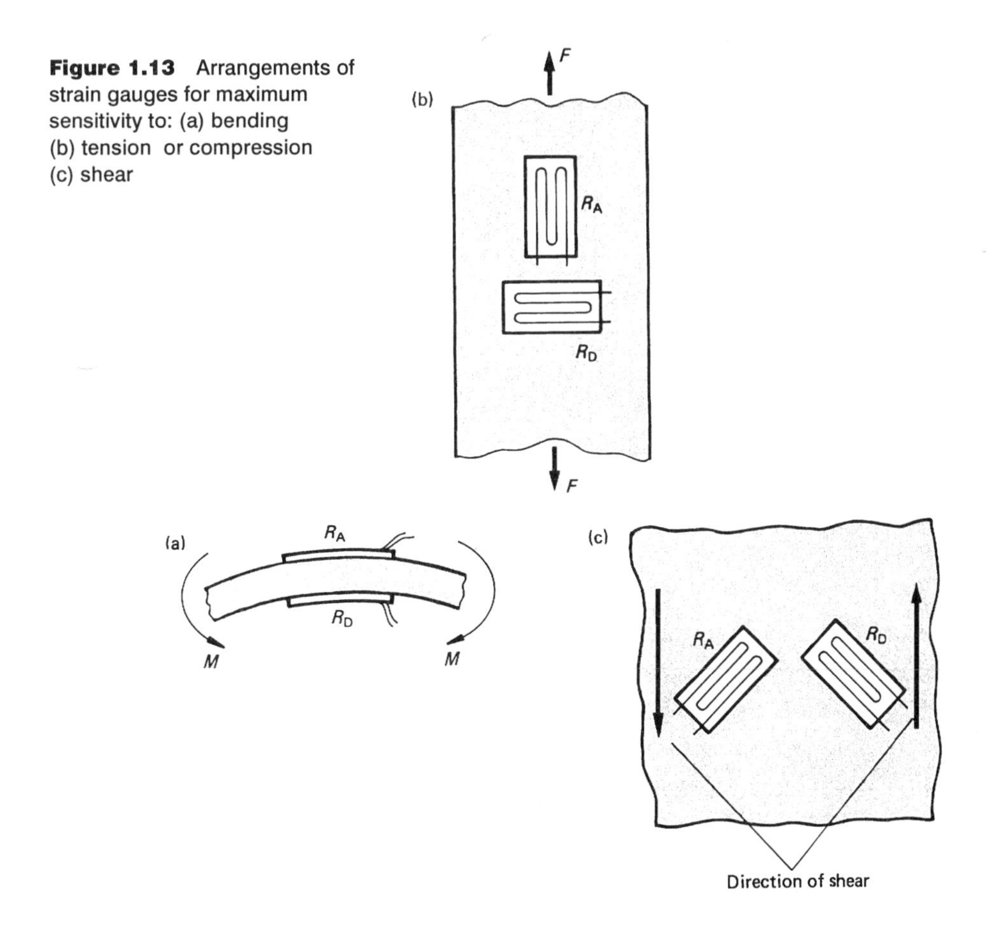


**Figure 1.13** Arrangements of strain gauges for maximum sensitivity to: (a) bending (b) tension or compression (c) shear

doubles the strain sensitivity and makes the arrangement insensitive to any end load which may be present, as the resistance change in the two gauges due to end load is of the same sign, and so cancels out.

Where the gauges are required to measure end load (tension or compression), the 'dummy' gauge is positioned where it will be acted upon by the *Poisson's ratio effect*, as shown in Figure 1.3(b). Poisson's ratio is the ratio of lateral strain to longitudinal strain. The two strains are always of opposite sign, and the actual value of Poisson's ratio depends on the material which is being strained. It has a value between 0.2 and 0.4 for most materials, so that making the 'dummy' gauge active in this way multiplies the sensitivity of the bridge by a factor of 1.2 to 1.4.

Where the gauges are required to measure shear force, the active and 'dummy' gauges are positioned at 45° to the direction of the shear and at 90° to each other, as shown in Figure 1.3(c), so that one is acted on by the tensile, and the other by the compressive strain set up by the shear. A 90° rosette as shown in Figure 1.7 could be used for this purpose.

## The all-active-gauge bridge

Carrying the principle of the twin-active-gauge bridge to its logical conclusion, we can make the complete Wheatstone bridge out of four strain gauges, all active, and so obtain the maximum in terms of sensitivity and immunity to temperature variation. Referring to Figure 1.5, all we have to do is to arrange the gauges so that gauges on opposite sides of the square are subjected to strain in the same sense (tensile strain on one pair of opposite sides, compressive strain on the other pair).

Thus in Figure 1.13a gauge $R_C$ would be situated alongside $R_A$ so that both increase, and gauge $R_B$ alongside $R_D$ so that both decrease in resistance. This has the effect of decreasing the voltage at P and increasing that at Q, giving four times the output compared with a single-active-gauge, if the cross section is symmetrical about the neutral axis.

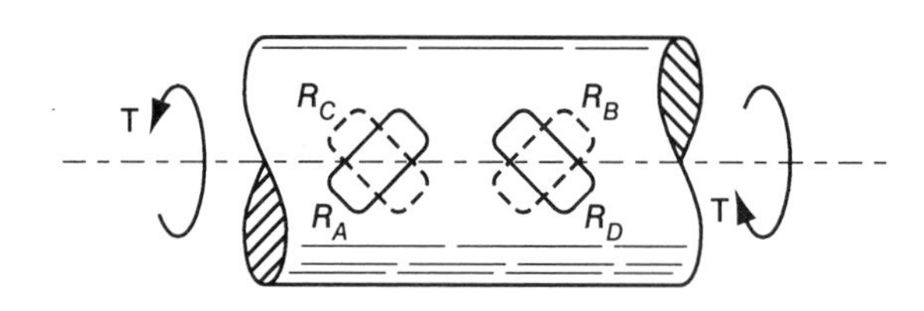


**Figure 1.14** After rotation, $R_C$ and $R_B$ will occupy positions of $R_A$ and $R_D$

Similarly, referring to Figure 1.13(b), a four-gauge bridge to measure tension or compression would have $R_C$ aligned with $R_A$, and $R_B$ aligned with $R_D$. $R_C$ and $R_B$ should be on the opposite face to $R_A$ and $R_D$ so that although the end load effect is additive, any bending effect cancels out. Torsion, applied to a shaft or torque tube, is reacted by shear stress in the material, with maximum shear stress at the outside surface. Consequently, a four-gauge bridge of strain gauges to measure torque should be arranged so as to measure shear stress, but with the gauges on opposite sides of the shaft, as shown in Figure 1.14, so that when the shaft has rotated 180°, $R_C$ and $R_B$ will occupy the positions that $R_A$ and $R_D$ occupied.

## Semiconductor strain gauges

These are also called *piezo-resistive strain gauges*, since they are made from crystalline materials in which large changes of electrical resistance occur when stress is applied to them. They have the advantage that they can be a hundred times as sensitive as a wire or foil gauge, so that they may not need any amplification of their output, or alternatively they can be used with amplification for measuring very small values of strain; dynamic strains can be measured down to 0.01 microstrain. Also they can be made very small, and so can be used for measuring highly localised strains, or in places inaccessible to foil or wire gauges. Their disadvantages are that their gauge factors may vary considerably as strain is applied, and also as temperature changes, so they are not quite as simple to use as 'conventional' gauges.

Semiconductor strain gauges are cut or etched from single crystals of silicon or germanium 'doped' with precisely regulated proportions of special impurities such as boron. Lightly doped gauges have gauge factors of 200 to 300 and comparatively high electrical resistance (200–1000 Ω) but their gauge factors vary considerably when either their strain or their temperature is altered. More heavily doped gauges have lower electrical resistance and lower gauge factors (50 would be a typical value). Their gauge factors, however, remain practically constant over wide ranges of temperature and strain.

Semiconductor gauges are manufactured in the form of filaments of rectangular cross-section, from 0.7 mm to 5 mm in length, with width about one tenth of their length, and thickness about 0.05 mm. They are bonded to a plastic backing material, together with solder tabs, to which they are connected by thin wires.

Strain measurement by semiconductor strain gauge is limited, by the elastic limit of the semiconductor material, to a maximum value of about 4000 microstrain. The gauges are bonded to surfaces by the same methods as are used for other types of strain gauge, but the material is very brittle, so care must be taken not to cause fracture by squeezing during the bonding process.

By doping the base crystals with different impurities, two types of material are produced; positive or p-type and negative or n-type. Tensile strain produces increase of resistance in p-type material and the opposite effect, decrease of resistance, in n-type material. However, with both types of material the gauge factor (and hence the resistance) decreases as the temperature is increased, because the temperature effect is a function of the base material only. Thus, by adjusting the proportions of the impurities, gauges can be produced in pairs, one p-type and one n-type, with (almost) equal and opposite gauge factors and with (almost) identical gauge factor variations due to temperature.

This enables temperature effects to be virtually cancelled out by a suitable arrangement of two such gauge pairs into a four-gauge bridge. The arrangement of gauges to measure tension in a flat tie-bar, for example, would be as shown in Figure 1.15. One pair of gauges would be wired as $R_A$ and $R_B$, while the other pair would be wired as $R_C$ and $R_D$, with the p-type gauges on either side of one output connection and the n-type gauges on either side of the other output connection. Then tensile strain would give increase of resistance in $R_A$ and decrease of resistance in $R_B$, while the associated lateral compressive strain (Poisson's ratio effect) would give increase of resistance in $R_C$ and decrease of resistance in $R_D$. The effect on the voltages at the bridge output connections would be as shown by the arrows in Figure 1.16. If the temperature increased, however, all four gauges would decrease in resistance equally, thus the temperature change would have no effect on the bridge output voltage.

The above principle is used to make miniaturised pressure transducers. The active element of the transducer is a disc of silicon which is thin enough to bulge very slightly when pressure is applied to it. The face of the disc is coated with masking material, leaving slots through which the impurity which produces a p-type gauge is diffused into the silicon. The disc is then remasked, this time with slots for the n-type material, which is similarly diffused into the

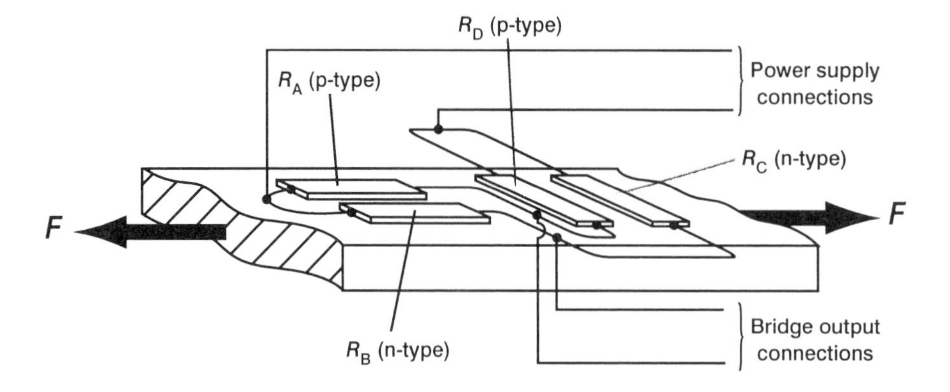


**Figure 1.15** Arrangement of p-type and n-type elements to measure tensile force $F$ in a four-gauge bridge

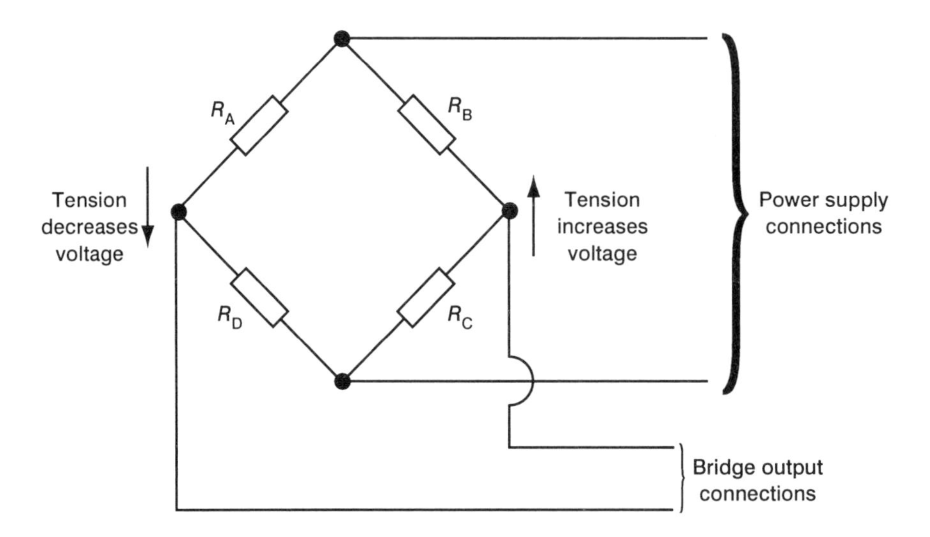


**Figure 1.16** Basic circuit diagram of the four-gauge bridge shown in Figure 1.15

silicon. The gauges thus produced interconnect to form a four-gauge bridge which can produce a maximum output signal of about 0.1 volt from a disc of about 2.5 mm diameter.

## Possible sources of error in strain gauge signals

### Cross-sensitivity

Because a strain gauge has width as well as length, a small proportion of the resistance element lies at right angles to the major axis of the gauge, at the points where the conductor reverses direction at the ends of the gauge. So as well as responding to strain in the direction of its major axis (YY in Figures 1.1, 1.2 and 1.3), the gauge will also be somewhat responsive to any strain there may be at right angles to YY.

In a flat grid gauge (Figure 1.2) a large part of each of the semicircular bends in the wire is effectively at right angles to the major axis, and its cross-sensitivity could be as much as 10% of its sensitivity in the direction of its major axis.

In an etched foil gauge (Figure 1.3) the portions of the conductor at right angles to the major axis are made much broader than the longitudinal portions, and hence of much lower resistance. So foil gauges have much less cross-sensitivity; 0.5% would be a typical value.

In a wrap-around wire gauge (Figure 1.1) the end loops lie almost perpendicular to the plane of the gauge, so that in this type of gauge there is very little cross-sensitivity. For this reason etched foil or wrap-around wire gauges should be used where it is necessary to reduce cross-sensitivity to a minimum.

**Bonding faults**

For perfect bonding, the particular manufacturers' recommendations as to suitable adhesives and procedures for bonding gauges to the strain-surface should be complied with. In general however the adhesive layer must be continuous and as thin as possible if the strain is to be fully transmitted to the gauge element. Before bonding, an area rather larger than the gauge must be cleared of any paint or corrosion, thoroughly degreased, and the surfaces thus prepared must be kept free from contamination from fingers until the gauge is applied and bonding is complete.

If the bonding is unsatisfactory, *creep* may occur. Creep is a gradual relaxing of the strain on the strain gauge, and it has the effect of decreasing the gauge factor, so that the output of the bridge becomes less than it should be. It is most likely to appear where the gauge is used to monitor a steady strain; short duration tests and dynamic tests are less likely to be affected. Temperatures above 50°C increase the probability of creep, and may require the use of special high temperature adhesives for bonding.

Creep may also occur where gauges have been used to measure dynamic strain, and have been subjected to many thousands of cycles of strain. In such a case, the bonding of the gauges, or the gauge base itself may begin to disintegrate due to fatigue, allowing the gauge element to relax.

**Hysteresis**

If a strain gauge installation is loaded to a high value of strain and then unloaded, it may be found that the gauge element appears to have acquired a permanent 'set', so that resistance values are all slightly higher when unloading than when 'loading-up'. The same effect continues when the direction of loading is reversed, so that the graph of resistance versus strain becomes a hysteresis loop, as shown in Figure 1.17. Repeating cycles of loading/unloading should cause the hysteresis loop to narrow to negligible proportions, so strain gauge installations are often cycled in this way to a higher strain level than their normal operational level to eliminate the effect. Severe hysteresis, which cannot be reduced in this way indicates faulty bonding of the gauge.

**Effects of moisture**

In damp conditions, the backing of the gauges or the bonding adhesive may absorb water. This can cause dimensional changes in the gauge which may appear as false strain values.

Another kind of fault occurs when moisture on gauges or soldered connections forms high resistance paths to the metal on which the gauge is bonded, so that in effect the strain gauge has a high resistance connected in parallel with it. As the example 1.3 shows, a parallel resistance of a few tens of kilohms can cause a considerable reduction in the equivalent resistance of the gauge, which would appear as a significant reduction in tensile strain or increase in compressive strain. To prevent this, gauges should be bonded only in dry conditions, and if condensation is a possibility, the installation should be damp-proofed with a suitable electrically insulating water repellent, such as a silicone rubber compound.

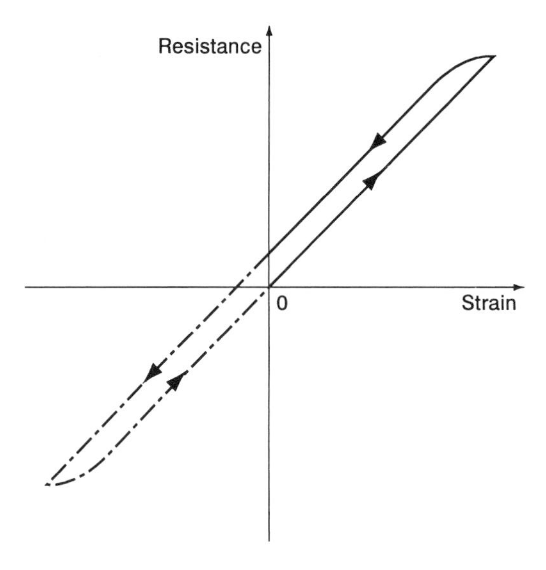


**Figure 1.17** Hysteresis loop given by a strain gauge

**Temperature change**

Throughout this chapter we have emphasised the importance of keeping all the gauges of a bridge as close together as possible, so that they all share the same temperature. Where this uniformity of temperature is in doubt, temperature compensated gauges must be used.

One possible source of temperature difference is the heat produced by the current through a strain gauge. When the bridge is first switched on, the gauges may warm up at different rates, so the bridge should not be used for measurement until sufficient time has been allowed for temperatures to stabilise. Foil gauges are better than other types of gauge at limiting such 'warm up' effects, as they have a much larger surface area for conducting heat away.

## Exercises on chapter 1

1 Sketch and describe three different types of strain gauge. Give two advantages of foil gauges compared with wire gauges.

2 A strain gauge having a resistance of 180 Ω and a gauge factor of 2.05 experiences a strain of 700 microstrain. What will be the corresponding change in its resistance?

3 A strain gauge is bonded to a short column which is then loaded in compression. The resistance of the strain gauge is 240.27 Ω before loading and 239.71 Ω while the load is applied. The gauge factor of the strain gauge is 2.15. The column has a cross-section 28 mm × 22 mm, and its material has a Young's modulus of 70 $GN/m^2$. Determine a) the strain, b) the stress and c) the load carried by the column.

4 A tensile load of 160 kN is applied to a 20 mm diameter tie-rod to which a strain gauge is bonded. Young's modulus for the material of the rod is 200 $GN/m^2$ and the gauge factor of the strain gauge is 1.95. Determine the percentage change in the resistance of the gauge. (Work in terms of $R$.)

5 a) What are the advantages of using a Wheatstone bridge circuit instead of a sensitive multimeter to determine the change of resistance of a strain gauge?

b) In a bridge circuit similar to Figure 1.6, $R_A$ was bonded to a cylindrical test specimen of diameter 25.08 mm. $R_D$ was 160.1 Ω, $R_C$ was 2502.7 Ω and $R_B$, at balance, was 2505.8 Ω when the specimen was unloaded and 2511.7 Ω when it was loaded with 50 kN. The gauge factor of the strain gauges was 2.10. Determine the value of Young's modulus for the material of the specimen.

6 Explain how a temperature-compensated strain gauge keeps its resistance constant in spite of temperature changes. Show, also, how the bridge circuit is modified to cancel out the effect of any temperature gradient in the gauge's connecting leads.

7 In a direct-reading bridge with calibrating resistors, using the circuit of Figure 1.12, the three calibrating resistances were 150 kΩ, 68 kΩ, and 47 kΩ. While $R_A$ was unstrained it had a resistance of 120.6 Ω and the output voltage of the bridge was +0.7 mV. Connecting each of the calibrating resistors in turn across $R_A$ gave bridge output voltages of +3.1 mV, +6.0 mV and +8.4 mV, respectively.

Dynamic loading of the structure to which $R_A$ was bonded caused the bridge output voltage to vary between –5.3 mV and +4.4 mV. Determine the corresponding range of stress, given that the gauge factor was 2.10 and Young's modulus for the material of the structure is 200 GN/m$^2$.

8 In a test to determine the gauge factor of a batch of strain gauges, two specimens each of 160.2 Ω resistance, were bonded to the upper and lower faces of a beam, and connected as $R_A$ and $R_D$ in Figure 1.13(a), to make a *twin-active-gauge direct-reading bridge*. The cross-section of the beam was a rectangle 30 mm (measured horizontally) by 5 mm deep and the modulus of elasticity of its material was 72 GN/m$^2$. When equal and opposite bending moments of 15 Nm were applied at the ends of the beam, the bridge output voltage changed from 0 to 26.5 mV. The power supply voltage was 15.0 V. Calculate (a) the change of resistance per gauge, (b) the maximum bending stress, (c) the maximum bending strain, (d) the gauge factor.

9 A load cell is constructed in which the transducer is a Wheatstone bridge made up of four active strain gauges. The first time it is tested the following results are obtained:

| Load (kN) | 0 | 0.5 | 1.0 | 1.5 | 2.0 | 2.5 | 2.3 | 1.8 | 1.3 | 0.8 | 0.3 | 0 |
|---|---|---|---|---|---|---|---|---|---|---|---|---|
| Output voltage (mV) | 0 | 3.8 | 8.1 | 12.0 | 15.7 | 19.9 | 18.7 | 15.3 | 10.8 | 7.2 | 3.0 | 0.7 |

Plot the results and determine the transfer function (output/input) of the load cell. What is the name of the fault shown up by the graph and how may it be eliminated?

10 Discuss the advantages and disadvantages of semiconductor strain gauges compared with gauges which use a wire or etched foil conductor. How are p- and n-type gauges produced, and why are they particularly useful in a combined form?

11 A strain gauge installation is suspected of giving inaccurate readings. Under what headings would you list the possible causes of error?

Assuming that the gauges are to be scraped off and replaced with new ones, what precautions would you take to ensure that, as far as possible, signals from the new installation will be free from error?

# 2
# Temperature transducers

## What is temperature?

The temperature of a substance is a function of the intensity of vibration of its atoms and molecules.

The Celsius scale of temperature (formerly called the centigrade scale) uses the *Celsius degree* as its unit of temperature. The Celsius degree is one-hundredth of the temperature change between the freezing and boiling points of water at standard atmospheric pressure, the freezing point being 0°C and the boiling point 100°C. The *kelvin* is the same unit of temperature as the Celsius degree. However zero on the kelvin scale is *absolute zero*, the temperature at which atoms and molecules are absolutely vibrationless (a temperature of –273.15°C). Thus the freezing and boiling points of water at standard atmospheric pressure are 273.15 K and 373.15 K respectively.

The standard for temperature measurement is the International Practical Temperature Scale (IPTS). This defines a number of accurately reproducible temperatures, such as the triple point* of water, the freezing point of zinc, the freezing point of silver, and so on. These points are measured by an accurate temperature-measuring instrument and are interpolated for temperatures between the IPTS points by assuming that the instrument reading varies linearly with temperature between the points. The interpolating instruments are a platinum resistance thermometer for temperatures up to 630.74°C, a 10% Rh-Pt/Pt thermocouple for temperatures between 630.74°C and 1064.43°C, and a radiation pyrometer for temperatures above 1064.43°C.

## Methods of temperature measurement

To measure temperature, we need a transducer which will convert intensity of atomic or molecular vibration into some kind of proportional signal. There are four ways in which this can be done:

- **Expansion** Temperature change can be converted into change of volume, as in a liquid-in-glass thermometer, or into change of shape, as in a bimetallic strip.
- **Thermoelectricity** A small voltage is generated where two different metals are joined together. The magnitude of the voltage generated depends on the temperature of the junction.
- **Electrical resistance** Resistance of metals and of semiconductors varies with temperature.
- **Radiation** The heat energy radiated by a body increases in quantity and includes radiation at shorter and shorter wavelengths as temperature increases.

For a temperature transducer to respond to a change of temperature it must gain or lose heat

---

*The triple point of a substance is the one temperature at which the solid, liquid and vapour phases can exist together. The triple point of water is 273.16 K (0.01°C).

energy. The three ways in which this can take place are by *conduction*, *convection*, and *radiation*.

For **conduction** of heat to occur, there must be physical contact between the transducer and the substance whose temperature is to be measured. The atoms and molecules of the transducer then gain kinetic energy of vibration from (or lose kinetic energy of vibration to) the atoms and molecules of the material with which it is in contact.

**Convection** occurs when the surrounding material is a liquid or gas. Temperature changes in the fluid in contact with the transducer (caused by conduction of heat) result in part of the fluid expanding or contracting. This causes a change in density so that fluid around the transducer rises or sinks, bringing fresh fluid into contact with the transducer, and so continuing the heat transfer process. (One of the effects of weightlesness in artificial satellites is that density changes do not result in convection so the circulation of air and other fluids must be forced by fans or pumps.)

Transfer of heat energy by **radiation** between bodies occurs all the time. The radiation is in the form of electromagnetic waves (radio waves) in the infra-red and, if the temperature is high enough, in the visible part of the electromagnetic spectrum (see Figure 2.1). The range of wavelengths radiated, and the power density (the power emitted at a given wavelength) depend on the temperature of the body, as shown in Figure 2.2.

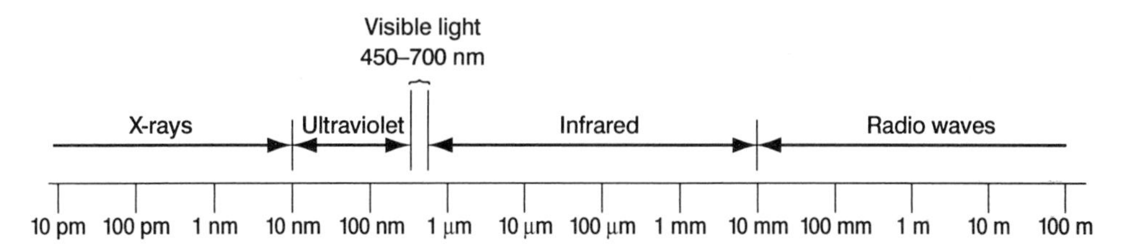


**Figure 2.1** The electromagnetic spectrum, on a logarithmic scale of wavelength

To see how these three methods of heat transfer affect the measurement of temperature, consider the apparently simple task of measuring the temperature of a few cubic centimetres of water in a test tube. It seems easy – let's suppose we use a thermometer.

If the thermometer is at room temperature, it immediately cools the liquid, the temperature of which we were trying to measure, as heat is given up by the water to raise the temperature of the thermometer bulb. Also the thermometer loses heat to the air by convection, though most mercury-in-glass thermometers have an inscription on the back stating the depth of immersion for which they are calibrated. So, provided we immerse the thermometer to the correct depth the reduction in temperature due to the continuous slight heat loss will have been allowed for, though in time that heat loss will further cool the contents of the test tube. A third source of error is radiation. If we were taking our temperature measurement in a cold-storage room, for example, we could expect our thermometer to give a slightly low reading because it would be radiating more heat to the surroundings than it was receiving in radiation from them. Similarly, if we were taking our measurement out-of-doors in bright sunshine, we could expect the thermometer reading to be slightly higher than the actual temperature, because it would be receiving more heat radiation from the sun than it would be giving out to its surroundings. Actually, the heat loss or gain due to radiation would be less for a mercury thermometer than for one employing coloured alcohol, because it depends on the emissivity of the liquid. The emissivity, $\varepsilon$, of a surface is the proportion of the radiation it receives or emits, compared with the radiation which would be received or emitted by a *black body* under the same conditions.

A 'black body' (which, incidentally, would not be seen as black if it were hot enough to be radiating any visible light!) is one having a surface which absorbs all the radiation falling on it, reflecting none. Such a surface would also be a perfect *emitter* of radiation, since none of the radiation of the body would be held back by internal reflection from the surface. A perfect black body, with an emissivity of 1 over the whole range of radiation wavelengths, does not exist in this imperfect world. The sun's radiation is very nearly that of a black body at 6000 K, thus approximating to the top curve of Figure 2.2. The emissivity of other surfaces may vary with the temperature of the object and also with the wavelength of the radiation; it cannot exceed a value of 1 at any temperature or wavelength.

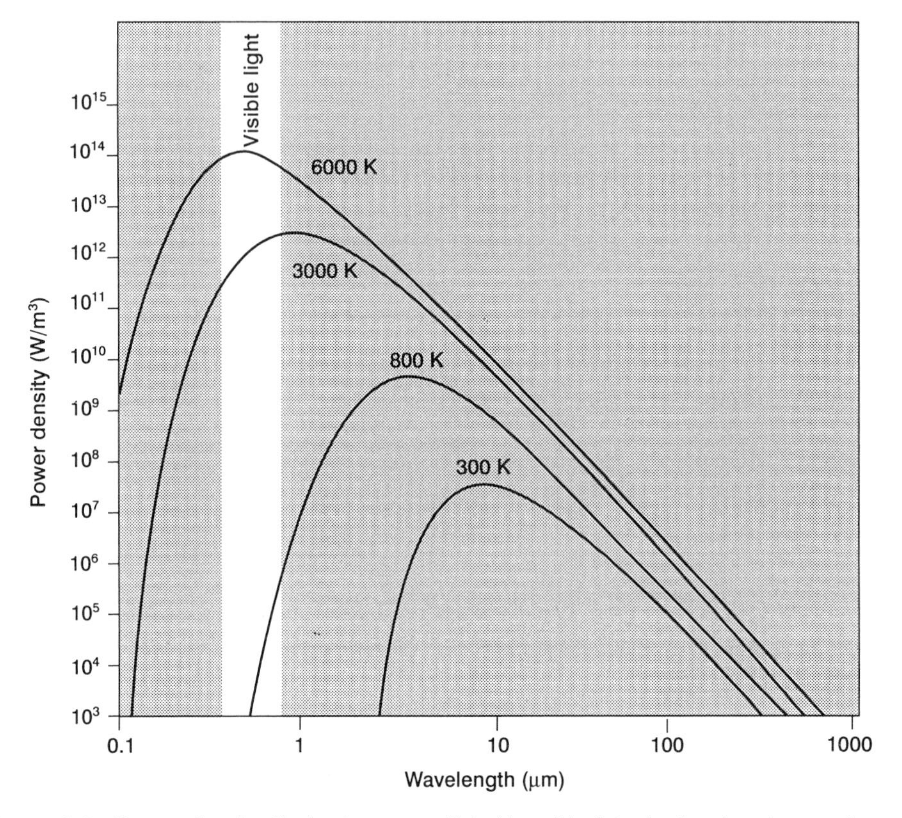


**Figure 2.2** Curves showing the heat power radiated by a black body at various temperatures

## Time constant

If we take a thermometer at room temperature and suddenly plunge it into boiling water it does not instantly indicate the temperature of the water. The mercury rises very quickly at first, but slows down as it rises, and we must wait a few seconds to be sure that it has reached its final value.

The rate of heat transfer to the thermometer is proportional to the temperature difference between the water and the thermometer; the smaller the remaining temperature difference, the more slowly does the temperature rise. If we plot indicated temperature against time, we get the type of curve shown in Figure 2.3.

This kind of curve is seen whenever a *step input* is applied to *any* system in which the rate of approach to the final value is proportional to the change which still has to be made. For

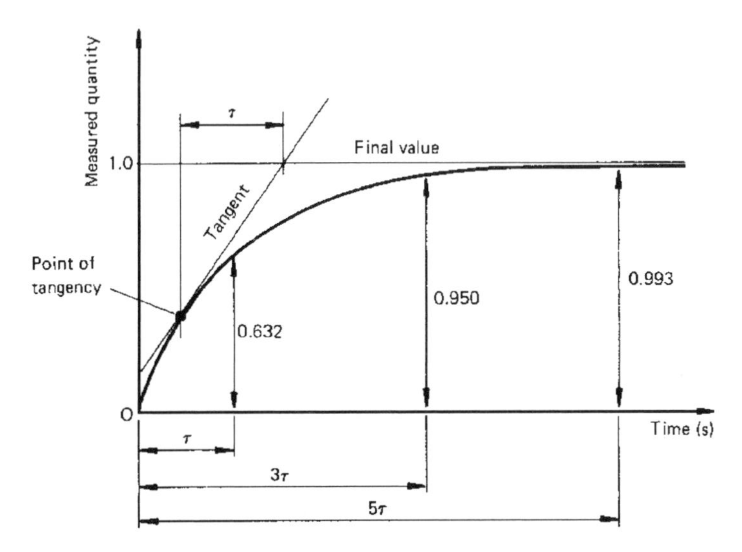


**Figure 2.3** The main features of time constant

instance, if a DC voltage is suddenly applied to a resistor and a capacitor in series, the voltage/time graph of the capacitor is exactly the same curve as in Figure 2.3. Such systems are called *first-order linear systems*, because they obey a first-order differential equation of the form:

$$\text{(Change which still has to be made)} = \text{(step input)} \times e^{\frac{-t}{\tau}}$$

where:

$e$ is the mathematical constant 2.718 ...,
$t$ is the time which has elapsed since the step was applied,
$\tau$ is the *time constant.*
($t$ and $\tau$ are both measured in seconds.)

The importance of the time constant of a transducer is that it enables us to estimate how much of the response to a step input has been completed in a given time, and hence how long we have to wait before taking a reading – that is before there is negligible difference between instrument reading and final value. The time constant can be found by taking a tangent to the curve at any point, as shown in Figure 2.3. It can also be found, more accurately, as the value of $t$ at which the ordinate of the response curve is 63.2% of the step input. This is shown by the example below.

*Example 2.1*
A step input is applied to a first-order linear system. Calculate and plot the percentage of final response which has been completed at the following times, measured from the instant the step was applied: 0, 0.15τ, 0.3τ, 0.6τ, τ, 1.5τ, 2τ, 3τ, 4τ and 5τ.

*Solution*
This question can be answered simply, using a scientific calculator. For example:

$$e^{\frac{-0.15\tau}{\tau}} \text{ is } e^{-0.15}$$

Enter 0.15 on the keyboard, press the ± key and then the $e^x$ key and we have 0.861. The other values in the table on the following page are obtained in the same way. The required percentages are obtained by multiplying the values in the bottom line of the Table by 100.

| Multiple of $\tau$ | 0 | 0.15 | 0.3 | 0.6 | 1.0 | 1.5 | 2.0 | 3.0 | 4.0 | 5.0 |
|---|---|---|---|---|---|---|---|---|---|---|
| $e^{\frac{-t}{\tau}}$ | 1 | 0.861 | 0.741 | 0.549 | 0.368 | 0.223 | 0.135 | 0.050 | 0.018 | 0.007 |
| $1-e^{\frac{-t}{\tau}}$ | 0 | 0.139 | 0.259 | 0.451 | 0.632 | 0.777 | 0.865 | 0.950 | 0.982 | 0.993 |

The graph should be similar to Figure 2.3. From the table and graph, we can note the following points. Up to about $0.15t$, the proportion of $t$ and the proportion of step input response completed are practically equal.

At $t = \tau$ the response has reached 63.2% of its final value.
At $t = 3\tau$ it has reached 95% of its final value.
At $t = 5\tau$ it exceeds 99% of its final value.

Thus if we are taking a reading, we should wait for three time constants if we want to be within 5% of the step, or five time constants if we want to be within 1%. The time for the response to be within any other limit of accuracy can be calculated by rearranging the equation given into the form:

$$e^{-\frac{t}{\tau}} = \frac{\text{Change which still has to be made}}{\text{Step input}}$$

and finding the value of $-t/\tau$ by entering the value of the right-hand side of the equation into the calculator and taking its log-to-the-base $e$ (by pressing the 'ln' key).

*Example 2.2*
A probe, which consists of a thermocouple in a protective sheath, has a time constant of 4.0 seconds. It is to be used to measure the temperature of a molten metal. If its initial temperature is 20°C, and the temperature of the metal will not be greater than 650°C, determine the minimum time that the probe should remain in the molten metal, to obtain a reading within two degrees of the true value.

*Solution*

Step input = 650° – 20° = 630°C

$$\therefore e^{-\frac{t}{\tau}} = \frac{2}{630} = 0.00317$$

$$\therefore -\frac{t}{\tau} = -5.75$$

$$\therefore t = 5.75\tau = 5.75 \times 4.0 = 23.0 \text{ s}$$

Therefore the probe should remain in the molten metal for a minimum time of 23 seconds.

## Sinusoidally varying temperatures

If the input to a first-order transducer, such as a thermometer, varies continuously, its time constant causes the corresponding variation in its output to lag behind that of its input. Also, the time constant diminishes the amplitude of the output variation, because before it can reach the required maximum or minimum the output is diverted in the opposite direction. The more rapid the variation in input, the more serious the errors from these two causes.

If the input variation takes the form of a sinewave (or can be approximated to one) the output variation is determined by the following two equations:

$$\text{Output amplitude} = \frac{(\text{input amplitude})}{\sqrt{\omega^2\tau^2 + 1}}$$

and

$$\text{Phase lag} = \arctan \omega\tau$$

where

$\omega$ is the angular frequency of the sinewave in radians/second.

Figure 2.4 shows how output amplitude and phase lag vary when $\omega\tau$ is varied.

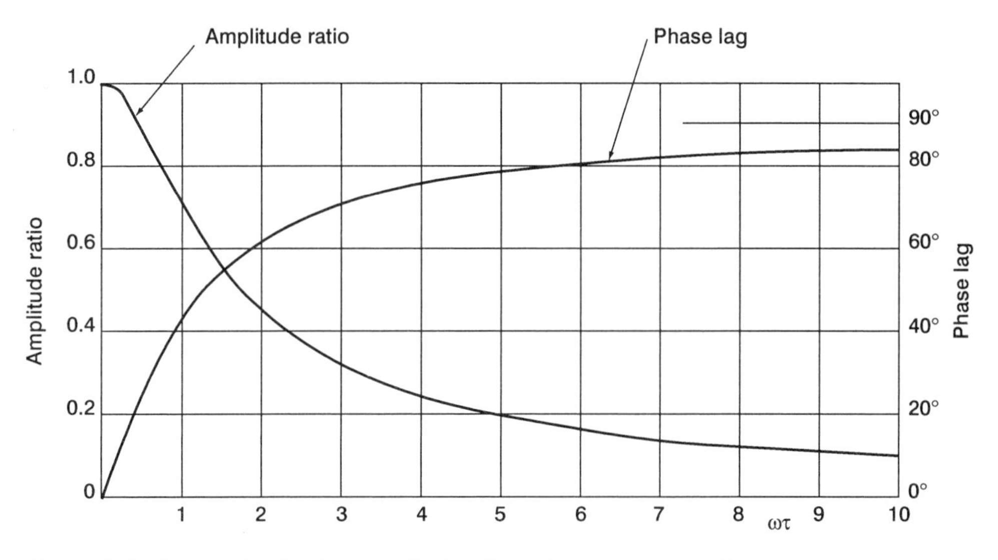


**Figure 2.4** Curves showing how amplitude ratio and phase lag vary with $\omega\tau$

*Example 2.2*
A transducer with a time constant of 5.2 seconds measures the temperature of a liquid flowing at a constant rate into a chemical process. The recorded temperature is found to vary approximately sinusoidally between 216°C and 238°C, with a periodic time of 18 seconds. Determine the actual temperature variation and the time delay between corresponding points in the actual cycle and in the recorded cycle.

*Solution*

$$\text{Frequency} = \frac{1}{(\text{period})} = \frac{1}{18} = 0.0556 \text{ Hz}$$

$$\therefore \text{ Angular frequency} = 2\pi \times 0.0556 = 0.349 \text{ rad/s}$$

$$\therefore \ \omega\tau = 0.349 \times 5.2 = 1.815$$

$$\text{Midpoint of temperature range} = \frac{238 + 216}{2} = 227°\text{C}$$

Amplitude of recorded temperature = 238 – 227 = 11°C

$$\frac{\text{(recorded amplitude)}}{\text{(true amplitude)}} = \frac{1}{\sqrt{1.815^2 + 1}} = \frac{1}{2.07}$$

True amplitude = 11 × 2.07 = 22.8°C

True temperature range is from 227 – 22.8 = 204°C to 227 + 22.8 = 250°C

Phase lag = arctan 1.815 = 61.1°

$$\text{Delay} = \frac{61.1}{360} \times 18 = 3.06 \text{ seconds}$$

## Thermocouples

Where two different metals are joined together, either mechanically or by welding, there is a continuous small voltage generated at the point where they meet (the junction). This voltage is generated by what is known as the *Seebeck effect* or *thermoelectric effect*. It depends on the temperature of the junction and on the metals used. The junction can therefore be used as a temperature transducer, converting temperature into voltage. This type of temperature transducer is called a thermocouple.

If a thermocouple is part of a complete electrical circuit, the voltage can drive a small current around the circuit. This principle is used in gas-fire safety devices in which a thermocouple projecting into the flame gives enough current to energise a solenoid holding a gas valve open against the force of a weak spring – if the flame goes out, the thermocouple ceases to generate current and the spring closes the gas valve.

Of course every point where dissimilar metals meet in the circuit is a thermocouple to some extent, and voltages generated by these other thermocouples will oppose and partly cancel out the voltage of the main thermocouple. The safety device mentioned above only works because the flame thermocouple junction is much hotter than all other junctions in the circuit.

The simplest circuit in which current can circulate is shown in Figure 2.5. The voltages $V_1$ and $V_2$ at the two junctions oppose each other and any current in the circuit is due to the voltage difference $V_1 - V_2$, which will be due to the temperature difference between the two junctions.

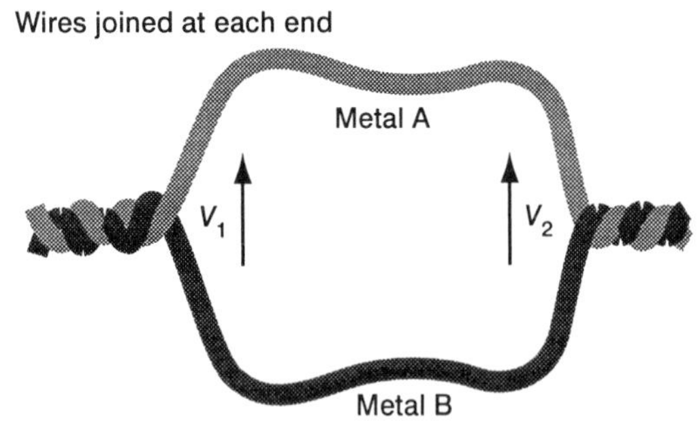


**Figure 2.5** Basic thermocouple circuit

To measure the voltage difference, we must introduce a voltmeter or the input stage of an amplifier into the circuit, and this will introduce a third metal, usually copper, as shown in Figures 2.6 and 2.7. We can assume that both of the junctions with the copper are at 'room temperature'. In Figure 2.6 the two additional voltages generated, $V_3$ and $V_4$, will be equal and opposite, and will therefore have no effect on the voltage difference ($V_1 - V_2$) which we have

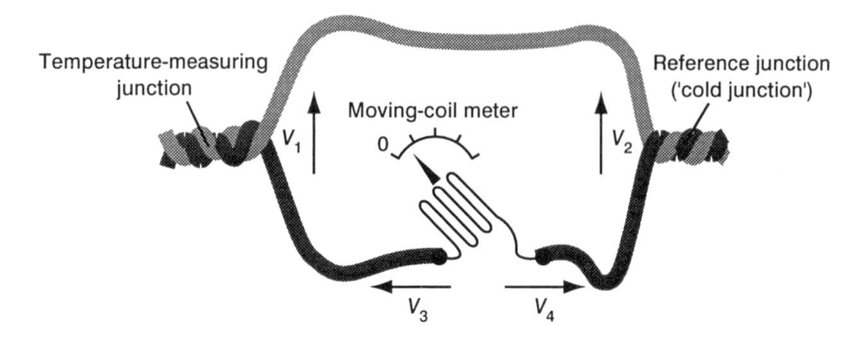


**Figure 2.6** Insertion of a third metal (e.g. the moving coil of a meter) into one of the thermocouple wires

to measure. In Figure 2.7, however, the two additional voltages will not be equal, because $V_5$ is generated by the junction of copper with metal A, while $V_6$ is generated by the junction of copper with metal B, but their difference ($V_5 - V_6$) will be the same as $V_2$, the voltage which would be generated by joining metals A and B together directly, at that temperature.

In general, therefore, we can say that introducing a third metal into a thermocouple circuit does not affect the output voltage, provided that both of the extra junctions are at the same temperature. It follows also from this rule that we can make up thermocouples by brazing or soldering the two metals together, or even by inserting them separately into a molten metal to measure its temperature.

In order to measure temperature with a thermocouple, the reference junction in Figure 2.6 must be kept at a constant, known temperature. For very accurate measurements, we should keep it in a 'triple-point apparatus', in which the temperature is maintained at the triple-point of water. For normal accuracy, a mixture of ice and water, giving 0°C for practical purposes, would be used instead. More conveniently, the reference junction may be installed in a tiny electrically heated oven thermostatically controlled to a temperature of, say, 45°C, so that it will always be at a higher temperature than its surroundings.

British Standard 4937 gives tables of EMF in microvolts versus temperature in °C, for industrial thermocouples. The EMF values are the voltage difference $V_1 - V_2$ when the reference junction is at 0°C. If the reference junction is at any other temperature, the EMF value for that temperature must be added to the EMF value in the table.

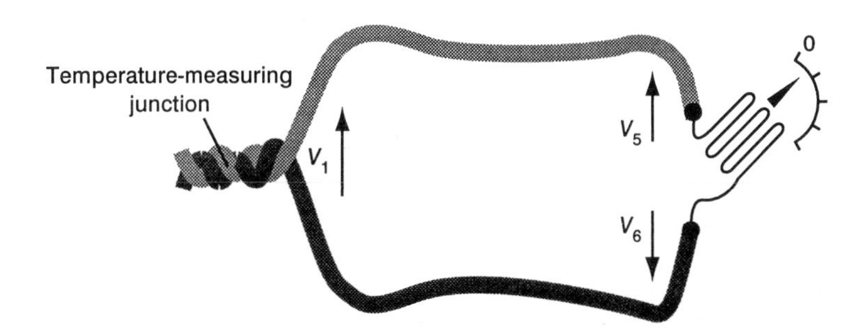


**Figure 2.7** Replacing the reference junction by separate junctions with the third metal

*Example 2.3*
A type N thermocouple (B.S. 4937 Part 8) gave a reading for $V_1 - V_2$ of 14.58 mV. Determine the corresponding temperature if the reference junction was at a) 0°C, b) 45°C, c) –12°C. Extracts from B.S. 4937 Part 8 (Nickel-chromium-silicon/nickel-silicon thermocouples) are given in Figure 2.8.

Table 2 Temperature in °C: EMF in mV — Reference junction at 0°C

| Temp. | 0 | –1 | –2 | –3 | –4 | –5 | –6 | –7 | –8 | –9 |
|---|---|---|---|---|---|---|---|---|---|---|
| –10 | –260.30 | –286.20 | –312.10 | –337.90 | –363.70 | –389.50 | –415.30 | –441.00 | –466.70 | –492.30 |
| 0 | 0.0 | –26.10 | –52.30 | –78.40 | –104.40 | –130.50 | –156.50 | –182.50 | –208.50 | –234.40 |
| **Temp.** | **0** | **1** | **2** | **3** | **4** | **5** | **6** | **7** | **8** | **9** |
| 40 | 1064.20 | 1091.50 | 1118.90 | 1146.30 | 1173.80 | 1201.30 | 1228.90 | 1256.50 | 1284.10 | 1311.80 |
| 50 | 1339.50 | 1367.20 | 1395.00 | 1422.80 | 1450.70 | 1478.60 | 1506.50 | 1534.50 | 1562.50 | 1590.50 |

Table 3 EMF in mV: temperature in °C — Reference junction at 0°C

| EMF | 0 | 50 | 100 | 150 | 200 | 250 | 300 | 350 | 400 | 450 |
|---|---|---|---|---|---|---|---|---|---|---|
| 14 000 | 427.6 | 428.9 | 430.2 | 431.6 | 432.9 | 434.2 | 435.6 | 436.9 | 438.2 | 439.5 |
| 14 500 | 440.9 | 442.2 | 443.5 | 444.9 | 446.2 | 447.5 | 448.8 | 450.2 | 451.5 | 452.8 |
| 15 000 | 454.1 | 455.5 | 456.8 | 458.1 | 459.4 | 460.7 | 462.1 | 463.4 | 464.7 | 466.0 |
| 15 500 | 467.3 | 468.7 | 470.0 | 471.3 | 472.6 | 473.9 | 475.2 | 476.6 | 477.9 | 479.2 |
| 16 000 | 480.5 | 481.8 | 483.1 | 484.4 | 485.7 | 487.1 | 488.4 | 489.7 | 491.0 | 492.3 |

**Figure 2.8** Extracts from BS 4937: Part 8, which gives values of EMF for type N (Nicrosil/Nisil) thermocouples

*Solution*
14.58 mV is 14 580 μV. Then, from Table 3 of Figure 2.8:

a) 14 580 μV corresponds to 443.0°C.

b) From Table 2 of Figure 2.8, the EMF at 45°C is 1201.3 μV.
14 580 + 1201.3 = 15 781.3 μV

From Table 3 this corresponds to 474.7°C.

c) From Table 2 of Figure 2.8 the EMF at –12°C is –312.1 μV.
14 580 + (–312.1) = 14 267.9 μV

From Table 3 this corresponds to 434.7°C.

The temperature values given above were obtained by interpolation (assuming that temperature varies linearly with EMF between the values given in the tables). Note that the answers to parts b) and c) are not 443.0 + 45 = 488°C and 443.0 – 12 = 431°C.

In the case of the circuit shown in Figure 2.7, we obviously cannot keep the meter in melting ice or an oven. In such cases we could add or subtract a correction to the reading as in the example above. But a modern electronic thermometer using a thermocouple will probably have an automatic correction for ambient temperature built into it, using a thermistor (see p.26) to obtain a voltage for cold junction compensation.

## Thermocouple materials

The operating temperature range, composition and accuracy of industrial thermocouples are defined in British Standard 1041 part 4. Tables of values of $V_1 - V_2$ versus temperature, when the reference junction is at a temperature of 0°C, are given in B.S. 4937, as illustrated in the previous example. Figure 2.9 shows how EMF/temperature characteristics compare for some of the more common types of industrial thermocouple. The curves shown relate to thermocouples in Table 2.1.

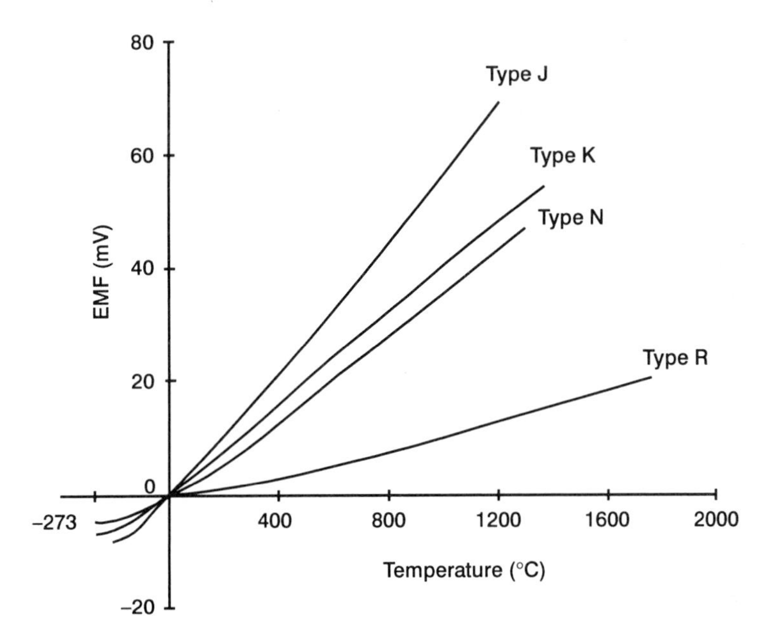


**Figure 2.9** EMF/temperature characteristics of some industrial thermocouples

**Table 2.1**

| Type | B.S. 4937 Part | Positive arm | Negative arm | Characteristics |
|---|---|---|---|---|
| J | 3 | Iron | Copper-nickel | Inexpensive. Widely used. High output EMF |
| K | 4 | Nickel-chromium | Nickel-aluminium | Higher max./ lower min. temperatures than type J. |
| N | 8 | Nickel-chromium-silicon | Nickel-silicon | Long life at high temps. No curie point step in EMF Neutron radiation immune. |
| R | 2 | Platinum-13% rhodium | Platinum | Expensive. High temp. applications in glass and ceramics manufacture. |

### Thermocouple compensating cable

To reduce the expense of running wires in thermocouple materials over long distances between the temperature-measuring junction and the reference junction or meter, *thermocouple compensating cables* are used. These are made from cheaper materials which satisfy the following requirements over the limited temperature range to which they will be subjected (typically 0°C to 80°C):

1 There must be no voltage generated where each compensating cable wire is joined to the corresponding thermocouple wire.
2 The voltage generated at the reference junction must be the same as would be generated by the thermocouple wires themselves at that temperature.

Thermocouple compensating cables have twin cores of stranded wire, each core being colour coded to ensure that it will be connected to the correct thermocouple wire.

### Sheathing

The temperature-measuring junction can be made of very fine wire so that it has a time constant of only a few milliseconds; but if so, it will be very fragile. Most thermocouples are enclosed in a metal or ceramic tube to protect them from physical damage or chemical action. A mineral filling such as magnesium oxide is used to insulate the wires from each other and from the tube. The wires may be welded or brazed to the end of the tube (observing the 'third metal rule') to reduce the time constant of the temperature probe, but in such a case it will be impossible to check the insulation electrically.

## Thermistors

Thermistors are made from materials in which a large change of electrical resistance is produced by a small change of temperature. The usual type of thermistor has a negative temperature coefficient (denoted *n.t.c.*) and a resistance-temperature relationship which is defined by the equation:

$$R = Ae^{B/T}$$

where $A$ and $B$ are constants for the particular material, and $T$ is the *absolute* temperature of the thermistor in kelvins. Writing this equation for a particular pair of values of resistance and temperature, $R_1$ and $T_1$, we get:

$$R_1 = Ae^{B/T_1}$$

and dividing the general equation by the particular, gives:

$$R = R_1 e^{B\left(\frac{1}{T} - \frac{1}{T_1}\right)}$$

from which the resistance $R$ can be calculated at any absolute temperature $T$, if $R_1$, $B$ and $T_1$ are known. $B$ is a temperature in kelvins, called the *characteristic temperature* of the thermistor.

*Example 2.4*
A thermistor has a resistance of 1500 Ω at 25°C and a characteristic temperature of 3975 K. Plot its resistance over the temperature range –40°C to +140°C.

*Solution*

$T_1 = 25 + 273 = 298$ K $\therefore \dfrac{1}{T_1} = 0.0033557$

by calculator or computer, using

$$R = 1500e^{3975\left(\frac{1}{T} - 0.0033557\right)}$$

the following values are obtained:

| °C | –40 | –20 | 0 | 20 | 40 | 60 | 80 | 100 | 120 | 140 |
|---|---|---|---|---|---|---|---|---|---|---|
| Ohms | 62 000 | 16 100 | 5090 | 1880 | 792 | 369 | 188 | 103 | 59.7 | 36.6 |

If these are plotted, the typical exponential curve for the resistance of a thermistor with negative temperature coefficient is obtained. The curve has a steep negative gradient at the low temperature end of the range, with the gradient reducing, apparently to almost zero as the temperature increases. However, if the ranges 0 to 140°C, 40 to 140°C and 100 to 140°C are plotted on 10-times, 100-times and 1000-times the vertical scale respectively, the 'horizontal' portion of the curve will be found to have exactly the same shape as the original curve. This is, of course, what we should expect from an exponential function. It illustrates one of the difficulties of using a thermistor for *precise* temperature measurement: the limited range of temperature over which we can use it without the change of resistance becoming excessive.

To measure temperature by means of the change in resistance of a thermistor, we have to pass a current through it. Figure 2.10 shows a simple circuit which uses a thermistor to indicate the temperature of the cooling water in a motor vehicle engine. The current meter can consist of just a bimetal strip carrying a pointer, the current being passed through a length of resistance wire wound round the bimetal strip to act as a heater. As the water temperature increases, the resistance of the thermistor decreases; thus the current passing through the heating coil increases, causing it to heat up the bimetal strip and so deflect the pointer. It is very crude, very cheap, but rugged, and immune to vibration and 'transients'. And anyway, we don't want to worry the driver by giving more information than 'cold', 'normal' or 'hot'.

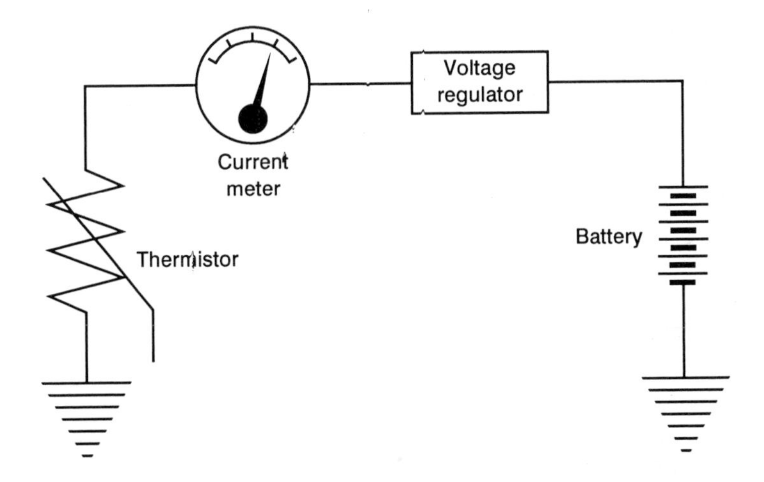


**Figure 2.10** Circuit diagram for a motor car water temperature indicator system

Where more precise temperature measurement is necessary, we must allow for the fact that when we pass current through the thermistor it will heat up until its temperature is such that its output power (in the form of heat) is equal to the electrical power being dissipated in it.

*Example 2.5*
The dissipation factor of the thermistor in the previous example is given as 8.5 mW/°C. It is to be used for temperature measurement over the range 0°C to 100°C. If its temperature is not to differ by more than 1.5°C from the temperature of its surroundings, calculate:

a) the maximum constant voltage which can be applied to it
b) the resulting current range corresponding to the temperature range.

*Solution*
a) Maximum allowable power = 1.5 × 8.5 = 12.75 mW = 0.01275 W.

Electrical power $P = \frac{V^2}{R} \quad \therefore V = \sqrt{P \times R}$

Therefore for a given value of power transfer, the maximum constant voltage is limited by the least possible resistance the thermistor can have within the given temperature range, that is, 103 Ω (from the table on p.27)

Then maximum constant voltage $= \sqrt{0.01275 \times 103} = 1.15\ \text{V}$

b) At this voltage, by Ohm's law, the current through the thermistor will vary from:

$$\frac{1.15}{103} = 0.0112\ \text{A} = 11.2\ \text{mA at } 100°\text{C}$$

$$\text{to } \frac{1.15}{5090} = 0.000226\ \text{A} = 0.226\ \text{mA at } 0°\text{C}$$

If the thermistor is to be enclosed in a protective coating or sheath however, its dissipation factor, and therefore its voltage and current limits, will be considerably reduced. It would then be better to put up with the unknown temperature error and calibrate the system by applying known temperatures to it and plotting the corresponding current values.

## Positive temperature coefficient (PTC) thermistors

These are thermistors in which increase of temperature *increases* their resistance, instead of decreasing it. They are not normally used for temperature measurement, but they are very useful for over-temperature protection of industrial equipment, and for over-current protection of electrical circuits.

The *over-temperature protection type* has a low resistance (typically 100 Ω) until the temperature reaches a specified reference temperature. Within a few degrees above that temperature the resistance increases about a hundred-fold, so that in effect it is a fail-safe device, switching off a circuit at temperatures around 100°C. Several of these devices connected in series can continuously monitor temperatures at sensitive places in an installation and switch off if any one of them becomes too hot.

The *over-current protection type* is also a switch-off fail-safe device, operated by the self-heating effect which occurs when too much current is passed through the thermistor. Up to a specified current the self-heating effect is negligible, but the resistance of the thermistor increases rapidly if the current limit is exceeded, thus bringing the current down to a small fraction of the specified limit. A typical application is the protection of the primary windings of a mains transformer.

## Resistance thermometers

Resistance thermometers offer a reliable, precise method of measuring temperature. Platinum is the metal usually used for the temperature-sensing resistor because it can withstand high temperatures without deterioration, and because its resistance-temperature relationship is almost perfectly linear. It can be used for temperature measurement throughout the range –260°C to +800°C.

In the laboratory type of resistance thermometer, the platinum is in the form of a wire wound on a mica former and enclosed in a glass bulb which may be evacuated or filled with an inert gas to protect the platinum.

In industrial resistance thermometers, which do not have to cover such a wide range of temperature but must be more shock-proof, the platinum wire is wound on a ceramic former and enclosed within a glass or ceramic protective sheath. The winding is usually arranged to be non-inductive, by making the current travel down one helix and back up a parallel helix.

For temperatures in the range –50°C to +500°C, a compact, economic form of resistance thermometer is the *platinum film sensor*. This has a resistance element in the form of a long continuous line of platinum-based ink, which, with many reversals in the line, forms a compact pattern, similar to the pattern of a foil strain gauge. The pattern is deposited on a substrate of alumina, and trimmed by laser to an exact value of resistance. For use in a stainless steel thermometer pocket, it may be encased in a ceramic material.

The resistance element of a resistance thermometer is normally used as one of the resistances in a direct-reading Wheatstone bridge circuit (already mentioned in connection with strain gauges – see p.7). To avoid errors due to self-heating of the resistance element, the bridge circuit should be supplied with power from a programmable current source: that is, one which will keep the current constant to a set value. This will also enable the sensitivity of the system to be adjusted. To eliminate errors due to temperature gradients in the wires connecting the bridge circuit to the resistance element, a four-wire element can be used. In this arrangement, the two extra wires are put in series with the resistor on the other side of the bridge output connection to cancel out any resistance change due to temperature in the connecting wires. (See Figure 2.11.)

The bridge output voltage can be displayed on a voltmeter graduated in degrees of temperature as shown, or used as the feedback signal in an automatic temperature control system.

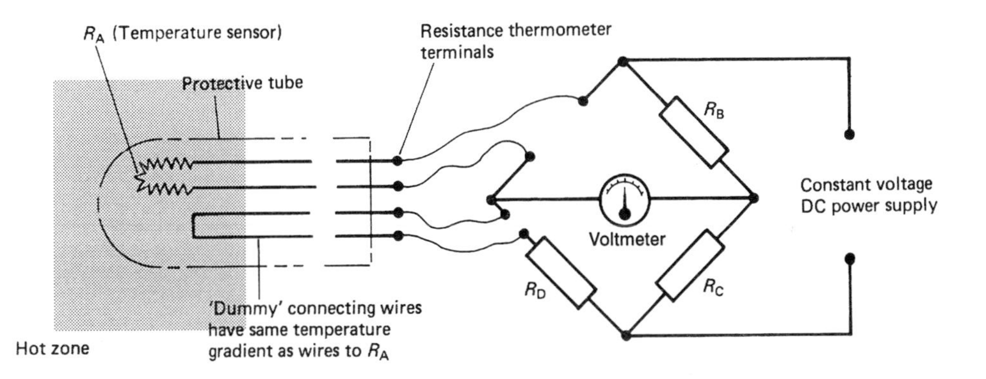


**Figure 2.11** Resistance thermometer circuit; four wire system

## Integrated circuit temperature sensor

This is a very cheap, simple means of temperature measurement or compensation, for temperatures in the range –40°C to +110°C. The integrated circuit, which looks like a small transistor, operates on a power supply between 4 and 30 volts, giving a linear output of 10 mVdegC$^{-1}$, with an accuracy of ±0.4°C.

## Radiation pyrometers

As the name indicates, these measure the temperature of a body from the electromagnetic radiation which it emits in the form of radiant heat, which is infra-red radiation and, if its temperature is high enough, visible light. Radiation pyrometers are normally used to measure very high temperatures, such as furnace temperatures, but, as the 300 K curve of Figure 2.2 shows, even an object at room temperature emits infra-red radiation from which its temperature can be determined, though we would normally use some simpler, cheaper instrument for this purpose, such as an ordinary thermometer. However, one of the virtues of a radiation pyrometer is that it does not make contact with the measured object, so if we have a situation where it is necessary to measure a 'low' temperature without making contact, a radiation pyrometer can be used.

In its simplest form a radiation pyrometer consists of a tube in which a temperature sensor is mounted. The internal surfaces of the tube have a matt black coating to minimise internal reflection. The tube may be open-ended, with the incoming radiation restricted to the area of the sensor by means of a disc with a central hole in it, like the 'stop' (or iris diaphragm) of a camera, as in Figure 2.12, or it may have a proper optical system with provision for sighting and focusing the radiation on to the sensor, as in the example shown in Figure 2.15.

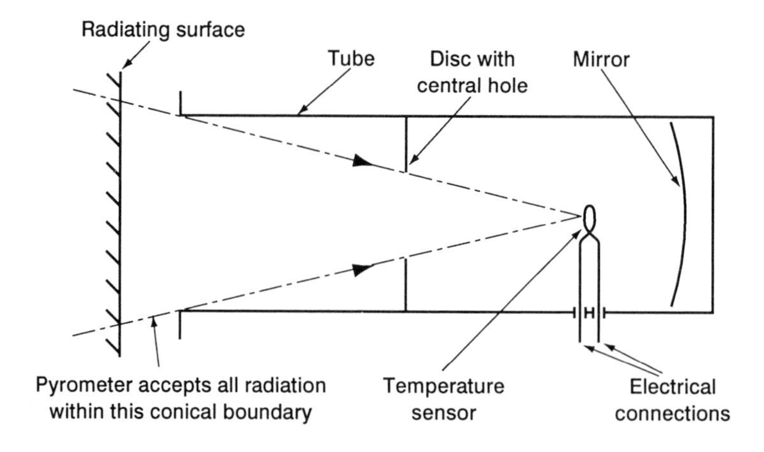


**Figure 2.12** Fixed-focus type of pyrometer

### Factors which affect radiation pyrometry

#### Emissivity of the radiating surface

A radiation pyrometer calibrated for black body temperatures will read low when used to measure the temperature of a surface with an emissivity less than 1. This may be overcome by recalibrating it for the emissivity of the given surface. Alternatively, the radiating surface may be covered by an enclosure with reflecting walls, with a hole facing the pyrometer, so it may

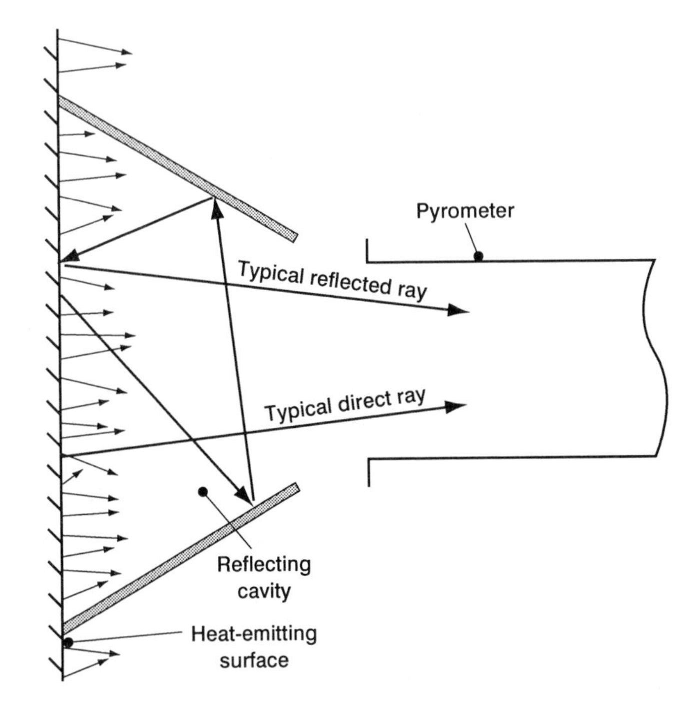


**Figure 2.13** Section through an internally reflecting enclosure used to increase the emissivity of a surface

'see' the surface through the hole. The pyrometer is placed where the hole completely fills its field of view, as shown in Figure 2.13. Most of the radiation emitted by the radiating surface is returned to it by reflection from the walls of the enclosure; some of that returning radiation is absorbed and the remainder is re-emitted. After many multiple reflections and re-emissions, the total radiation emitted through the hole is of such an intensity that the target has an emissivity within 1% of that of a black body. By the same principle, the emissivity of the inside of a furnace viewed through a small hole approximates to that of a black body.

**Attenuation of radiation by air or glass**

- Air and glass are transparent in the visible spectrum, and so absorb very little radiation, but over large parts of the infra-red spectrum they are relatively opaque, and thus reduce any radiation passing through them within those wavebands.
- Attenuation (weakening) of radiation by the optical system of a pyrometer can be allowed for in its calibration.
- Optical filters may be used to filter out those wavelengths at which air is opaque and the distance between the pyrometer and the target should always be as short as possible, so that radiation losses to the air may be kept to a minimum. This may make it necessary to water-cool the pyrometer if it is permanently installed close to a furnace.
- A smoky or dusty atmosphere will obviously attenuate the radiation. Care should be taken also to ensure that the optical system of the pyrometer is kept clean. Optical mirrors, which are usually silvered on the outside, must be cleaned very gently using alcohol on a fine camel-hair brush, so that the reflecting coating is not rubbed off.

**Failure of the radiating surface to completely fill the field of view of the pyrometer.** In this case, some radiation will be received from the background, and as this will be at a different temperature, a false reading will result.

For the simple pyrometer shown in Figure 2.12, the field of view is limited by the conical boundary shown. If the radiating surface is moved further away from the sensor the area filling the field of view increases as the square of the distance from the sensor, but because the radiation is in all directions, the proportion reaching the sensor *decreases* as the square of the distance. These two effects cancel out, so provided the radiating surface fills the field of view, and provided we can neglect the attenuation caused by the air in the path of the radiation, we can say that temperature readings are independent of the distance from the radiating surface to the sensor. If the pyrometer has an optical system which reduces its field of view to a narrower cone or even to a parallel beam of radiation, the above rule still applies.

## Temperature sensors for radiation pyrometers

The temperature sensor loses heat to its surroundings, and so has a lower temperature than the object being measured. A compact device of low thermal capacity is required. Thermistors may be used for 'low' temperature measurements; thermocouples or *thermopiles* are used for higher temperatures. A thermopile is a number of thermocouples connected in series, the voltage generated by a single hot-junction/cold-junction pair being multiplied by the number of such pairs. Figure 2.14 shows how they are arranged in a radiation pyrometer. In the example shown there are 15 hot junctions and 15 cold junctions in series (counting the two junctions with copper wire as one cold junction, as in Figure 2.7) giving 17 times the EMF of a single thermocouple at any given temperature.

The cold junctions may be kept at a stabilised temperature by installing the thermopile in a small temperature-controlled box – the oven mentioned on p.23 – with a small hole in the side facing the thermopile so that only the hot junctions are exposed to the radiation.

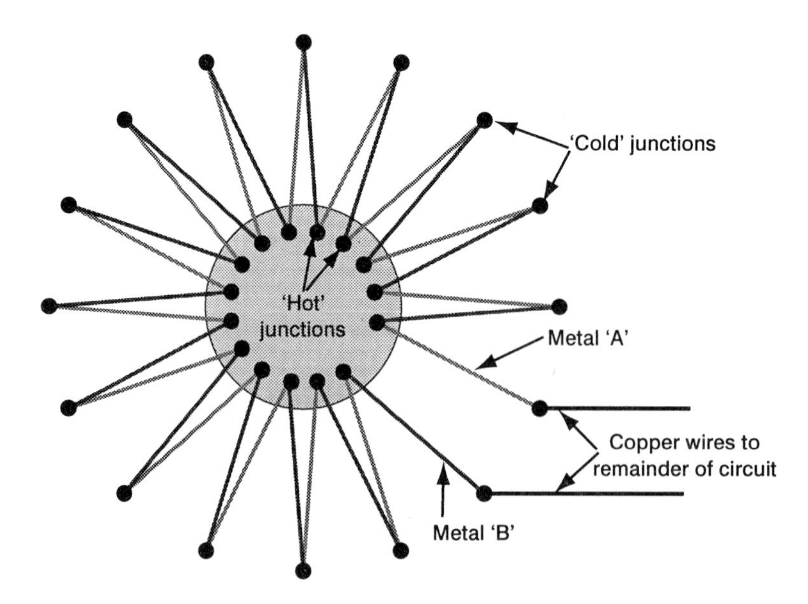


**Figure 2.14** Arrangement of junctions in a thermopile for a radiation pyrometer. Shading indicates area exposed to heat radiation

The thermopile behaves as a first-order linear system, with response as in Figure 2.3. To speed up its response and to prevent 'drift' in the output voltage due to gradual warming up of the pyrometer, the DC output may be converted to AC by means of a 'chopper'. The chopping is done mechanically, with a disc driven at constant speed by an electric motor. Sectors of the disc are cut away, so that the thermopile is alternately exposed to radiation and blanked-off, for equal periods of a few milliseconds. As the hot junctions of the thermopile alternately warm up and cool down, the output is a 'sawtooth' waveform. Increase or decrease in temperature of the radiating surface causes a corresponding increase or decrease in the peak-to-peak value of the output voltage.

A typical pyrometer incorporating these refinements is shown in Figure 2.15. The objective lens seals the pyrometer to prevent the entry of dust or smoke, and can be adjusted to focus on the radiating object. The beam splitter is a glass plate treated so that it reflects most of the radiation on to the thermopile but passes some through to the eyepiece, so that the operator can aim and focus the pyrometer. The chopper disc is gold-plated and polished, to make the blanking-off of the thermopile as effective as possible.

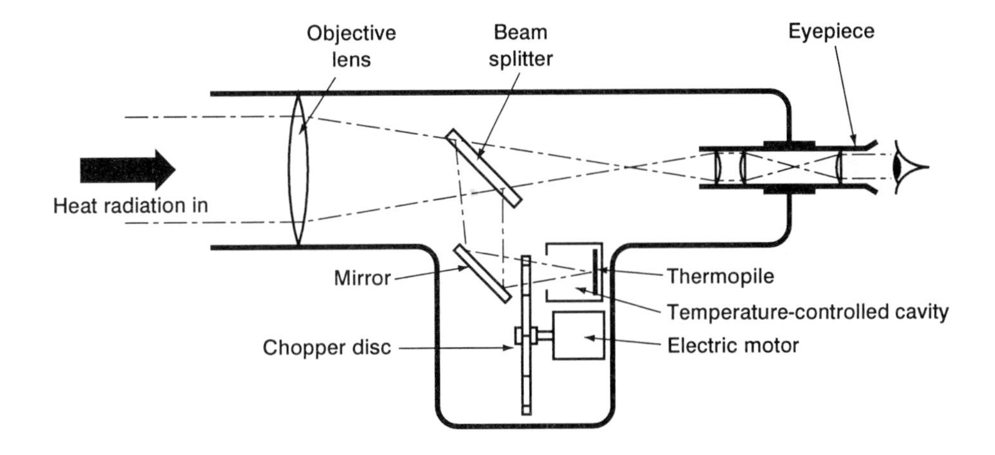


**Figure 2.15** Diagrammatic section through a radiation pyrometer with thermopile and chopper

## Calibration

Calibration may be *primary* or *secondary*. In a primary calibration an instrument is calibrated by reference to values obtained from first principles under laboratory conditions. In a secondary calibration the instrument is calibrated against values determined by a standard instrument which has itself received a primary calibration.

In the case of radiation pyrometers a primary calibration involves measuring the output at known temperatures on the International Practical Temperature Scale. The high temperature end of the IPTS is defined by the temperatures at which pure metals such as zinc, antimony, silver, gold and nickel solidify. As a molten metal cools and solidifies, there is a short period of time during which no change of temperature occurs because the heat lost in cooling is being replaced internally by the latent heat of solidification. The pyrometer output at this instant can be plotted against the known solidification temperature for that metal (given by the IPTS), to obtain one point on the calibration curve for the pyrometer. By repeating this process for each of the various metals and substances listed in the IPTS, a complete calibration curve is obtained.

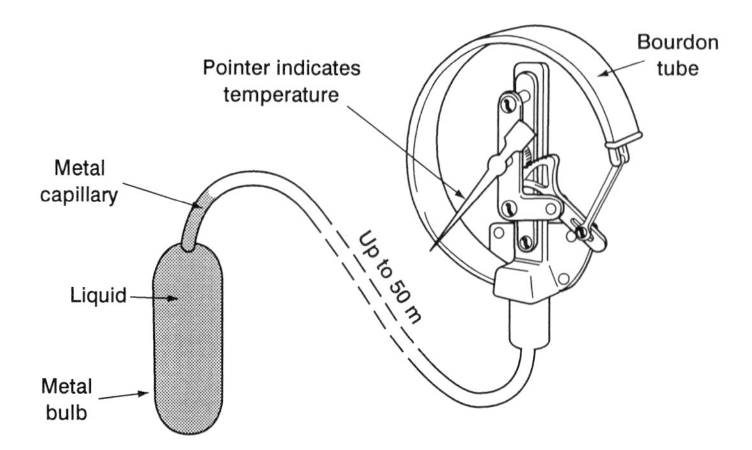


**Figure 2.16** A liquid-in metal thermometer

## Sealed-fluid thermometers

A *liquid-in-metal thermometer* consists of a metal bulb connected by means of a metal capillary tube to a Bourdon tube pressure gauge, as shown in Figure 2.16. The system is completely filled with a liquid under pressure, and sealed. Mercury is usually used for the liquid, with a steel bulb and capillary tube. Because the liquid cannot freely expand or contract, a change in its temperature causes a corresponding change in pressure, and this is indicated by the pressure gauge on a dial graduated in degrees of temperature. The capillary tube can be up to 50 metres in length, so this is a remote-reading device; any excess length in the capillary can be coiled up out of the way at a suitable point in its run. Provided the capillary is securely clamped down, it is unaffected by shock or vibration (except at the pressure gauge) and the only source of error is expansion or contraction of the capillary tube due to a change in ambient temperature. Where a significant change in ambient temperature is likely, this error may be eliminated by using the vapour-pressure form of the thermometer, shown in Figure 2.17. This uses the fact that for any vapour in contact with the liquid from which it has been evaporated, there can be only one pressure corresponding to a given temperature. Any change in temperature or volume causes liquid to evaporate or vapour to condense until the pressure appropriate to that temperature is regained.

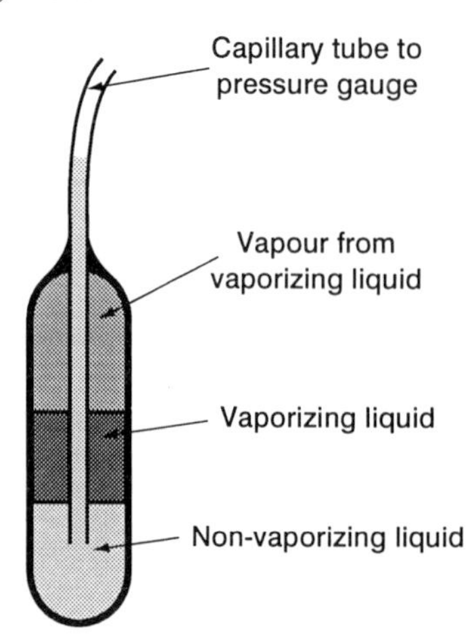


**Figure 2.17** Section through the bulb of a vapour pressure thermometer

As Figure 2.17 shows, the vapour pressure is transmitted through its liquid to a non-vaporizing liquid and hence to the pressure gauge. This neatly eliminates the effect of volume changes in bulb, capillary and Bourdon tube, and the liquids used are much cheaper than mercury, but there are some disadvantages:

1 The temperature range is very much reduced compared with that of liquid expansion thermometers.
2 The relationship between temperature and vapour pressure is non-linear.
3 The bulb must be kept upright or else the vaporizing liquid or its vapour will enter the capillary tube.

## The bimetallic strip

This consists of two metals with unequal coefficients of linear expansion, bonded together to form a single piece of material. A change in temperature causes them to expand or contract unequally, and this results in a proportionate increase or decrease in the curvature of the material. For maximum sensitivity to temperature change, one of the metals is usually *invar*, an alloy of iron and nickel which has a very low coefficient of linear expansion.

Simple dial-and-pointer type thermometers are made by forming the bimetallic strip into a spiral or a helix. This is fixed at one end to the body of the instrument and carries a pointer at the other end. Temperature change causes the bimetallic strip to curl or uncurl so that the pointer indicates temperature on a circular dial.

### Exercises on chapter 2

1 State four different effects of temperature change which may be used to convert temperature into a proportional signal. For each of these effects give an example of a transducer which makes use of the effect for the measurement of temperature.

2 a) What is meant by the term 'black body'?
b) 'An oxidised aluminium surface has an emissivity of 0.11.' Explain what this statement means, and its implication for the accuracy of your readings if you are trying to determine the temperature of an aluminium ingot by means of a radiation pyrometer. Sketch and describe a means of improving the accuracy of your readings.

3 The following readings were obtained from a clinical thermometer while it was being used to take a patient's temperature:

| Time (seconds) | 0 | 10 | 20 | 30 | 40 | 50 | 60 | 70 | 80 | 90 | 100 |
|---|---|---|---|---|---|---|---|---|---|---|---|
| Reading (°F) | 94.0 | 95.7 | 96.8 | 97.4 | 97.8 | 98.0 | 98.2 | 98.3 | 98.3 | 98.4 | 98.4 |

a) Plot these values and hence determine the time constant of the instrument (i) by the tangent method, (ii) by the 63.2% ordinate.
b) Use your answer to part (a) to determine the minimum time that the thermometer should remain in the patient's mouth to give a reading within 0.1 deg F of the patient's temperature, if it starts from a temperature 40 deg F below that of the patient. (Give your answer to the nearest whole number of time constants, converted into seconds.)

4 A casting is removed from its mould and allowed to cool in a current of air of temperature 20°C. The surface temperature of the casting is taken every five minutes by means of a radiation pyrometer and the following readings are obtained:

| Time (minutes) | 0 | 5 | 10 | 15 | 20 | 25 | 30 |
|---|---|---|---|---|---|---|---|
| Temperature (°C) | 420 | 344 | 288 | 240 | 200 | 168 | 140 |

Plot a graph of these results and determine the time constant of the cooling process. Hence determine how long it will take for the casting to cool to a surface temperature of 30°C.

5 A type N (Nicrosil/Nisil) thermocouple system gave a reading for $V_1 - V_2$ of 14.68 mV. Determine the corresponding temperature if the reference junction was at a) −15°C, b) 0°C, c) 40.5°C. (Take values from Figure 2.8.)

6 The temperature at the temperature-measuring junction of a type N (Nicrosil/Nisil) thermocouple system is 464°C. What is the output EMF of the system if the reference junction is at a) 0°C, b) −11°C, c) 43°C? (Take values from Figure 2.8.)

7 A thermistor has a resistance of 14.56 kΩ at 25°C and a characteristic temperature of 3740 K. Calculate its resistance at a) 0°C, b) 100°C.

8 In a test to determine the characteristic constants of a thermistor, a constant current of 20 μA was passed through it and the voltage drop across it was measured while it was immersed in water at various temperatures. The following results were obtained:

| Temperature (°C) | 22 | 26 | 32 | 40 | 51 | 67 | 82 | 100 |
|---|---|---|---|---|---|---|---|---|
| Voltage drop (mV) | 89.8 | 77.0 | 61.6 | 46.4 | 32.2 | 19.7 | 13.0 | 8.2 |

a) Plot thermistor resistance against temperature, assuming that any self-heating effect was negligible.

b) Calculate the characteristic resistance $A$ and the characteristic temperature $B$ in the equation $R = Ae^{B/T}$ by substituting into the equation the co-ordinates of two points on the curve, to obtain a pair of simultaneous equations. (Note: $T$ is absolute temperature.)

9 What is a 'positive temperature coefficient thermistor'? Give two different uses for this type of device, explaining, in each case, what function the thermistor performs.

10 a) Describe, with the aid of a sketch, a simple form of radiation pyrometer, indicating the essential features.

b) State three possible reasons why a radiation pyrometer might give a low reading when it is used to take the temperature of an object. In each case explain how the error may be reduced or eliminated.

11 a) Explain the difference between a primary calibration and a secondary calibration of an instrument.

b) Explain how a primary calibration of a radiation pyrometer would be carried out.

12 a) Explain, with the aid of a diagram, the principle of operation of a liquid-in-metal thermometer.

b) Draw a diagram to show how a vapour pressure thermometer differs from a liquid-in-metal thermometer.

c) Why in some circumstances is a vapour pressure thermometer likely to be the more accurate of the two?

d) List three *dis*advantages of a vapour pressure thermometer in comparison with a liquid-in-metal thermometer.

# 3
# Displacement transducers

## Mechanical devices

These usually consist of various combinations of rack-and-pinion, gear train, cable-and-drum, and lever. Their purpose is to amplify small displacements. The dial gauge is an example: a small displacement of the plunger is amplified mechanically, by means of a rack-and-pinion and gear train, to give a large displacement of a pointer around a dial. Descriptions of this and many similar examples will be found in textbooks on metrology. Such devices have the advantage that they are positive in action – that is, the amplification factor (the *gain*), is set by the design of the instrument, and cannot vary. They have some serious disadvantages however:

a) The input displacement must be applied with sufficient force to overcome friction in the mechanism, and this force may distort the displacement being measured.
b) The inertia of the components of the mechanism is magnified by the gain of the mechanism, so very large input forces are needed if the device is accurately to follow rapidly varying displacements.
c) Purely mechanical devices are not remote reading. To enable their output to be transmitted over any but the shortest distances some kind of secondary transducer must be used to convert their mechanical output into an electrical signal.

## The potentiometer

The potentiometer (often shortened to *pot*) is shown in principle in Figure 3.1. A resistance element (the *track*) is connected across a source of voltage, and a wiper, which can be

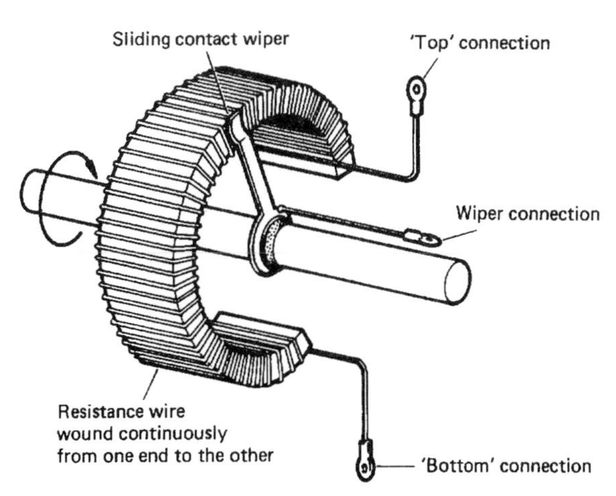


**Figure 3.1** Potentiometer

positioned anywhere along the track, picks off a corresponding proportion of the source voltage.

Figure 3.1 shows a *wire-wound* potentiometer. Wirewound potentiometers are robust and wear-resistant, but their resolution is limited because their output voltage increases or decreases in steps corresponding to the successive turns of wire contacted by the wiper.

Alternative types of resistance element are *carbon track*, *cermet*, and *conductive plastic*. **Carbon track** is the cheapest and most common type of element. **Cermet** elements are made by coating a ceramic base with a metal oxide resistance track. They have excellent electrical and thermal stability, wear resistance and low 'noise'. **Conductive plastic** elements can be moulded and trimmed so that the relationship between wiper angle and resistance is linear to within very close tolerances, so that the potentiometer can be used as a position transducer in electrical servos.

Some resistance elements are made deliberately non-linear, to cancel out an opposite non-linearity in the remainder of the circuit. A common example is the *logarithmic track* or *log* potentiometer, in which the resistance of the element is so arranged that, at any point, the logarithm of the resistance from one end of the track is proportional to the angle turned by the wiper. Potentiometers of this type are used as volume controls in radio receivers where they counteract the fact that at the 'low' end of the track, a small increment of resistance gives a much greater percentage increase in volume than a similar increment at the 'high' end of the track.

Other types of potentiometer are the *slide potentiometer*, and the *multiturn potentiometer*. In the **slide potentiometer**, the resistance element is straight, the wiper travelling in a straight line along its length instead of being rotated by a spindle. In the **multiturn potentiometer**, resistance wire is wound on a helical former of (usually) ten turns. The wiper is made to travel along a helical path with the same pitch as the resistance element, so that it is always in contact with the resistance. Thus the input range of this type of potentiometer is increased from about three-quarters of a revolution to ten (or whatever) revolutions.

Whatever the design of a potentiometer the input displacement of the wiper has to be limited by the stops at either end of the track. So potentiometers cannot be used as displacement transducers where the displacement may continue indefinitely. In fact the usable input displacement is further limited by a short length of zero-resistance track at each end, where the electrical connections are made. The specification of a potentiometer therefore usually states a *mechanical* rotation angle and a slightly smaller *electrical* rotation angle.

*Example 3.1*

A 1 watt wirewound potentiometer to be used as a displacement transducer has a resistance of 500 Ω, formed by 150 turns of wire. It has a mechanical rotation of 285° and an electrical rotation of 265°. It is to be connected to a 20 V power supply:

a) check that its power consumption rating is not exceeded
b) determine
   i) its resolution
   ii) its transfer function
   iii) its output voltage (assuming negligible current is drawn from the wiper) when the input spindle has rotated 120° from the stop at the 0 V end of the track.

*Solution*

a) Power consumption =

$$\frac{V^2}{R} = \frac{20^2}{500} = 0.8 \text{ W}$$

This is (just) acceptable though it might be advisable to substitute a potentiometer with a higher power rating if it is to be used in high temperature surroundings.

b) i) *Resolution* is the greatest change of input value which can occur without change of output.

$$\therefore \text{resolution of transducer} = \frac{265°}{150} = 1.767°$$

ii)

$$\text{Transfer function} = \frac{\text{output change}}{\text{corresponding input change}}$$

$$= \frac{20}{265} = 0.0755 \frac{V}{\text{degree}}$$

iii) Angle of 'dead band' at each end $= \dfrac{285 - 265}{2} = 10°$

$\therefore$ wiper is positioned 120° – 10° = 110° from 0 V end of track.
$\therefore$ output voltage = 110 × 0.0755 = 8.30 V.

## Potentiometer loading

It was assumed in the above example that negligible current was drawn from the wiper of the potentiometer. But what is negligible current, and what happens if we exceed it?

To measure or to make use of the output voltage of a potentiometer we must apply a voltmeter to it or connect it to the input of an amplifier. Whatever we connect to it (the *load*) will have a resistance, so we have the circuit shown in Figure 3.2. In Figure 3.2, $R_L$ is the resistance of the voltmeter or the input resistance of the amplifier. The portion of potentiometer

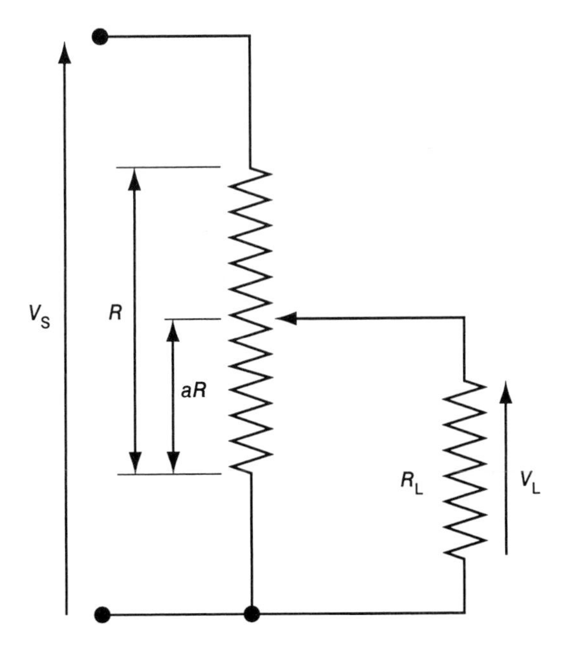


**Figure 3.2** Potentiometer with output load $R_L$

resistance between the wiper and its lower end will have some value $aR$, where $R$ is the total resistance of the potentiometer and $a$ is a number between 0 and 1. Then $aR$ and $R_L$ are in parallel, and their effective resistance is

$$R_e = \frac{aR \times R_L}{aR + R_L}$$

The portion of potentiometer resistance above the wiper has resistance $(1 - a)R$, and so the voltage applied to the load is

$$V_L = \frac{R_e}{R_e + (1 - a)R} \times V_S$$

Substituting for $R_e$ gives

$$V_L = \frac{a}{1 + a(1 - a)(R/R_L)} \times V_S$$

The curves in Figure 3.3 show how the output voltage $V_L$ deviates from the linear relationship $V_L = aV_S$ as the ratio $R/R_L$ is increased from 0 to 1 (that is, as $R_L$ is decreased from infinity).

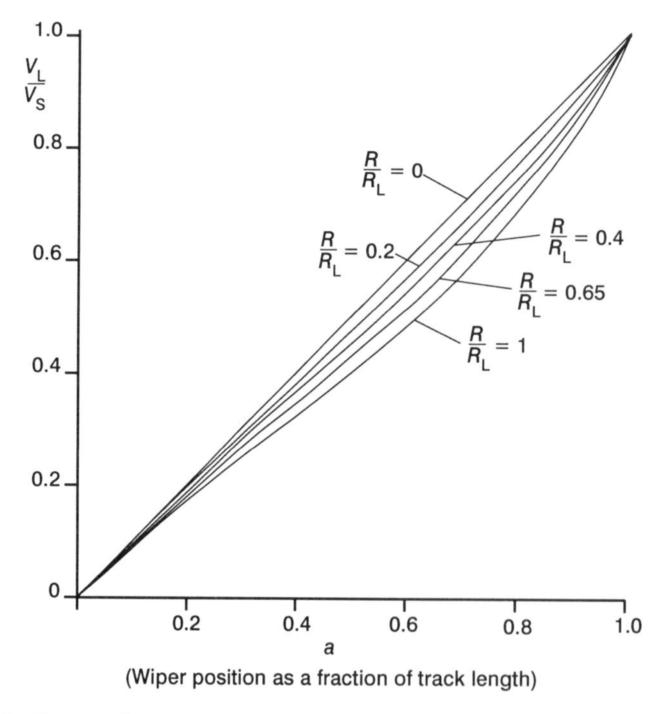


**Figure 3.3** Curves of potentiometer output for various resistance ratios of potentiometer/load

*Example 3.2*

A displacement measurement system is to be made up of a 1 kΩ wirewound potentiometer with its output connected to a moving coil voltmeter. The voltmeter has a range of 0 to 15 V and a resistance stated as 1 kΩ/V. The power supply is 15 V and the potentiometer is rated at 1.5 W.

a) Check that the power rating of the potentiometer will not be exceeded.

b) Calculate the percentage error in the output of the potentiometer at mid-range due to potentiometer loading.
c) Suggest how the accuracy of the system could be improved.

*Solution*

a) Power dissipated in the potentiometer =

$$\frac{V^2}{R} = \frac{15^2}{1000} = 0.225 \text{ W}$$

which is well within the rating of 1.5 W.

b) We could use the final equation above, but it is quite easy to work from first principles: 1 kW/V on a range of 0 to 15 V gives a voltmeter resistance of $1000 \times 15 = 15\,000\ \Omega$. Referring to Figure 3.2, at mid-range of the potentiometer the wiper divides the potentiometer resistance into two equal portions of 500 Ω each. The voltmeter resistance is in parallel with the lower portion. The effective resistance of 500 Ω and 15 000 Ω in parallel is:

$$\frac{500 \times 15\ 000}{500 + 15\ 000} = 483.9\ \Omega$$

Therefore the output voltage of the potentiometer at mid-range is:

$$\frac{483.9}{500 + 483.9} \times 15 = 7.377\ V$$

At mid-range, the output voltage of the potentiometer would be 7.5 V if the system were linear. Therefore the percentage error is:

$$\frac{7.5 - 7.377}{7.5} \times 100 = 1.64\%$$

c) The following improvements could be made:
   1 Instead of a wirewound potentiometer, use one with a conductive plastic track. This would have infinite resolution, and the linearity of its resistance element would be better.
   2 Instead of a moving coil voltmeter use a digital voltmeter. This would have an input resistance of (typically) 20 MΩ, giving negligible non-linearity when used as the load on the potentiometer output, and its accuracy and resolution would be much better than that of an analogue meter.

## Potentiometer supply voltage

This must be kept constant within very close limits if the potentiometer is to be used for displacement measurements. Any variation will cause a corresponding variation in the output voltage and thus will cause an error in the output signal.

## Potentiometer noise

In time the resistance track and the wiper of a potentiometer acquire a coating of dirt and corrosion products. This introduces a varying resistance between the wiper and the resistance track, causing electrical 'noise' in the signal from the potentiometer, as the wiper travels along the track. It can sometimes be heard as audible noise when the volume control of an elderly radio is adjusted. The wiper and track should clean themselves to some extent as the wiper rubs along the track, so the effect is less apparent in potentiometers in frequent movement.

Noise can be reduced at the design stage by:

1 sealing the potentiometer casing against dust and moisture
2 using a multiple-contact wiper, or using a solid carbon brush as the wiper.

## The linear variable differential transformer (the LVDT)

This is a transducer which, when the signal has been processed, gives a DC voltage proportional to displacement in a straight line. Thus its function is the same as that of a slide potentiometer but as there is no contact between fixed and moving parts, there is no friction to be overcome and negligible electrical noise in the signal.

The transducer itself is shown in Figure 3.4. The fixed part is a transformer consisting of three coils: a primary winding and two matched secondary windings, symmetrical about the primary. The primary is supplied with AC of constant amplitude and constant frequency from an oscillator – a typical frequency would be 8 kHz. The core is a cylinder of ferromagnetic material (e.g. soft iron) which strengthens the alternating magnetic field of the primary, increasing the amplitudes of the AC voltages picked up by the secondaries. The input displacement is applied as an axial displacement of the core, by means of a rod of non-magnetic material. With the core in mid-position in the transformer as shown, the secondaries pick up AC voltages which are equal in magnitude but opposite in phase, so they cancel out. Displacement of the core towards one secondary causes an increase in the amplitude of the AC in that winding and a corresponding decrease in amplitude in the other winding. Because the secondaries are connected so that their AC voltages are added algebraically, the output is an AC voltage with amplitude proportional to the displacement, and either *like* or of *opposite* phase to the AC in the primary, depending on whether the displacement is in one direction or the other from the mid-position of the core. Thus as the core traverses through the mid-position, the AC output decreases to zero, then increases from zero again but with a phase change of 180°.

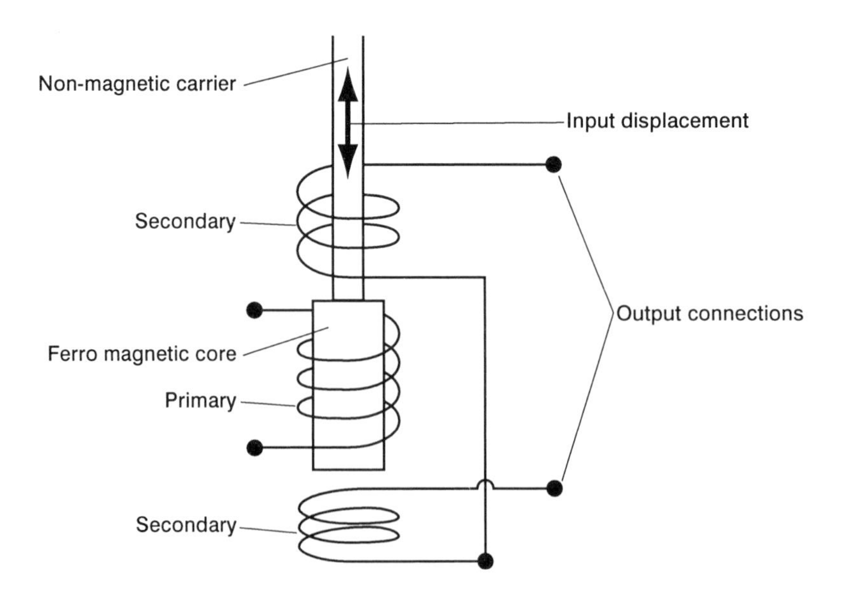


**Figure 3.4** Diagram of the LVDT transducer

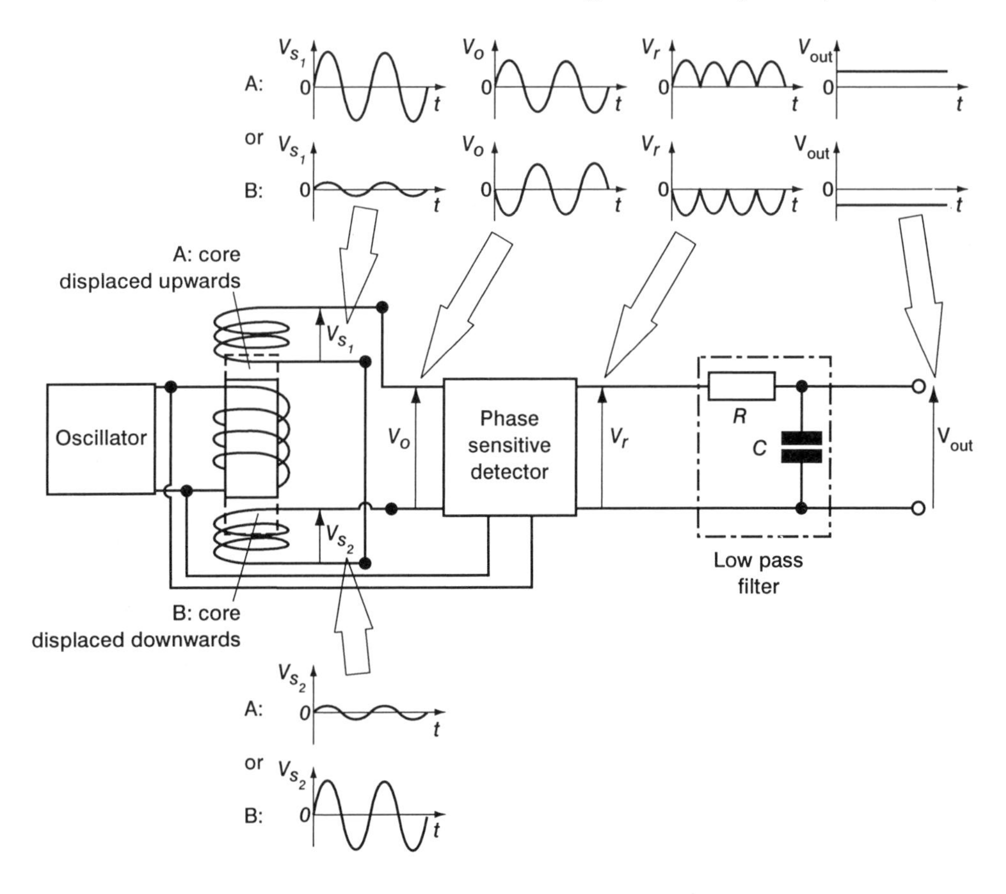


**Figure 3.5** A complete LVDT displacement measurement system

The LVDT output is passed through a *phase-sensitive detector* which converts it into full-wave rectified (but unsmoothed) DC of either positive or negative polarity, depending on whether it is in phase with or antiphase to the output of the oscillator. This DC is then smoothed by passing it through a low-pass filter. Thus a complete LVDT displacement measurement system is as shown in Figure 3.5.

## LVDT limitations

1. There is a practical limit to the dimensions of the core and windings in an LVDT hence the range of displacement which can be measured is also limited. A typical displacement range would be ± 15 mm.
2. There is a displacement frequency limit imposed by the smoothing circuit. Thus high-frequency mechanical vibrations may get smoothed out of the output signal and disappear.
3. No matter how carefully the two secondaries are matched during the winding of the coils there will tend to be a slight difference of capacitance effects when the inductance effects are in balance, and vice versa. This causes the phase difference between the AC voltages in the secondaries to slightly differ from 180°, which has the effect of obscuring the signal with noise as the core passes through the mid-position. The effect is significant only when the core is very close to mid-position, as the 'noise voltage' is usually less than 1% of the maximum output voltage of the transducer.

### Self-contained LVDTs

LVDT transducers may incorporate oscillator, phase-sensitive detector, and output filter as an integrated circuit within the transducer casing. The external connections to the casing may be reduced to two wires for the power supply and either one or two wires for the DC output voltage.

## Variable inductance transducers

The simplest form of inductive displacement transducer is shown in Figure 3.6. It consists of a coil mounted on an E-shaped iron core. The coil is supplied with AC from a constant voltage source and an AC current meter shows the current passing through the coil. The displacement alters the air gap between the end of the E-shaped core and a mild steel plate. The alternating current through the coil sets up an alternating magnetic field, indicated by the dotted lines representing the magnetic flux path. The *reluctance* of the magnetic flux to pass through air is about a thousand times greater than its reluctance to pass through iron, so a small reduction in the air gap causes a large increase in the magnetic field strength in the flux path. Higher magnetic field strength along the axis of the coil makes it more difficult for the current through the coil to change rapidly, thus increasing its resistance to the passage of AC (that is, increasing its *impedance*). Thus a reduction in the air gap causes a reduction in the reading on the AC meter. Similarly, an increase in the air gap causes an increase in the meter reading. The change in meter reading is proportional to the change in air gap over a limited range of gap (usually about 5 mm).

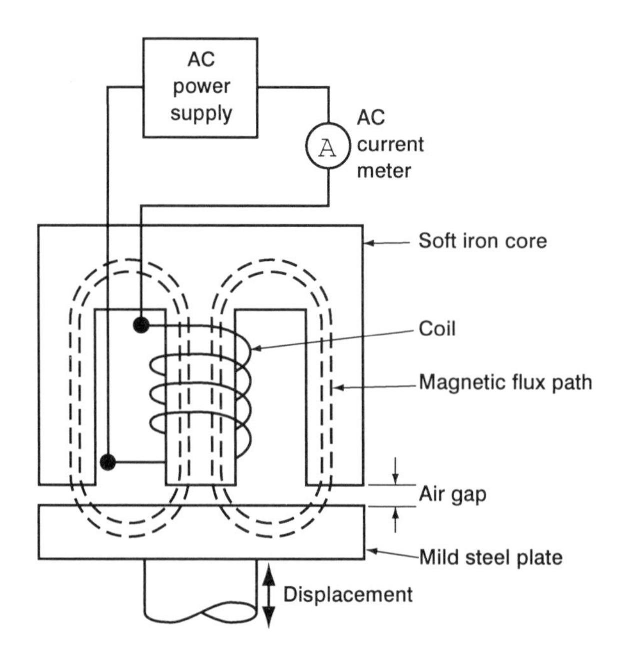


**Figure 3.6** An inductive displacement transducer

## Inductance bridges

The Wheatstone bridge circuit works on the principle that instead of trying to measure a small change in a large quantity, we connect it so that it is opposed by an almost equally large quantity, and just measure the difference between the two quantities. Strain gauge circuits, resistance thermometer circuits and even the LVDT, make use of this principle, and it is equally applicable to inductance transducers if they are used in an AC bridge circuit.

Figure 3.7 shows the inductance bridge circuit, and Figure 3.8 shows it applied to the measurement of pressure. Usually, the AC power supply has a frequency of 1 to 10 kHz. The pressure transducer is shown measuring a pressure difference, $P_1 - P_2$. (To measure a gauge pressure, the other pressure input is left open to the atmosphere.) The pressure difference causes the steel diaphragm to bulge slightly towards one of the coils (increasing its impedance to AC) and away from the other coil (reducing its impedance).
The output voltage of the bridge is:

(voltage at A) – (voltage at B)

and the voltage at B is virtually zero, since it is at the mid-point of the resistance chain $R + R$ across the AC supply.

The voltage at A will be AC with an amplitude proportional to the difference in impedance between $L_2$ and $L_1$ (and hence proportional to the pressure difference $P_1 - P_2$). It will be either

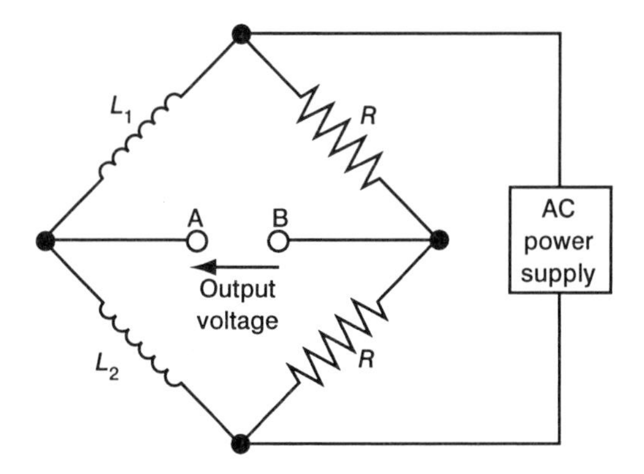


**Figure 3.7** An AC bridge circuit for inductive transducers

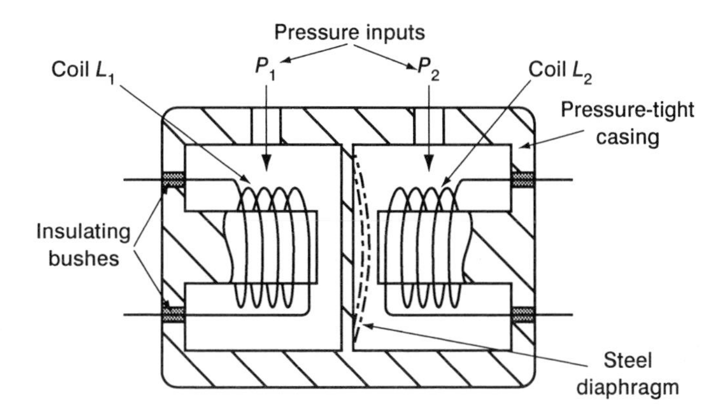


**Figure 3.8** Diagram of an inductance-bridge pressure transducer (shown in section)

in phase with, or of opposite phase to the supply, depending on whether $P_1 > P_2$ or $P_1 < P_2$.

If $P_1 - P_2$ is known to be always positive, the output voltage can be displayed on an AC voltmeter graduated in units of pressure, but if $P_1$ may on occasion be less than $P_2$, the output must be taken through a phase-sensitive detector and filter, as with the LVDT.

## Capacitive transducers

Capacitance occurs where two electrical conductors are separated by an insulator. The conductors are usually in the form of parallel 'plates', and the insulator, which may be air, empty space or any non-conductive material, is called the *dielectric*. Capacitance can be calculated in farads (abbreviation: F) from the formula:

$$C = \frac{\varepsilon_o \varepsilon_r A}{d}$$

where

$C$ is the capacitance in farads,
$\varepsilon_o$ is the *permittivity of free space* (= $8.85 \times 10^{-12}$ F/m),
$\varepsilon_r$ is the *relative permittivity* of the dielectric (it is also called the *dielectric constant*). This is a multiplying factor to take account of the increase in capacitance caused by the actual material of the dielectric.
$A$ is the area of overlap of the plates,
$d$ is the thickness of the dielectric.

From the above formula, it can be seen that capacitance is

i) proportional to the area of overlap of the plates
ii) inversely proportional to the thickness of the dielectric
iii) dependent on the material of the dielectric.

Any of these three relationships may form the basis of a capacitive transducer.

A capacitor prevents the passage of DC (except for a slight leakage current in electrolytic capacitors, to maintain the dielectric), but AC passes through by continuously repeating the cycle of:

charge → discharge → charge with opposite polarity → discharge→

A capacitor has a resistance to AC (a *reactance*) given by the formula:

$$X_c = \frac{1}{2\pi f C}$$

where

$X_c$ is the reactance in ohms,
$f$ is the frequency of the AC in hertz,
$C$ is the capacitance in farads.

From the formula it can be seen that reactance is inversely proportional to frequency. Thus small 'accidental' capacitances, for example between parallel wires which are close together, may cause significant leakage of AC at high frequencies.

The change of capacitance in a capacitive transducer is usually quite small, so it is normally used as part of an AC bridge circuit, as shown in Figure 3.9. Also, if long connecting leads are used, their self-capacitance may form quite a large proportion of the total capacitance in a circuit, so capacitive transducer manufacturers usually mount bridge and signal processing circuits as close to the transducer as possible. Phase-sensitive detection is used in the same way

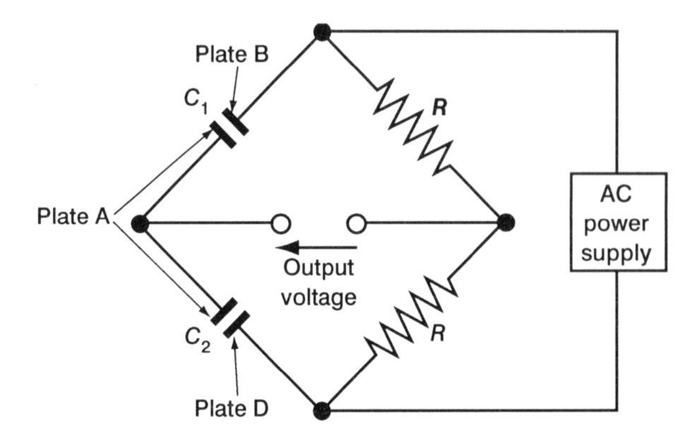


**Figure 3.9** An AC bridge circuit for capacitive transducers

as with the inductive bridge, and linearizing and temperature compensation circuits may also be included so that the output is a signal in the form of a DC voltage or current.

An example of a capacitive transducer is shown in Figure 3.10. This is a differential pressure transducer, fulfilling a similar function to that of the inductive pressure transducer in Figure 3.8. The two pressures, $P_1$ and $P_2$, are applied each to an isolating diaphragm at either end of the transducer body. The diaphragms deflect and transfer the pressures through silicone oil, which also acts as the dielectric, to the opposite faces of the sensing diaphragm. The sensing diaphragm is the central plate of a differential capacitor, corresponding to plates A in Figure 3.9. Plates B and D are fixed plates formed by a thin coating of metal on the surface of the solid insulating material which fills up most of the volume of the transducer body. Deflection of the sensing diaphragm towards one of the fixed plates reduces the thickness of dielectric on that side, increasing its capacitance, thus reducing its reactance, while the deflection away from the other fixed plate similarly increases the reactance on that side. The output of the bridge is an out-of-balance AC voltage proportional to $P_1 - P_2$, similar to that obtained from the inductive pressure transducer.

Another type of capacitive transducer is used to measure the contents of fuel tanks in aircraft. The 'plates' of the capacitor are two metal tubes, one inside the other, projecting from

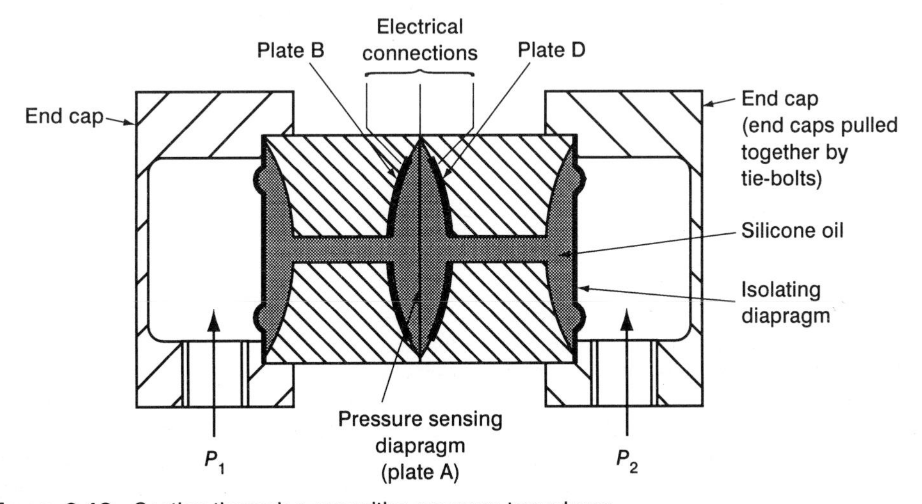


**Figure 3.10** Section through a capacitive pressure transducer

top to bottom, inside the tank. The annular space between the inner tube and the outer tube is filled partly by the fuel and partly by the air above the fuel. Since air and fuel have different relative permittivities, the capacitance depends on the quantity of fuel in the tank. The main advantage of this method of fuel measurement is that if the total capacitance is made up of a number of such pairs of tubes at suitable places in the tank, the reading is largely unaffected by the attitude of the aircraft in pitch or roll.

## Binary coded disc

This may be used to give a signal indicating the angle turned through by, for example, a weather vane, or it may be used to determine the linear displacement of an object such as the saddle of a lathe, by measuring the rotation of a threaded rod which positions it. The disc is read by sensors which convert the sectors of the disc into a voltage in the form of binary code; '0' representing 0 V and '1' usually 5 V. Figure 3.11 shows such a disc, converting one revolution into 16 parts identified by the binary sequence from 0000 to 1111.

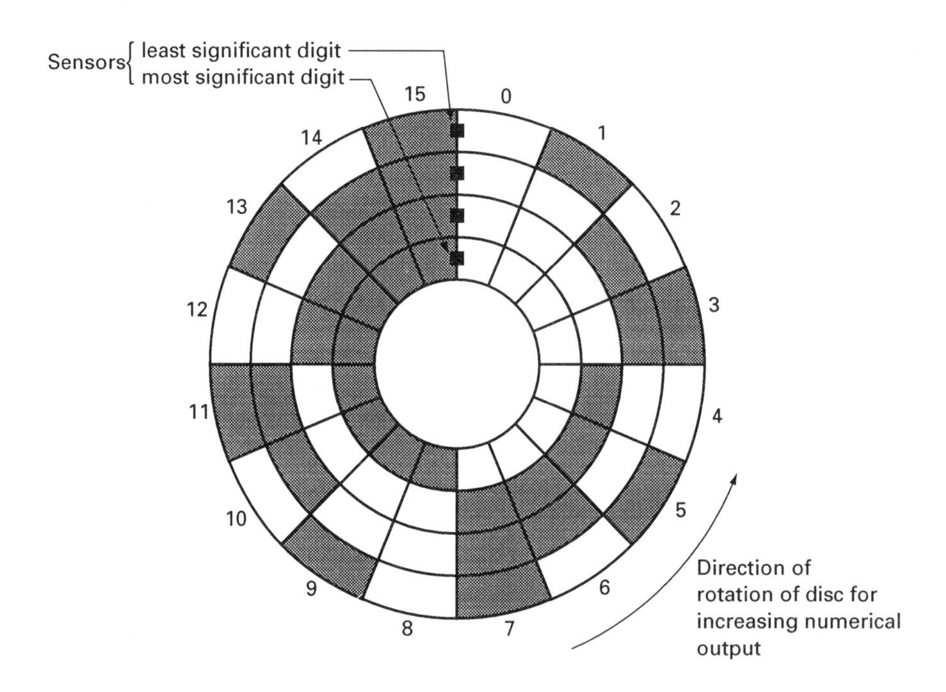


**Figure 3.11** Disc encoded with the natural binary sequence from 0 to 15 (0000 to 1111)

### Overcoming ambiguity

The trouble with the simple binary-coded disc of Figure 3.11 is that where more than one binary digit changes (for example in the position shown, between sector 15 and sector 0, where 1111 changes to 0000) the sensors do not necessarily record the change simultaneously. Thus the output at the instant shown might be, for example, 0110, indicating sector 6. One way of overcoming this is to add an anti-ambiguity track and an extra sensor, as shown in Figure 3.12. A '0' from this sensor prevents the computer from responding to any changes in the output of the other sensors, so output changes are only signalled when the disc is in positions where the sensors cannot make a mistake.

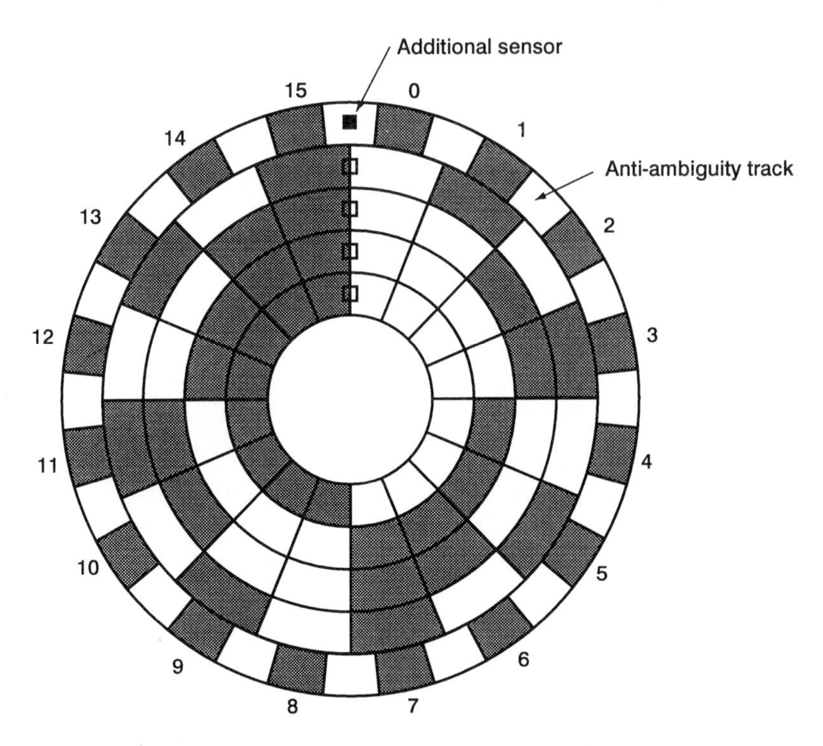


**Figure 3.12** The disc of Figure 3.11 with an anti-ambiguity track added

A better way of preventing ambiguity is by means of a Gray cyclic coded pattern on the disc, as shown in Figure 3.13, instead of the natural binary pattern. The principle of a Gray cyclic code is that only one digit at a time changes, so that the only possible error is a slightly early or slightly late change of output. Of course the sequence of binary output values is not the same

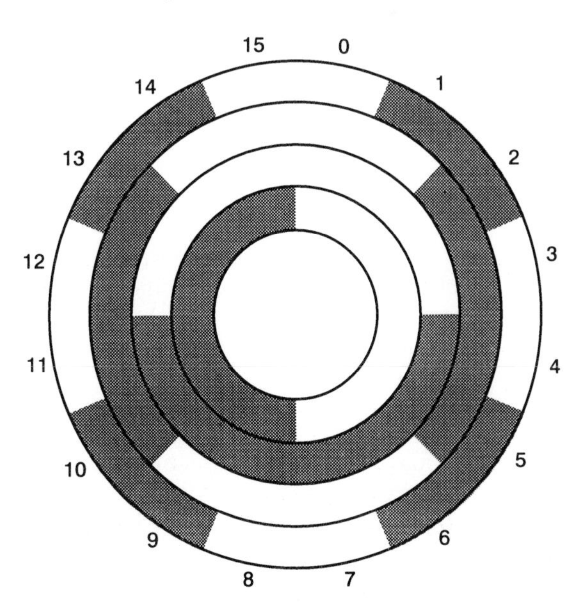


**Figure 3.13** Disc encoded with a simple Gray cyclic code for the numbers 0 to 15

as the natural sequence from 0000 to 1111, but it can be converted to that sequence by logic circuits or by a small subroutine in the computer program.

The discs shown in Figures 3.11 to 3.13 measure an angle of rotation to the nearest 1/16 of a revolution, that is, to the nearest 22.5°. Each disc has four tracks, because $16 = 2^4$. If we wanted to measure rotation with, for example, an error not greater than one degree, the disc would need to have nine tracks (because $2^8 < 360 < 2^9$).

### Sensors

For severe environmental conditions such as high temperature or severe mechanical shock, brushes may be used, contacting on a disc in which the 'dark' sectors are of electrically conducting material and the 'light' sectors are insulating. Alternatively, optical sensors may be used. The dark and light sectors may be formed on a transparent disc, through which beams of infra-red light may be projected on to phototransistors. Otherwise the disc may be opaque, with each track on the disc reflecting infra-red light to a phototransistor housed in the same package as the light source, as shown in Figure 3.14.

A *phototransistor* is a special form of transistor in which light falling on the base induces a base current proportional to the intensity of illumination. The transistor itself amplifies this base current in the usual way.

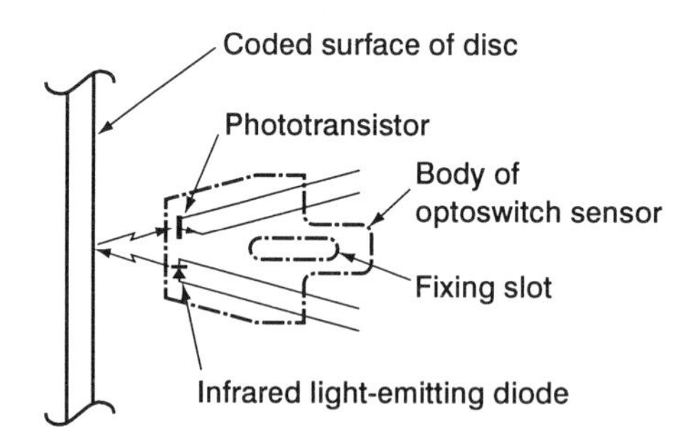


**Figure 3.14** A reflecting opto-switch sensor

The output of the sensors goes through signal conditioning circuits to produce the correct voltage levels corresponding to logic 1 and logic 0, and to give a sharper change-over from one logic level to the other. The output then goes to logic circuits or to the controlling computer to be counted up (or down) so that the displacement of a body positioned by a drive screw at any instant can be determined from the number of complete revolutions made by the drive screw plus the odd fraction of a revolution.

Whichever type of system is used, sensor and disc must be totally enclosed in a housing which gives them complete protection from the environment in which they have to work.

## Exercises on chapter 3

1 State the disadvantages of using mechanical displacement transducers. What are the main advantages?
2 Potentiometers may be classified by three alternative 'shapes', four types of track material, and two ways in which resistance may be distributed between one end of the track and the other. What are these classifications, and what is the main use of each type?
3 Sketch the internal construction of a rotary potentiometer indicating the wiper and showing its electrical connection in correct relationship to the other two connections.

4 A wirewound potentiometer to be used as a position transducer has the following specification:

Resistance: 250 Ω
Power rating: 3 W
Electrical rotation: 270°
Mechanical rotation: 300°

The resistance element consists of 120 turns of wire.
Calculate:
a) the resolution of the potentiometer,
b) the maximum allowable power supply voltage.
If a voltage of 24.0 V is applied across the ends of the track, calculate:
c) the transfer function
d) the output voltage when the wiper is 170° from the stop at the 0 V end of the track:
i) when negligible current is drawn from the wiper,
ii) when there is a 1 kΩ load between the wiper and 0 V.

5 a) Explain the operating principle of a differential transformer (LVDT) type of transducer.
b) How is the signal from the transducer converted into a proportional DC voltage so that its polarity indicates the direction of displacement of the core from the mid-position?
c) Explain how this signal processing circuit may be simplified if the core can only move in one direction from the mid-position.

6 a) Draw a diagram of a simple inductive displacement transducer with an ammeter to indicate the displacement measurement.
b) Will the meter show an increase or a decrease of current if displacement increases the air gap? Give the reasoning which determines your answer.
c) What changes should be made to the circuit if a voltmeter is to be used as the display device instead of an ammeter?

7 a) Sketch a section through an inductive transducer for measuring the difference between two pressures. Show the coils of the transducer connected into an inductance bridge.
b) Name the other circuits which would be needed between the bridge output terminals and a centre-zero voltmeter to display positive and negative pressure differences.

8 A pressure transducer has a diaphragm of 50 mm diameter. The diaphragm acts as the middle plate in a differential capacitor, pressure deflecting it towards the plate on one side of it and away from the plate on the other side. As an approximation the outer plates can be taken to have the same area as the diaphragm and to be spaced 1 mm away from it. The space between the plates is filled with a liquid having a relative permittivity of 2.2. Calculate:
a) the capacitance between the diaphragm and one of the outer plates,
b) the reactance due to that capacitance, at an oscillator frequency of 1000 Hz.
The permittivity of free space is $8.85 \times 10^{-12}$ F/m.

9 A capacitive transducer, for measuring the quantity of fuel in a tank, consists of two concentric tubes 300 mm long, one having an outside diameter of 20 mm and the other an inside diameter of 21 mm. Take the plates of the capacitor as each having an area equal to the mean of those two cylindrical areas and calculate, for an oscillator frequency of 1.5 kHz, the change in reactance as the fuel between the tubes is replaced by air. The relative permittivities of fuel and air are 2.2 and 1.0006 respectively. The permittivity of free space is $8.85 \times 10^{-12}$ F/m.

10 a) Explain the essential difference between a Gray coded disc and a disc coded with the natural binary sequence.
b) The spindle of the wind direction vane of an automatic weather station carries a binary-coded disc to convert wind direction into a binary signal. The specification requires wind direction to be measured with an error of not more than 5°. How many tracks are required on the binary coded disc to comply with this?

# 4
# Force, torque and pressure transducers

## Elastic sensors

These are components, usually made of steel, which are designed to deflect proportionately to the force, torque or pressure applied to them. This deflection may either be used to indicate directly the magnitude of the load, as in a spring balance, or else some side effect such as strain in the material may be used for this purpose, as in a strain-gauged load cell.

As far as possible, the material should be *perfectly elastic*; that is, there should not be the slightest tendency for any measurable deflection to remain when the load is removed. The material should also be free from *creep*. Creep is the slow plastic deformation of metals which occurs when they are under constant stress, especially at high temperatures.

Elastic force sensors include coil springs for the measurement of tensile and compressive forces, spiral springs and torque rods for the measurement of torque, and beams and cantilevers which measure force by bending. Another type of elastic sensor is the *proving ring*, in which the load is applied, and the corresponding deflection measured, along a diameter of the ring. The deflection may be measured by a dial gauge or converted into an electrical signal as in Figure 4.7 on p.58.

## Hydraulic load cell

Figure 4.1 shows the principle of this transducer. The force to be measured is applied to a diaphragm which forms one side of a capsule containing oil. The resulting pressure is shown

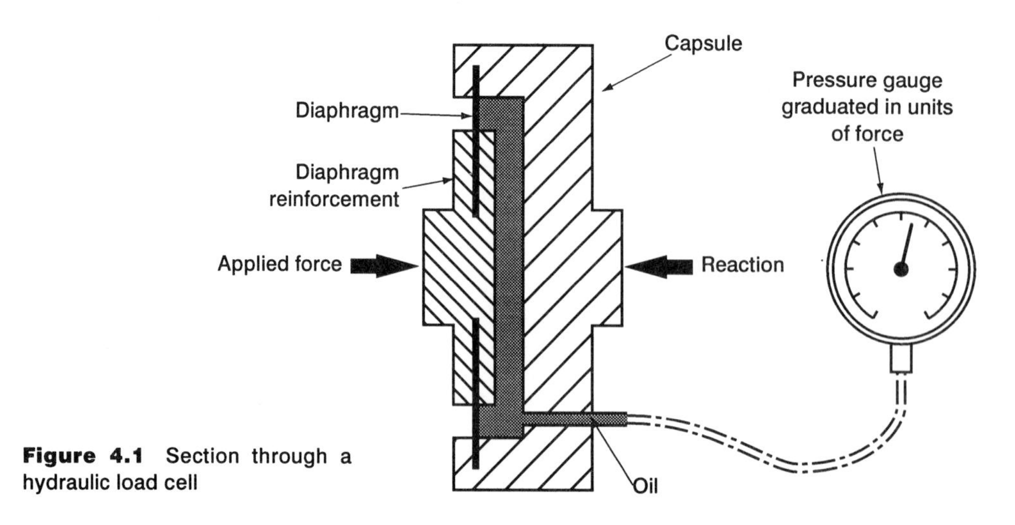


**Figure 4.1** Section through a hydraulic load cell

on a pressure gauge graduated in units of force. The capsule can be very thin, as its deflection under load is negligible, so it can often be inserted into a load path where other types of force transducer would take up too much room.

## Piezoelectricity

Figure 4.2 shows the principle of the piezoelectric transducer. If a crystal of quartz is cut so that its crystallographic $x$-axis is normal to the two faces it acquires electrostatic charge proportional to the load when it is loaded in compression and a corresponding voltage appears between the electrodes. It can thus be used as a force transducer.

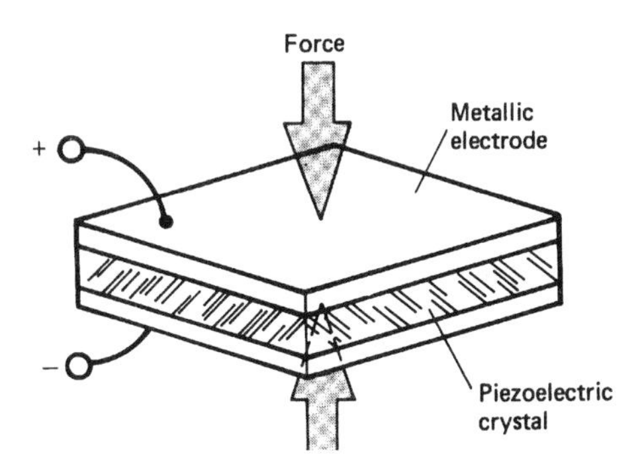


**Figure 4.2** The piezoelectric effect

**Quartz** is a naturally occurring mineral which has the advantage of very high electrical insulation, so that the electrostatic charge it acquires leaks away more slowly than it would in other piezoelectric materials. However, because it has to be machined it can only be produced in a limited number of configurations. Also, it cannot transmit as much power as synthetic piezoelectric materials when it is used as an active vibrating element. Other natural piezoelectric crystals such as tourmaline and Rochelle salt have similar limitations.

**Synthetic piezoelectric materials** are ceramics such as lead zirconate titanate. The material consists of crystals with their axes in random directions, so to make it piezoelectric it must be polarized by allowing it to cool slowly from a temperature above the Curie temperature to room temperature while a high DC voltage is applied by means of electrodes. It loses its piezoelectric property if it is reheated above its Curie temperature; this limits its use in high-temperature situations.

Although not much work is done when piezoelectric materials are compressed, hence not very much charge is produced, nevertheless the capacitance between the electrodes can be so small that quite a high voltage can result:

$$\text{voltage} = \frac{\text{charge}}{\text{capacitance}}$$

Piezoelectric material is therefore used in gas lighters, the voltage produced by compressing or striking it being high enough to flash over as a spark.

The piezoelectric effect can also be used 'in reverse' – that is, if a voltage is applied to the electrodes of a piezoelectric crystal it expands or contracts as if the corresponding force had been applied. Thus an AC voltage causes it to emit sound at the frequency of the AC; this may be in either the audio range at frequencies from about 750 Hz upwards, or in the ultrasonic range. Sound generators of this type, usually in the form of thin discs, provide the 'squeaks' in toys, watches, calculators and electronic games.

Quartz piezoelectric crystals are also used to stabilize the frequency of oscillators in radio transmitters, clocks and computers. Piezoelectric materials have a very high modulus of elasticity and an individual crystal can be cut to dimensions which will give it a particular natural frequency of mechanical vibration. This frequency is virtually constant. When the crystal is incorporated into an oscillator it maintains the frequency of the oscillator at the natural frequency of vibration of the crystal.

So besides being used as force transducers piezoelectric materials are also used as spark generators, sound and vibration generators, and as frequency stabilizers. They are also used as microphones, converting the pressures of sound waves into AC voltages.

## A piezoelectric force transducer

Figure 4.3 shows a practical form of piezoelectric force transducer; a Kistler-Swiss force link. The piezoelectric material is in the form of two quartz discs, one on either side of a metal disc which acts as one of the electrodes and which is connected to the central conductor of the coaxial cable socket. The outer faces of the quartz discs are earthed to the metal body and to the outer conductor of the coaxial socket. Because the quartz discs cannot be loaded in tension they are given a compressive preload by tightening the nuts against each other. Then tension is measured as a decrease in the preload on the discs, compression as an increase.

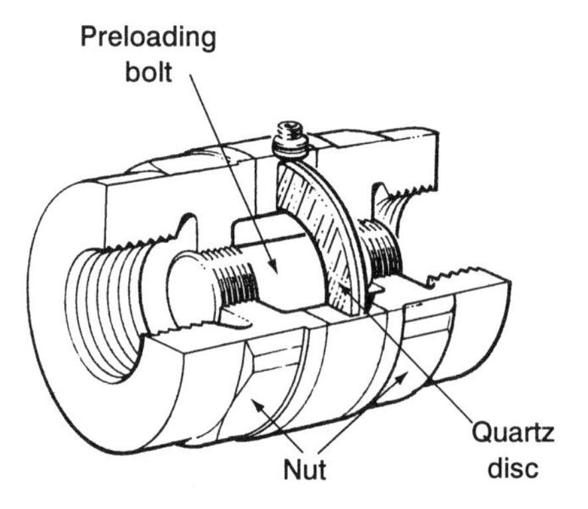


**Figure 4.3** A Kistler-Swiss type 9351 piezoelectric force link

## The step response of a piezoelectric transducer

When a load is applied to a piezoelectric force transducer, the tiny amount of work done in compressing the crystal becomes converted into electrostatic charge. This immediately begins to leak away at a rate proportional to the voltage between the electrodes. We thus have a similar situation to that of the thermometer described in Figure 2.3: a first order linear system, which will have a time constant. The only difference is that whereas the thermometer's output was rising, in this case the output is falling. So the exponential curve in Figure 2.3 will be inverted. It follows the law:

$$v = Ve^{-t/\tau}$$

where $V$ is the initial voltage, $t$ is the elapsed time after the step, and $\tau$ is the time constant.

*Example 4.1*

The following output voltages were obtained from a piezoelectric transducer connected to a storage oscilloscope, when a weight was suddenly lifted from the transducer.

| Time (s) | 0 | 0.2 | 0.4 | 0.6 | 0.8 | 1.0 | 1.40 | 1.80 | 2.20 |
|---|---|---|---|---|---|---|---|---|---|
| Voltage (V) | 16.2 | 10.8 | 7.2 | 4.8 | 3.2 | 2.2 | 0.98 | 0.44 | 0.20 |

Plot these results with the time axis horizontal, and hence determine the time constant of the transducer-oscilloscope combination.

*Solution*
From a graph of the results, a voltage of $(1 - 0.632) \times 16.2$ V is reached at 0.49 seconds from the step input. Therefore the time constant is 0.49 seconds. Readers might use this value in the above equation to calculate $V$ for one or two values of $t$, to check that the curve obeys the law.

The step input response depends on the following factors:

1 The initial voltage is determined by the sensitivity (transfer function) of the piezoelectric transducer in volts/newton. This is given by the sensitivity of the crystal divided by the capacitance

$$\frac{\cancel{\text{coloumb}}}{\text{newton}} \times \frac{\text{volts}}{\cancel{\text{coloumb}}} = \frac{\text{volts}}{\text{newton}}$$

Thus a long cable connecting the transducer to the 'voltmeter' can considerably reduce the output voltage because the capacitance between the conductors of the cable is in parallel with, and therefore adds to, the capacitance of the transducer.

2 The time constant depends on the resistance of the path joining the electrodes of the piezo transducer. To give the longest possible time constant, whatever is connected to the transducer should have the highest possible input resistance.

Because of these two considerations, a piezoelectric force transducer is normally connected to a special kind of amplifier, known as a *charge amplifier*.

## The charge amplifier

This is shown in principle in Figure 4.4. The piezoelectric crystal is represented by the capacitor $C$ which acquires a charge $q$ in proportion to the force applied to it. $C$ is connected to the inputs of an operational amplifier, and the inverting input is also connected to the output by a feedback capacitor $C_f$.

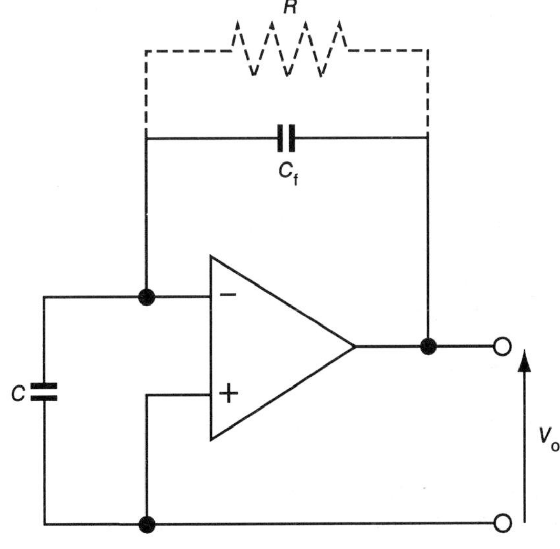


**Figure 4.4** The principle of the charge amplifier

If the amplifier's gain and input impedance are both very high (as they normally are in an operational amplifier), it can be shown that:

$$V_o = -\frac{q}{C_f}$$

This result means that the total capacitance of crystal plus cable is now irrelevant to the output voltage of the charge amplifier, which depends only on the value of the feedback capacitor. A charge amplifier usually includes a series of alternative values of $C_f$ which are switched into circuit by means of a range switch.

Figure 4.4 also shows a resistance $R$ in parallel with $C_f$. This is usually necessary to stabilise the amplifier, by providing a DC feedback path. With a high quality operational amplifier $R$ is usually in the order of 100 MΩ but it inevitably decreases the time constant of the decaying exponential output voltage. However, if very high quality components are used throughout, time constants of hundreds of seconds are possible.

*Example 4.2*

a) A piezoelectric force transducer has a sensitivity of 28 pC/N. It is connected to a charge amplifier which has a feedback capacitor of 22 nF. Calculate the output voltage of the charge amplifier at the instant when a step input of 5 kN is applied to the transducer.
b) The system of piezo transducer and charge amplifier has a time constant of 90 seconds. How long will it take to lose the first 5% of the output step?

*Solution*

a) The load of 5 kN charges the transducer with a charge of:

$$q = 28 \times 5000 = 140\,000 \text{ pC} = 140 \text{ nC}$$

$$V_o = -\frac{q}{C_f} = \frac{140 \times 10^{-9}}{22 \times 10^{-9}} = 6.36 \text{ V}$$

(The negative sign in the original equation merely means that the output voltage of the charge amplifier will be of opposite polarity to its input voltage.)

b) Figure 4.5 shows the exponential decay of the output voltage of the charge amplifier after the step.

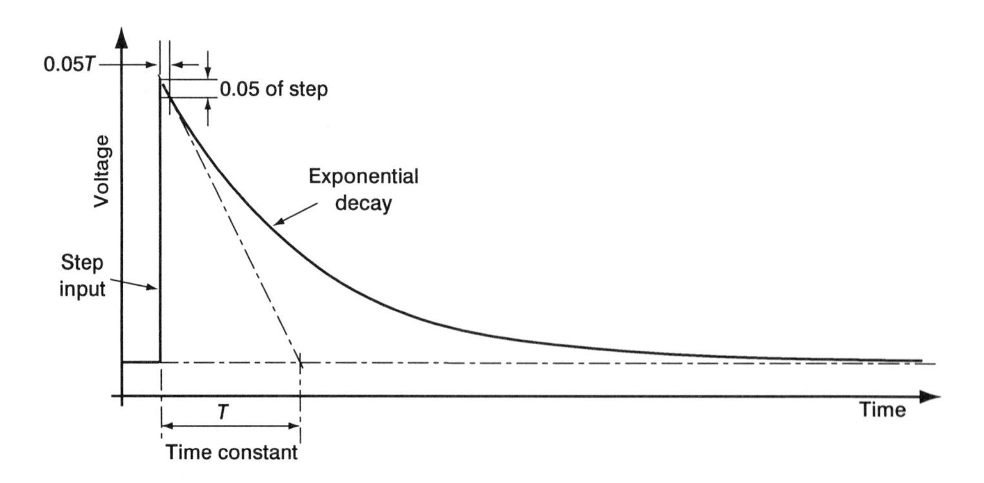


**Figure 4.5** Exponential decay of the output of a charge amplifier

By the approximation on p.20, 0.05 of the step will have been lost in a time of 0.05τ, that is, in 0.05 × 90 = 4.5 seconds. Or, more accurately, using

$$v = Ve^{-t/\tau}$$

$$0.95V = Ve^{-t/90}$$

$$\therefore \ln(0.95) = -\frac{t}{90}$$

$$\therefore t = -90 \times \ln(0.95) = 4.61 \text{ seconds}$$

## The electromagnetic force balance

This is used for measuring very small forces, of the order of thousandths of a newton. It is a form of automatic control system (a servo), the mechanical part of which is based on the principle of the moving-coil current meter. Figure 4.6 shows the complete force balance system. The moment of the applied force $P$ is opposed by the torque due to the current through the moving coil, so the coil and arm are kept in equilibrium. Any movement of the arm alters the ratio of the capacitances of the differential capacitor. This alters the output voltage from the capacitance bridge and phase-sensitive detector. This output voltage is applied to one input of an operational amplifier, changing its output current, which flows through the moving coil altering the torque it exerts and so restoring the arm to its original position.

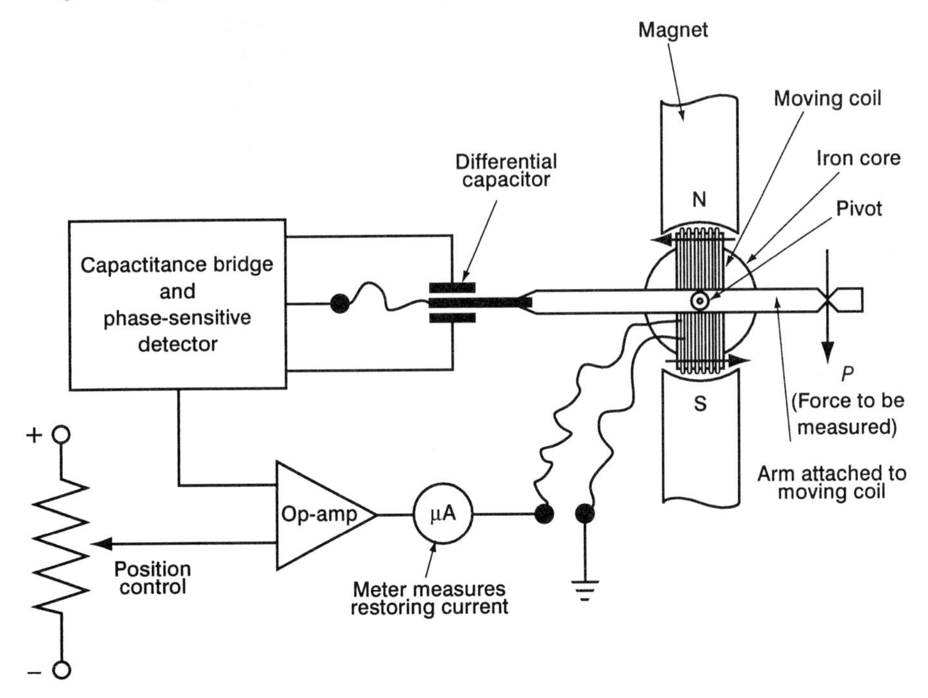


**Figure 4.6** The principle of the electromagnetic force balance

The displacement of the point of application of the force can be varied by varying the voltage applied to the other input of the operational amplifier, by means of a position control potentiometer. This enables the relationship between displacement and force to be plotted over short ranges of displacement, as in the determination of the surface tension of a liquid, for example.

## Combinations of two or more transducers

An example is the combination of a proving ring and an LVDT shown in Figure 4.7. The proving ring converts force into deflection along a diameter, and this is normally measured by a dial gauge. An LVDT may be used instead of the dial gauge, converting the deflection into an electrical signal suitable for use in an automatic control system.

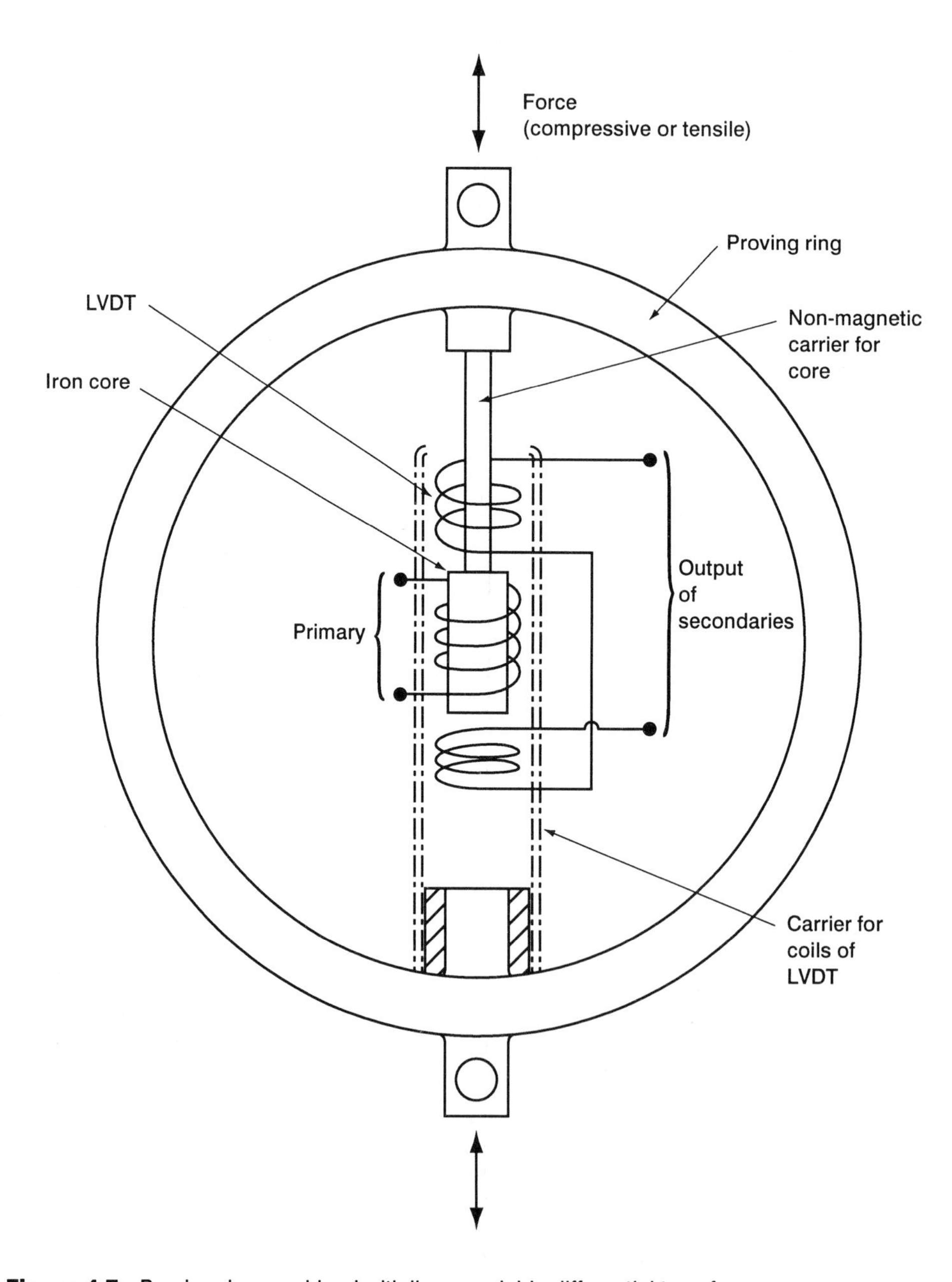


**Figure 4.7** Proving ring combined with linear variable differential transformer

## Torque measurement

Where torque measurement is required for determining the power exerted by a motor driving a machine, the simplest method is to measure the torque *reaction*, by mounting one side of the motor or machine on a hinge and the other side on a load cell. Then if $F_1$ is the load indicated by the load cell when the motor is driving the machine, and $F_2$ the corresponding load when it is at rest, the torque exerted is

$$(F_1 - F_2) \times \text{(moment arm from hinge to load cell)}$$

### Measurement of torque in a rotating shaft

The measurement of torque by means of strain gauges has already been dealt with in chapter 1. The difficulty of measuring torque in a rotating shaft is the difficulty of making electrical connections to strain gauges which are rotating. The usual method of making electrical connections to a rotating circuit is by slip rings and brushes, but if we use this method to connect to strain gauges, electrical 'noise' (caused by slight variations in the resistance between brush and slip ring as the shaft rotates) is liable to swamp the small changes in resistance of the strain gauges themselves. The signal/noise ratio can be improved by using special materials for the brushes and slip rings and by using high-resistance strain gauges so that brush/slip ring resistance variations are proportionately less. A better technique is to make up a complete four-gauge bridge circuit on the surface of the shaft and amplify the output voltage by means of an integrated circuit operational amplifier (see p.110) before passing it to the slip rings, so that electrical noise voltages will then be negligible in comparison with the amplified output signal. Three slip rings are needed: two for the power supply and one for the amplifier output voltage. All wiring, together with the integrated circuit chip, must of course be insulated from the shaft, and the arrangement must be designed to withstand the maximum centrifugal force expected. Conductors may be painted on to the insulated surface using a conducting paint, in which the conductor is finely divided silver. The chip may be held down by a winding of glass fibre tape.

Instead of slip rings and brushes, electrical connections to a rotating circuit may be made by *rotary transformer* or by *telemetry*. The circuit of the **rotary transformer** is shown in Figure 4.8. AC power is supplied from an oscillator to the strain gauge bridge by means of the

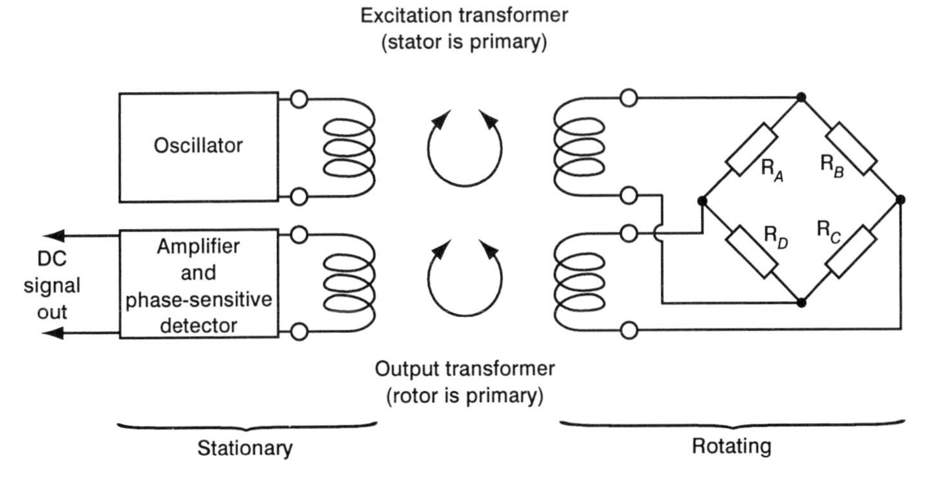


**Figure 4.8** Rotary transformer circuit for supplying power to strain gauges on a rotating shaft and receiving their output.

excitation transformer. The bridge output, in the form of an AC voltage, is passed to an amplifier and phase-sensitive detector by means of the output transformer. The rotating windings of the two transformers are wound on the shaft. The fixed windings are mounted inside a cylindrical metal casing concentric with the shaft; each is in the form of a coil surrounding its rotating winding with a small air gap to provide a clearance from it. Because the rotary transformer requires careful manufacture and setting up it may be supplied complete with a short strain-gauged shaft, as a *torque cell,* which can be inserted into a gap in existing shafting.

Rotary transformers are more suitable than slip rings for use in hostile environments since there are no sliding electrical contacts to become dirty and wear, so electrical noise is eliminated. In general, rotary transformers are better suited to permanent installations and slip rings to temporary measurements.

**Radio telemetry** is an alternative method of obtaining an output signal from a strain gauge bridge on a rotating shaft. In this system a small battery and a radio transmitter are strapped to the shaft, the battery supplying the power for both the bridge circuit and the transmitter. The transmitter need be no larger than the battery since it only has to transmit its signal a few millimetres from its aerial to a receiver mounted alongside. The transmitter modulates its carrier wave with the bridge output signal; the receiver demodulates the carrier wave to recover the signal. The system is calibrated before use by applying known values of torque to the shaft and measuring the corresponding receiver output voltages.

Unlike slip rings or the rotary transformer, radio telemetry permits the shaft to move freely in any direction relative to the receiver. It is less expensive than slip rings, has a higher signal-to-noise ratio, and requires only about 100 mm of unobstructed shaft length for its installation.

**Torque measurement from torsional deflection** makes use of the fact that one end of a shaft twists relative to the other in proportion to the torque applied. Figure 4.9 shows how the angle of twist may be measured over a short length of the shaft. The difficulty of reading a

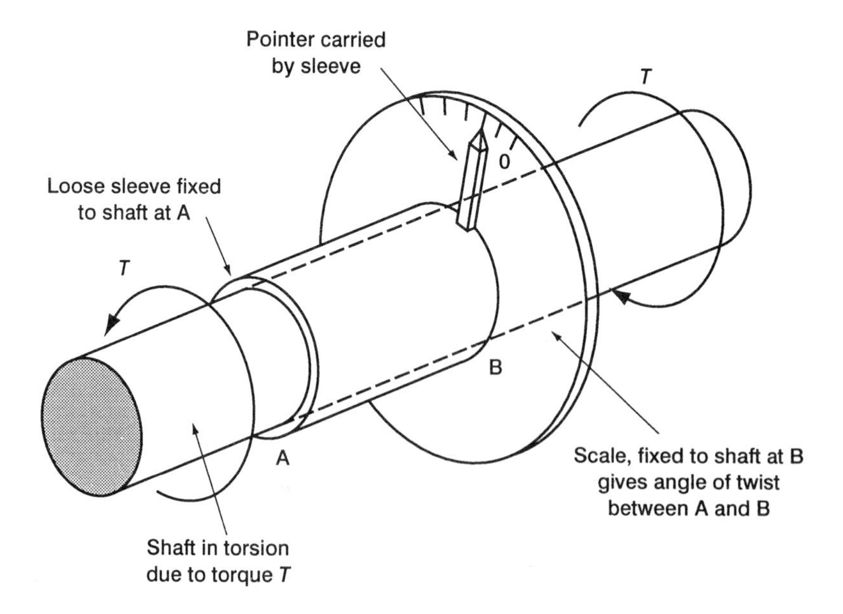


**Figure 4.9** Torque measurement from angle twist of shaft

rotating scale is overcome by illuminating it with a stroboscope flashing at the same frequency as the rotational speed of the shaft, so that the scale and pointer appear stationary.

Another method of using torsional deflection is shown in Figure 4.10. Two radially slotted discs are attached to the shaft, a small distance apart. A non-rotating light source and photodetector are used to measure the illumination received through the slots in the discs. The slots are exactly aligned when the shaft is under no torque. As torque is applied, the shaft twists, causing one disc to rotate slightly relative to the other. The light received by the photodetector is reduced in the same proportion as the torque applied. Provided that a photodetector with a short rise time is used (e.g. a phototransistor), its output when the slots pass through the beam of light is the same as when the shaft is stationary.

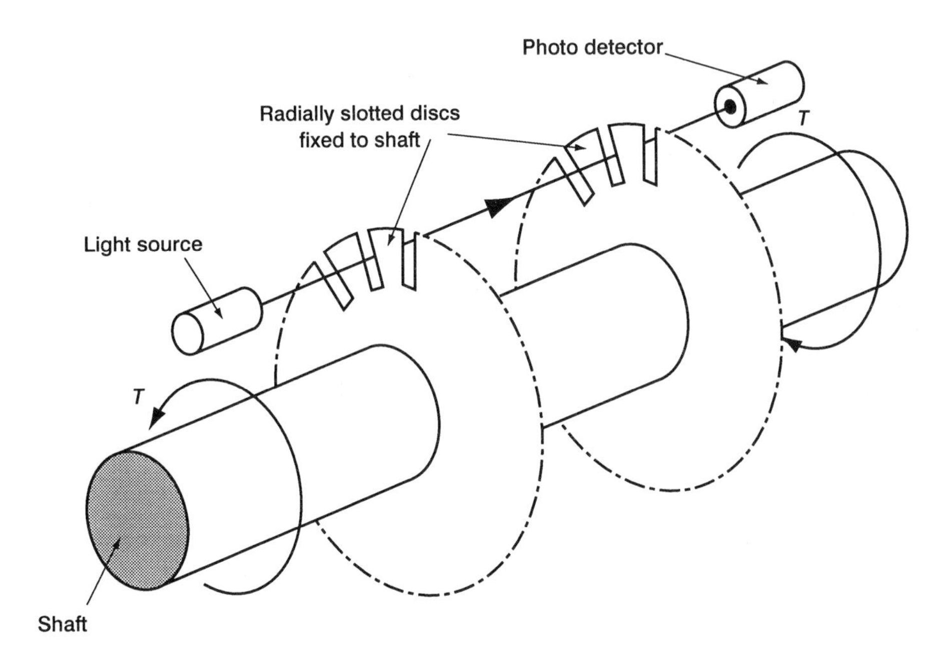


**Figure 4.10** Torque measurement by measuring the illumination transmitted through the slots of a pair of radially slotted discs

## Dynamometers

These devices apply a braking torque to an engine or motor to measure its torque output when it is undergoing a performance test. They absorb the full mechanical power output of the engine. In the simplest types of dynamometer the braking torque is applied to a brake drum by friction, using brake blocks or a rope, and measured by means of a spring balance or a pair of spring balances. The brake drum has to be cooled by a continuous inflow of water to dissipate the heat produced by the continuous mechanical energy input. Slight variations in frictional torque as the drum rotates cause spring balance readings to oscillate, making precise reading difficult, and any cooling water spin-off from the brake drum is a potential source of trouble for electronic instrumentation which may be present.

## The Froude hydraulic dynamometer

This is illustrated in Figure 4.11. It uses a continuous flow of water for the dual purpose of providing the braking torque and carrying away the heat which results from the energy conversion. The shaft of the dynamometer carries a rotor in the form of a thick disc with radial pockets in each face, shaped to throw the water in the direction of rotation. The rotor is enclosed in a watertight casing with similar radial pockets on the inside of the casing, shaped to receive the water from the rotor. Water is conducted to the casing through a flexible pipe, and is discharged from the casing to a drain by means of an outflow pipe restricted by a wheel valve. The casing is carried on trunnions, so that it is free to rotate through a small angle.

Centrifugal force acts on the water being carried round by the pockets of the rotor, causing it to flow outwards until it is forced on to the pockets on the inside of the casing. It flows down these and back on to the rotor in a continuous circulation. Its inertia as it leaves one set of pockets and strikes the other set applies a torque to the casing, which is measured by a spring balance, and applies an equal and opposite braking torque to the engine.

The outflow from the casing is restricted, to pressurise the water inside the casing and so prevent cavitation (bubbles formed when below-atmospheric pressures occur in eddies). The braking torque is controlled by two pairs of semicircular blanking plates which can move in the gap between the casing and the rotor. They are carried on threaded rods each of which has a right-hand and a left-hand thread, so that when the rods are rotated by the operation of a handwheel, the pairs of blanking plates either approach each other to blank off more of the rotor from the casing, and reduce the applied torque, or move apart to increase it.

The spring balance has a weight equal to its full-load reading suspended from it, and its dial is graduated so that it reads in reverse – that is it shows how much of the weight is being lifted by the torque from the engine. If, for example, the weight has a value of 200 N, the spring balance shows 0 when it is carrying all of the weight, and 200 N when the torque is lifting all of the weight and the spring balance is carrying none of it. This allows the range of the dynamometer to be extended by adding extra weights, each equal to the first weight, as required.

To enable the braking torque to be measured as accurately as possible, a 'dashpot' is fitted to damp out cyclic torque variations. Also the spring balance can be raised or lowered by means of a handwheel so that the reading may be taken with the torque arm exactly horizontal – this is indicated by the alignment of two pointers, one fixed and the other carried by the torque arm. The torque output of the engine is given by:

(spring balance reading) × (torque arm $l$).

## The electric dynamometer

Basically, this is a DC generator driven by the engine under test. The generator is mounted on trunnions so that its casing is free to rotate through a small angle. The torque exerted by the engine on the armature is transmitted through the internal magnetic field to the casing, and measured by a spring balance or load cell at the end of a torque arm, as in the hydraulic dynamometer. Figure 4.12 is a diagram showing the internal construction of the generator. The field coils are supplied with current from an external source (the electrical power required for this is small in comparison with the power being absorbed from the engine), and the braking torque on the engine is controlled by varying this current by means of a variable resistance in series with the field coils. The mechanical power output of the engine is converted into electrical power in the armature windings, which must be connected by means of the commutator and brushes into a complete electrical circuit in which the power will be dissipated. Ideally the electrical power output of the dynamometer could be made use of, but this is usually impracticable because of the wide variations of voltage and current output which

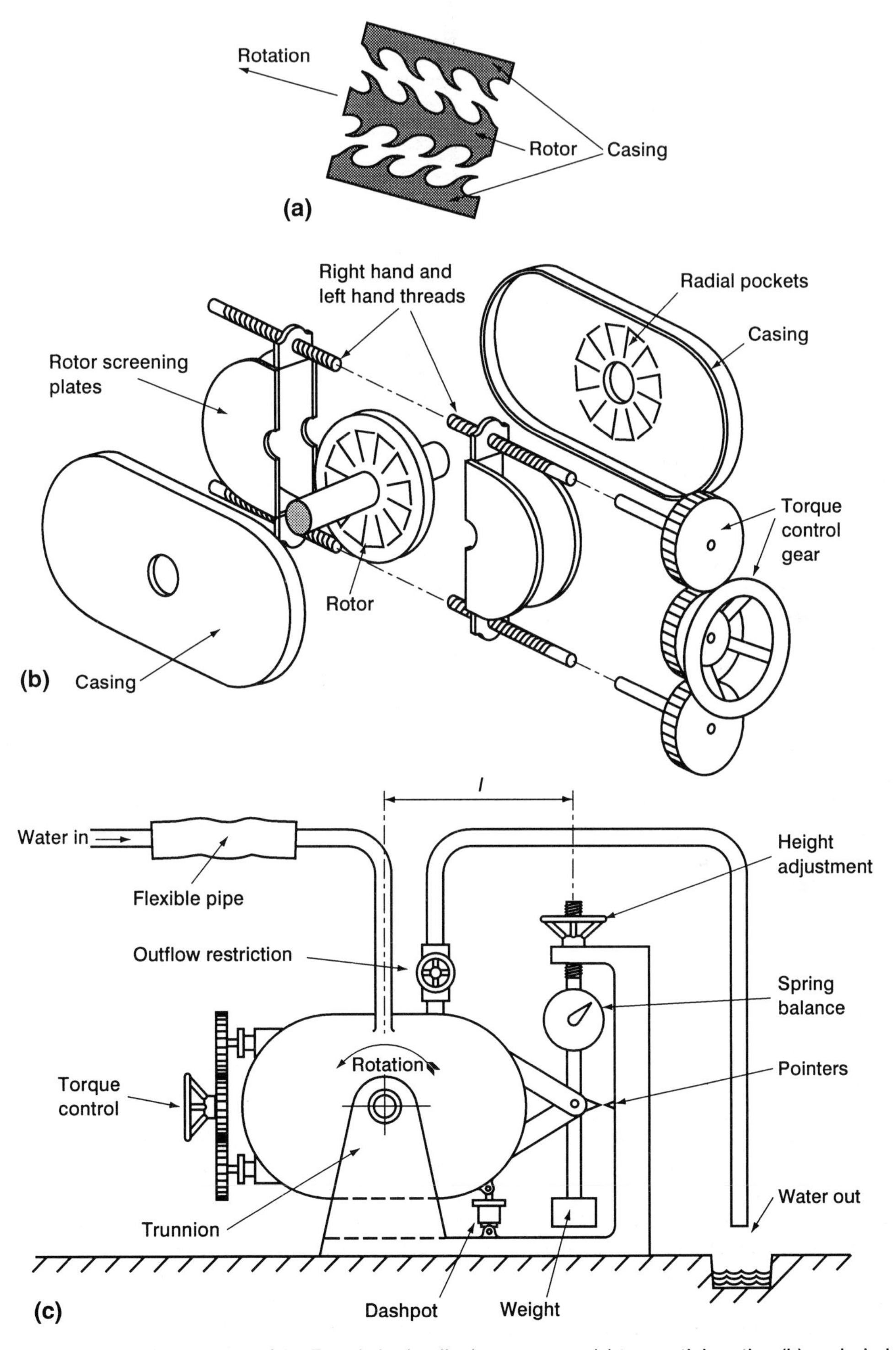


**Figure 4.11** Three views of the Froude hydraulic dynamometer (a) tangential section (b) exploded diagram (c) schematic diagram

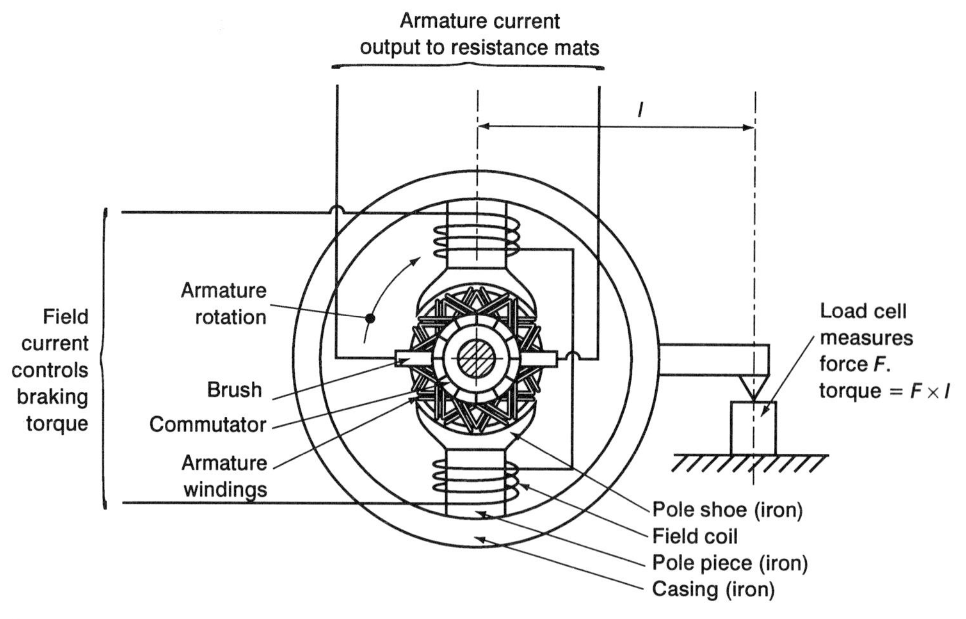


**Figure 4.12** Section through an electric dynamometer

may occur during a test, and the random availability of such power. Instead the power is usually dissipated as heat to the atmosphere by passing the current through resistance 'mats' consisting of large numbers of resistance heating elements connected in parallel, assembled as a stack through which air is convected or blown.

Because the same electrical machine is a generator when supplied with mechanical power and an electric motor when supplied with electrical power, an engine can be started by switching the field and armature windings of its dynamometer to a suitable electrical power supply. The power losses due to frictional effects also can be measured in this way, by using the dynamometer to drive a 'dead' engine.

## The eddy current dynamometer

Figure 4.13 shows an example of this type of dynamometer. It consists of a toothed disc (the rotor) driven by the engine under test, rotating in a cylindrical casing. Both disc and casing are made of iron of high magnetic permeability. The casing is supported by trunnions which allow it limited freedom of rotation, and it is magnetised by a continuous DC current passed through a coil surrounding the rotor. The magnetic field across the gap inside the casing generates eddy currents in the teeth of the disc as they rotate through it. These eddy currents set up their own magnetic fields, which interact with the applied magnetic field, causing a braking torque on the toothed disc and an equal and opposite torque on the casing tending to drag it round with the disc. This torque is measured by a load cell in the usual way. The braking torque is controlled by controlling the current through the magnetising coil. The mechanical energy input from the engine is converted into electrical energy in the form of eddy currents, and this energy in turn is converted into heat by the electrical resistance of the iron disc and casing. The heat produced in the disc is radiated to the casing across the very narrow air gap (less than 0.5 mm) between casing and disc; some heat is also carried away by air flowing outwards through the air gap. The faces of the casing adjacent to the disc are cooled by a continuous flow of water through annular passages in the casing.

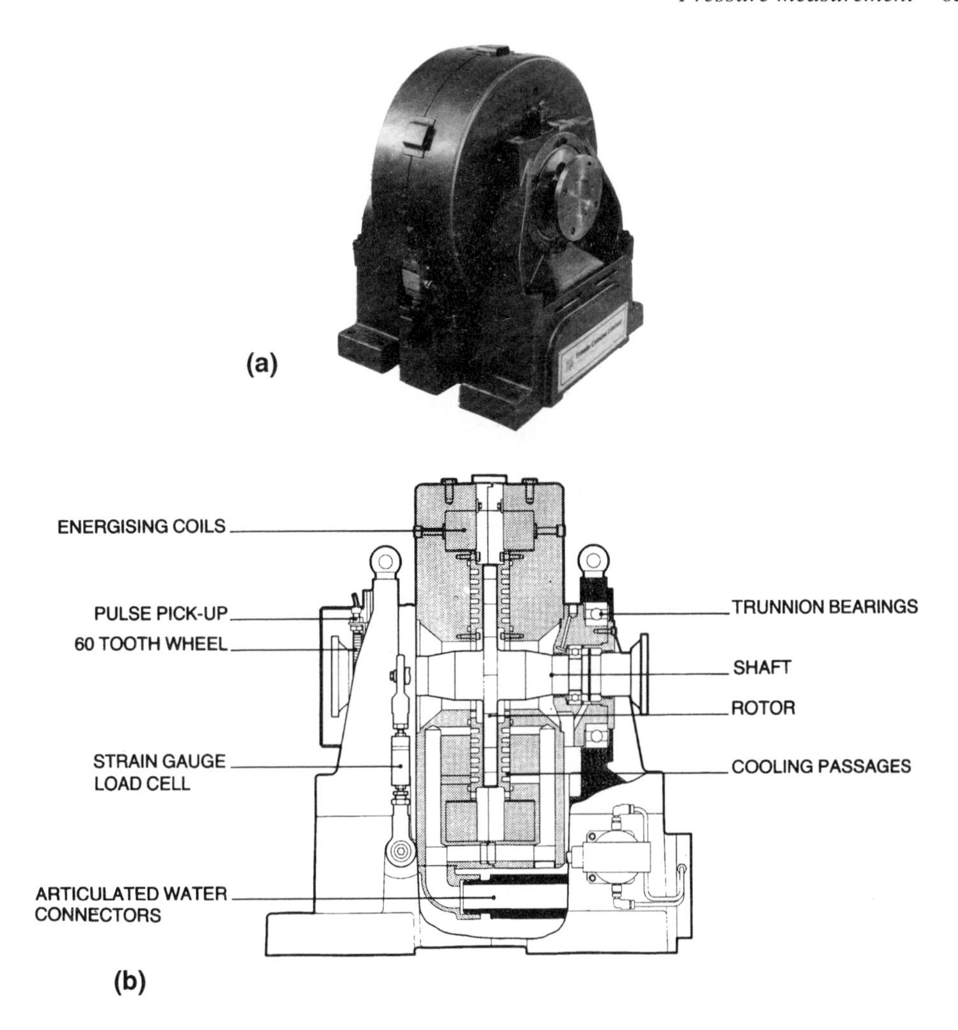


**Figure 4.13** The Froude Consine eddy current dynamometer (a) external view (b) cross section

Because of its compactness and simplicity the eddy current dynamometer is well suited to the routine testing of engines. Where a precise measurement of torque is not required, for instance in the testing of engines coming off a production line, the load cell can be dispensed with as the braking torque may be inferred from the value of magnetising current being applied and a knowledge of the braking torque which is produced by a given current at a given speed. This same simplicity of operation makes it easy to incorporate the eddy current dynamometer into an automatically controlled engine testing system.

## Pressure measurement

The 'standard' way of measuring pressure is to use a pressure gauge. In this type of instrument, the pressure is applied to the inside of a **Bourdon tube** (a metal tube bent into a circular arc and flattened so that its cross-section is oval). Internal pressure causes the flattened cross-section to open out to a more circular form, and this causes the tube itself to tend to straighten out. The

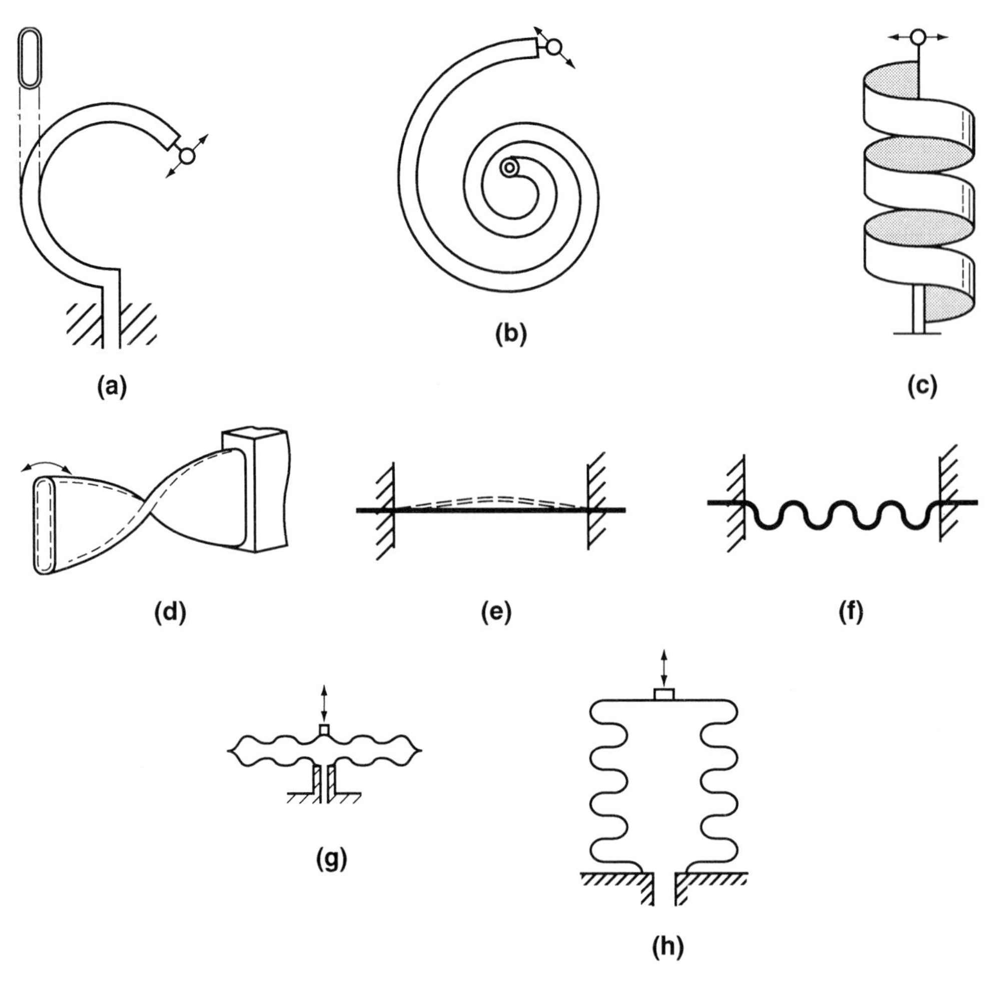


**Figure 4.14** Transducers which give displacement proportional to pressure. Bourdon tubes: (a) C-type, (b) spiral, (c) helical, (d) twisted. Diaphragms: (e) flat, (f) corrugated. Capsule: (g). Bellows: (h)

free end of the tube is connected to a toothed quadrant which drives a pinion carrying the pointer. Pointer rotation is proportional to the pressure inside the Bourdon tube. Figure 4.14 shows a C-type Bourdon tube (the usual form) at (a), together with other ways of obtaining a deflection proportional to pressure. Instead of a Bourdon tube, some low-pressure gauges have a diaphragm (e) which deflects under pressure. Or to make the Bourdon tube more resistant to shock and vibration it may be twisted (d) instead of curved, so that as the flattened tube opens out under pressure, one end rotates relative to the other. To increase its sensitivity as a transducer a Bourdon tube may be lengthened and formed into a spiral (b) or a helix (c) to accommodate this extra length. A diaphragm may be corrugated into concentric rings (f) to increase its sensitivity. Two such corrugated diaphragms, joined face to face form a capsule (g); an aneroid barometer has its atmospheric pressure transducer in this form. Inside the capsule there is a complete vacuum, the capsule being prevented from collapsing by a spring which holds the two faces apart.

Another form of pressure transducer is the hydraulic bellows (h). This device is formed out of thin-walled metal tube by very high hydraulic pressure, which produces corrugations in the cylindrical surface. On its own it has little stiffness, so as a pressure transducer it must usually be used in conjunction with a spring, which supports most of the load on the end face.

### U-tube manometers

These work on the principle that at the lower of the two levels in the U-tube, the pressures in the two limbs are equal, so the difference between the pressures acting on the two surfaces is given by $p_1 - p_2 = \rho g h$, where $\rho$ is the density of the liquid, $g$ is gravitational acceleration ($9.8\ m/s^2$) and $h$ is the height of liquid above the lower level.

While a U-tube gives a simple direct measurement of pressure, which 'can't go wrong', the liquid is apt to surge up and down the tube under sudden variations of pressure, and in practice its use is limited to measuring pressure differences of less than two atmospheres. Also, as it requires a human being to read it, it cannot be incorporated into an automatic control or recording system.

### Inductive and capacitive pressure transducers

Examples of these have already been given in the preceding sections on inductive transducers and capacitive transducers.

### Vacuum measurement

Very low absolute pressures may be stated in pascals (1 Pa is 1 $N/m^2$), in millimetres of mercury (1 mm Hg = 133.322 Pa), or in torr (1 torr is 1 mm Hg).

Bourdon tube pressure gauges can be used to measure absolute pressures down to 10 mm Hg; diaphragm-operated gauges can measure down to about 1 mm Hg absolute. Pressures lower than this may be measured by the *McLeod gauge*, by the *Pirani gauge*, or by gauges which measure the ionisation current produced in the gas (a) by electrons from a hot cathode in a form of thermionic valve, or (b) by alpha particles from a radioactive source. For detailed descriptions of these devices the reader is referred to more specialised texts, such as *Measurement Systems* by Doebelin (see Suggested further reading on p.203).

## Exercises on chapter 4

1 a) Sketch a section through a hydraulic load cell and name the component parts.
  b) Give reasons why a hydraulic load cell may be more suitable than a spring balance as a force measuring device in some situations.

2 The following results were obtained when a step input of force was applied to a piezoelectric force transducer connected to a charge amplifier:

| Time since step (seconds) | 0.0 | 2.5 | 5.0 | 10.0 | 15.0 | 25.0 | 35.0 | 50.0 |
|---|---|---|---|---|---|---|---|---|
| Output voltage (volts) | 14.2 | 11.4 | 9.2 | 5.9 | 3.8 | 1.6 | 0.7 | 0.2 |

Plot a graph of these results and hence determine the time constant of the system.

3 A piezoelectric force measurement system has a time constant of 20 seconds.
  a) 2.5 seconds after a step input of force was applied the output voltage was 16.3 V. Determine the instantaneous voltage produced by the step.
  b) How many seconds after the input step will the voltage be down to 60% of the instantaneous voltage?

4 a) A piezoelectric force transducer has a sensitivity of 12 pC/N. It is connected to a charge amplifier which has a feedback capacitor of 4.7 nF. Calculate the output voltage of the charge amplifier at the instant when a step input of 2 kN is applied to the transducer.

b) The system of piezo transducer and charge amplifier has a time constant of 48 seconds. How long will it take to lose one-eighth of the output step?

5 In each of the following cases suggest a suitable measurement system to give an electrical output signal, and explain how the system works.
a) For continuous measurement of the load on a crane hook which carries a ladle dispensing molten metal in a foundry.
b) For measuring the force variation as a probe passes through a very thin membrane. The maximum force will not be greater than 0.01 N.

6 a) Draw a rotary transformer circuit for measuring the torque carried through a rotating shaft.
b) State, and briefly describe, two other methods of measuring the torque output of an engine without absorbing any of the power.

7 Briefly explain the operating principle of each of the following:
a) a hydraulic dynamometer
b) an electric dynamometer
e) an eddy-current dynamometer.
In each case explain how the energy input from the engine is dissipated.

8 The difference in height of the levels of mercury in the two arms of a U-tube is 958 mm. Calculate the corresponding difference in the pressures applied to the two mercury levels if the pressures are applied by:
a) air
b) oil of relative density 0.82.
The relative density of mercury is 13.6.

9 Sketch, in section, an inductive pressure transducer and a capacitive pressure transducer, and explain the difference between them.

# 5
# Velocity transducers

## Linear velocity measurement

This usually means measuring the speed of a body, such as a wheeled vehicle, a ship or an aircraft. In the case of a wheeled vehicle the simplest method is to convert the linear velocity into a rotational velocity by driving a speedometer from the output shaft of the gearbox. The speedometer is really a *tachometer* (an instrument for measuring rotational velocity) graduated in units of speed; mile/h or km/h. The most commonly used type of tachometer for this purpose is the eddy-current drag cup (see p.73).

In a ship or an aircraft, speed transducers measure speed through the water or the air. This is *not* the same as speed over the ground unless the medium in which the transducer operates, water or air, is stationary. If the medium is in motion the ground speed can only be obtained from these transducers by vector addition.

The transducer used to measure speed through the air is a *pitot tube*. To measure speed through water, the transducer (the log) can be a submerged paddle wheel projecting through the hull or a 'spinner' with skewed vanes which rotates as it is towed through the water on a flexible cable. The output of either type of transducer is a series of voltage pulses from a proximity pickup (see p.72), which detects the passage of a magnetic insert in the paddle wheel or spinner. The speed of a vehicle may also be measured by Doppler effect or from

$$\frac{\text{distance}}{\text{time}}.$$

### Velocity from distance/time

If two positions of the vehicle are known, velocity may be calculated by dividing the distance between them by the time taken to travel that distance. If the rate of change of velocity is small enough and the time interval is short enough, the average speed thus obtained will be a good approximation to the true velocity. Decca, Loran and GPS navigational radio receivers continuously calculate velocity in this way, giving both magnitude and direction.

### Doppler effect

This is the change in the frequency of a sound wave or radio wave reaching a receiver when there is relative velocity between the receiver and the transmitter. A common example is the apparent alteration in pitch of a police siren as it approaches, passes and recedes from an observer. It is caused by the changing distance which the waves have to travel between transmitter and receiver. As the source of the waves approaches, the distance is decreasing, so each wave has a shorter distance to travel than the previous wave, and thus arrives earlier than it would if the source were stationary. The number of waves reaching the observer in a given time (i.e. the frequency) is increased. Conversely for a receding source, an increasing path length gives a decrease in the received frequency. The relationship between the emitted and received frequencies is:

$$\text{frequency received} = \frac{c}{c - v} \times \text{frequency emitted}$$

where

$c$ is the velocity of transmission through the medium
$c \approx 330$ m/s for sound in air
$c = 2.997925 \times 10^8$ m/s for light and radio waves
$v$ is the relative velocity of the source towards the receiver.

$v$ is positive if the distance between receiver and source is decreasing, negative if it is increasing. If the relative velocity of the source is at an angle $\theta$ to a line joining source and receiver, then $v$ = (relative velocity) $\times \cos\theta$.

## The pitot-static tube

The principle is shown in Figure 5.1. The pitot tube is open-ended and faces into the flow. Air flowing past comes to rest at the mouth of the tube, and its kinetic energy is thereby converted into pressure energy, causing a *stagnation pressure* which is higher than the pressure of the undisturbed air. The static tube, which surrounds the pitot tube, has a front end which is closed and rounded, to cause minimum disturbance to the flow. It senses the pressure of the undisturbed air (the *static pressure*) through holes or slits in the wall of the tube. The velocity of the air flowing past the aircraft may be calculated from the following formula, which is derived from Bernoulli's equation:

$$v = \sqrt{\frac{2(p_1 - p_2)}{\rho}}$$

where

$v$ is the flow velocity
$\rho$ is the density of the fluid (in this case, air),
$p_1$ is the stagnation or total pressure,
$p_2$ is the static pressure.

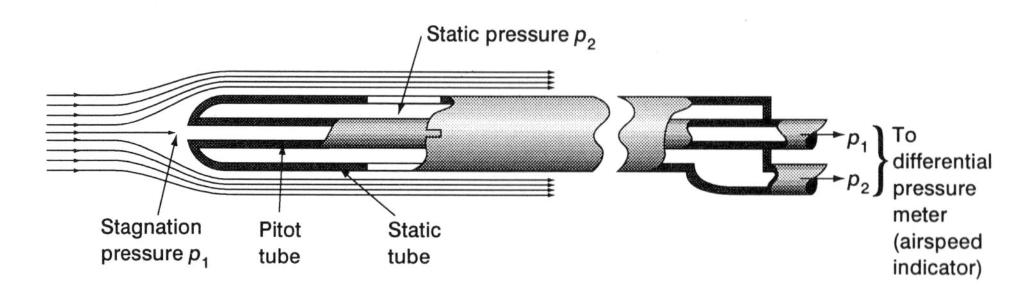


**Figure 5.1** Pitot-static tube used as an airspeed transducer on aircraft

In an aircraft, the pitot-static tube is called the pressure head. It usually contains a small electric heater to prevent the pressure-sensing orifices from being blocked by ice. The two pressures are piped to a differential pressure meter graduated in units of airspeed (knots). Because air density decreases with altitude the indicated airspeed reads low by a factor equal to the square root of the relative air density at any particular altitude. Other possible sources of error are (1) the misalignment of pitot-static tube with the airflow at high angles of incidence, and (2) the compressibility of the air as the speed of sound is approached.

As well as being used on aircraft, pitot tubes are used to measure the velocity of airflow in ducts, and the velocity of other gases and liquids in pipes. In these applications, an inclined manometer may be used to measure the pressure difference, and the flow velocity needs to be high enough to give a measurable reading. A combined pitot-static tube may be used, as on an aircraft, or a static pressure tapping may be made in the wall of the duct or pipe. The advantage of the pitot tube over other methods of flow measurement is that it causes minimal obstruction to the flow. Its disadvantages are (1) the pressure sensing orifices may become blocked if the fluid carries any particles in suspension, and (2) it can only measure the velocity at one point in the cross-section of the pipe. (This is not necessarily a disadvantage as the pitot tube can be set at various radii in the pipe, to enable the velocity gradient from the wall to the centre of the pipe to be plotted.)

## Rotational velocity measurement

### The stroboscopic lamp

This generates flashes of light of extremely short duration, at a set frequency, by means of a gas discharge tube. The flash rate is indicated on a dial by a pointer on the control knob. Alternatively it may be shown by a frequency meter incorporated in the lamp assembly.

The stroboscope has the property of apparently 'freezing' the rotation of a shaft or the motion of a mechanism, so that a stationary image is seen when the flash frequency coincides with the frequency of rotation of the shaft or the cyclic frequency of the mechanism. The stationary image is also seen when the flash occurs exactly once every $n$ revolutions or cycles (where $n$ is an integer) so when using the stroboscope to measure rotational speed, care must be taken to get the true speed and not some submultiple of it.

To measure the speed of a rotating shaft the stroboscope is aimed at the end face of the shaft and the flash frequency is gradually decreased from maximum until a single stationary image of some feature on the end of the shaft (such as a keyway or a radial line) is seen. The flash frequency is then (presumably) the speed of the shaft in rev/min – stroboscopes are usually graduated in flashes/min. However, a check should be made by decreasing the flash frequency further until a single stationary image is again obtained. This should be at about half the original frequency (the shaft being illuminated once every two revolutions). If the second frequency is significantly greater than half the first frequency, the true speed is above the range of the stroboscope and must be deduced by comparing the ratio of frequencies with the ratio of possible fractions of the true speed. Alternatively, the gradual decrease of flash frequency should be continued so that as well as the first two frequencies, $f_1$ and $f_2$, further frequencies, $f_3$, $f_4$, ... $f_n$ which give single stationary images are obtained. The true speed can then be calculated from the formula

$$speed = \frac{f_1 f_n (n-1)}{f_1 - f_n}$$

By this means speeds can be measured up to about ten times the maximum flash rate of the stroboscope, but the accuracy of the result depends on the accuracy of the first and $n$th readings.

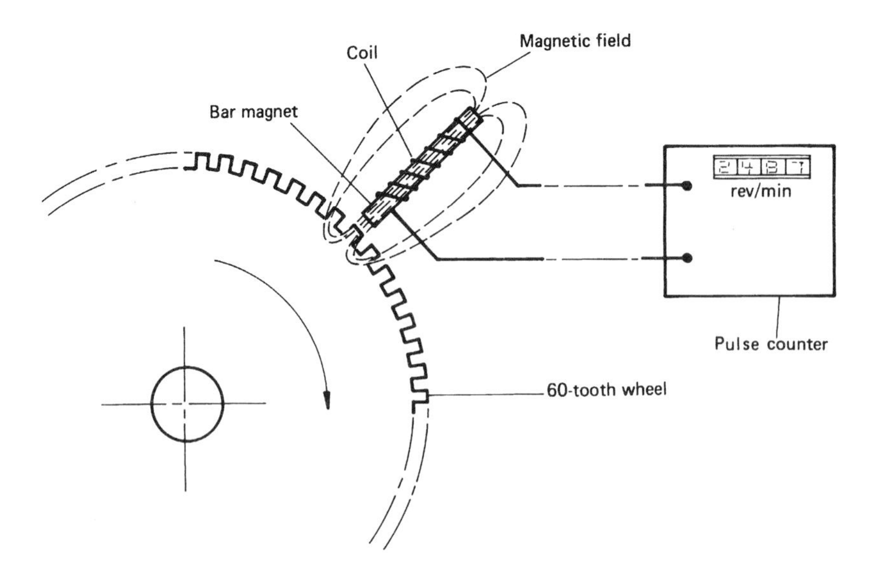


**Figure 5.2** Rotational speed measurement by magnetic transducer and pulse counter

## Rotational velocity from toothed wheel and proximity pickup

This is the most usual way of obtaining a continuous measurement of rotational speed for the control of industrial processes. The principle is shown in Figure 5.2. The magnetic proximity pickup – a coil wound on a bar magnet – generates a pulse of voltage each time a tooth moves past it. These pulses are fed to the pulse counter which counts them for a set period of time (usually one second), displays the result for a few seconds to enable it to be read, and then repeats the process. The toothed wheel must be made of some form of iron. At its simplest it need only be one of the gear wheels in the gear drive to a machine. It is more usual however to use a 60-tooth wheel so that the pulse counter, counting for exactly one second, automatically shows the speed in rev/min; an example is seen in Figure 4.13(b) (left-hand side).

The accuracy of the result depends on the number of teeth passing the pickup in the set time interval and the accuracy with which the time interval is measured. The pulse count will be accurate to ±1 pulse, while the timebase of the counter usually has an accuracy of the order of 1 part per million. Thus the inaccuracy in the time interval is negligible compared with the possible error of 1 in the number of pulses counted. When measuring relatively slow rotational speeds, therefore, it may be necessary to drive the toothed wheel at a higher speed, through gearing, to obtain the required accuracy.

By its nature, this system of measurement can only give the average speed over the time interval, and so is unsuitable for measuring rapidly varying speeds. The pulse-counting principle can also be used with other methods of generating voltage pulses; for instance the light source, phototransistor and black-and-white sectored disc described under 'Sensors' in chapter 3. Another possible source is the *Hall-effect transducer*.

## Hall-effect transducers

Figure 5.3 shows the Hall-effect principle. If a magnetic field acts at right angles to the current in a conductor, a voltage, $v$, is generated at right angles to both current and field. The voltage is due to the variation in density of the charge carriers within the cross-section, as they are deflected in accordance with Fleming's left-hand rule for motors. In a metal conductor, the

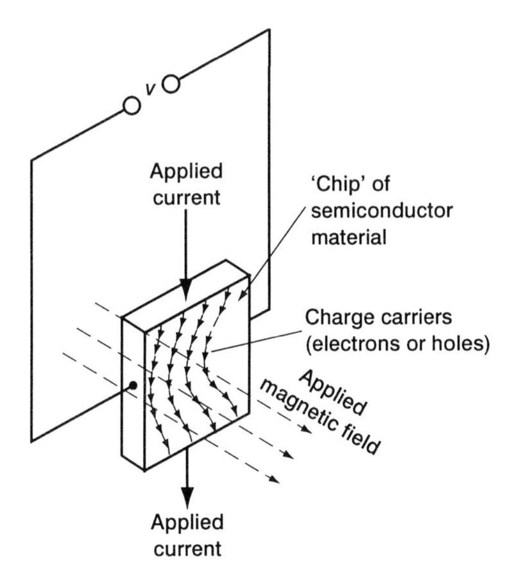


**Figure 5.3** Hall effect

charge carriers are electrons, and the generated voltage is virtually short-circuited by the low resistance of the metal. A Hall-effect transducer therefore uses semiconductor material; in this the charge carriers may be either electrons or holes. (An electron may be thought of as a negatively charged particle, and a hole as the absence of an electron, and therefore as a 'particle' of positive charge.) A constant current is passed through the chip, and $v$ is then proportional to the magnetic field, up to a limiting value of field strength.

Hall effect transducers (also called *Hall effect sensors* or *Hall probes*) are used i) for magnetic field measurement, ii) for current measurement (from the field strength around a conductor), and iii) as proximity pickups (replacing the contact-breaker in a car ignition system for example). The proximity pickups sense the presence of iron from the change it causes in the field strength of a small magnet positioned so that its field passes through the chip. They act as solid-state switches, switching $v$ on and off within a few microseconds, with the advantage that they are free from contact 'bounce' and are unaffected by dust and changes of ambient lighting which sometimes affect optical devices. The chip usually has an integrated circuit amplifier which amplifies $v$ and enables the transducer to supply a few milliamps of current.

## The eddy current drag-cup tachometer

Most analogue-display instruments which indicate rotational speed are eddy current drag-cup tachometers. These include motor vehicle speedometers, which are actually drag-cup tachometers gear-driven via a flexible drive cable from the output shaft of the vehicle's gearbox, their dials being graduated in miles per hour instead of rev/min. Figure 5.4 shows the construction of such an instrument. The input rotation is applied to a magnet which rotates inside an aluminium cup carried by the spindle on which the pointer is mounted. The field of the magnet passes through the aluminium cup to a fixed iron casing. As the magnet rotates, the magnetic field, travelling through the aluminium, generates eddy currents in it. These create their own magnetic field, which is attracted to that of the rotating magnet, and so exerts a torque on the cup proportional to the speed of the input rotation. The cup and pointer assembly therefore rotate to the position at which this torque is exactly balanced by the opposing torque reaction from the torsion spring.

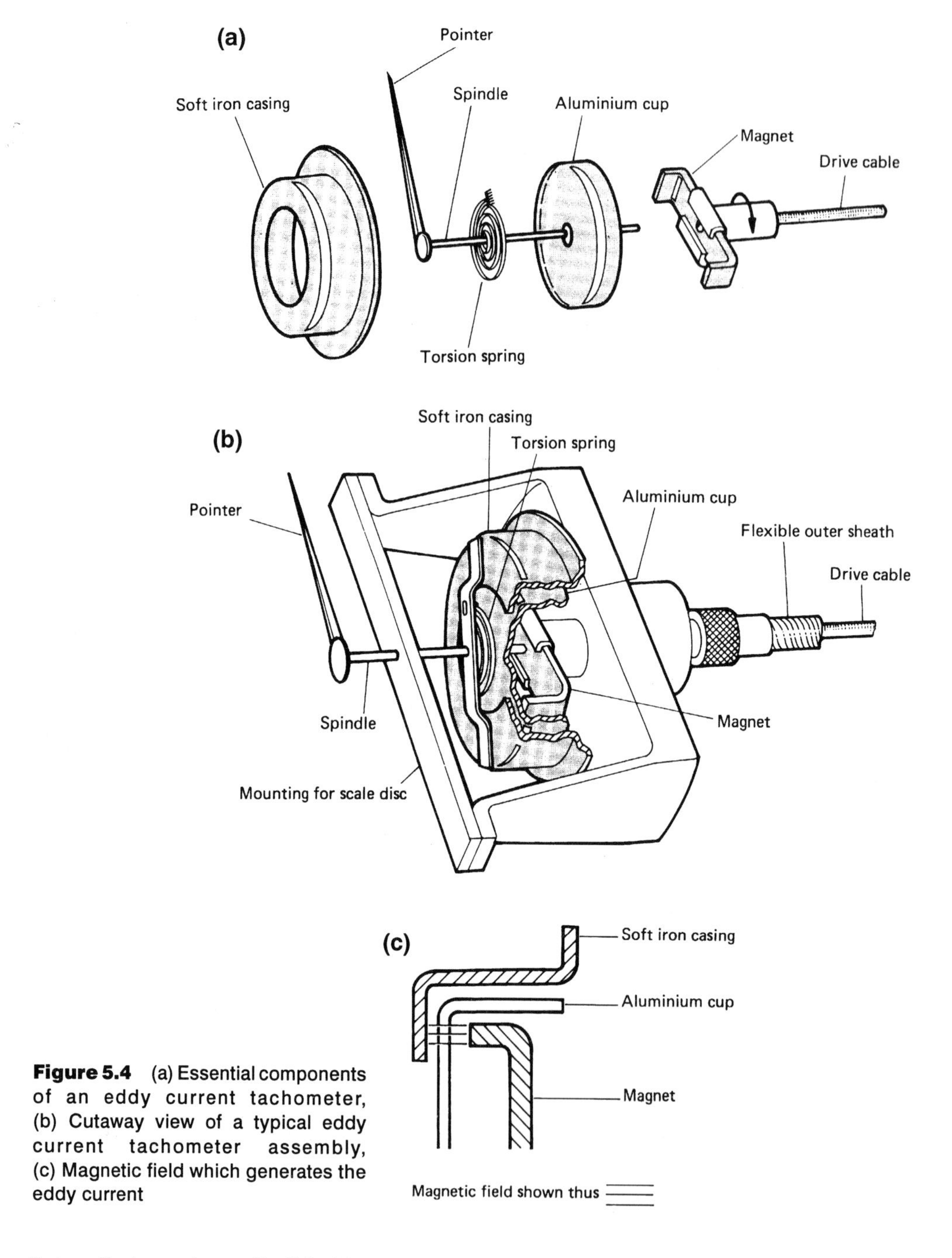


**Figure 5.4** (a) Essential components of an eddy current tachometer, (b) Cutaway view of a typical eddy current tachometer assembly, (c) Magnetic field which generates the eddy current

In installations where a flexible drive cable cannot be used because the cable drive point on the machine is too far away from the instrument, a small 3-phase AC generator is used as a transducer to convert engine speed to AC frequency. The AC output of the generator is connected to a small 3-phase synchronous motor mounted on the instrument, giving an input rotation to the tachometer which is exactly the same as that applied to the generator.

### DC tachogenerators

An ordinary DC generator with a permanent magnet field gives an output voltage approximately proportional to the speed at which it is driven. A DC tachogenerator is a miniaturised version of the same type of machine, with design features which give it a constant ratio of output voltage to rotational speed. A typical tachogenerator has a sensitivity of 5 V per 1000 rev/min, over a speed range of 0 to 6000 rev/min, with a non-linearity of the order of 0.01%. To obtain this kind of performance it has its permanent magnet in the form of a fixed central core surrounded by an outer steel ring, so that the magnetic field is radial in the gap between them. The armature is a non-metallic sleeve carrying the coils, which rotates in the gap. The output load must not be less than 250 kΩ (i.e. the current drawn from the generator must be negligible). The DC output voltage is taken from the coils through a commutator and brushes, and hence has a small ripple superimposed on it together with small voltage 'spikes' which occur when commutator segments meet or leave the brushes. The main AC component of the ripple is about 5% of the DC output. It has a frequency of:

$$2 \times (\text{number of coils}) \times (\text{rev/s}).$$

The ripple and spikes may be smoothed by passing the output through a low-pass filter, but at low speeds, and hence low frequencies of ripple, the ripple may still get through the filter. The polarity of the tachogenerator's output voltage reverses when the direction of rotation of the armature is reversed.

### AC tachogenerators

The simplest type of AC tachogenerator is a small brushless alternator in which a permanent magnet rotates inside a fixed winding generating the output AC, which has amplitude proportional to the speed of rotation. Unfortunately, the frequency of the AC produced is also proportional to speed, and this may affect the accuracy of the instrument.

The *AC drag-cup tachogenerator*, illustrated in Figure 5.5, is an improvement on this, as its output frequency remains constant at the frequency of excitation. The stator (the stationary

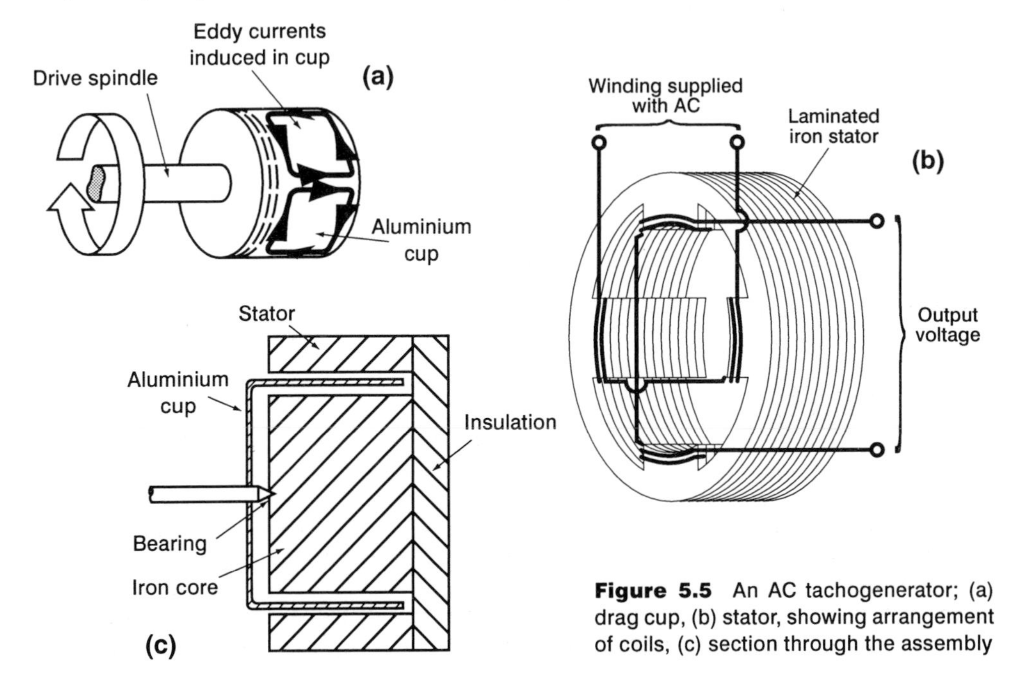


**Figure 5.5** An AC tachogenerator; (a) drag cup, (b) stator, showing arrangement of coils, (c) section through the assembly

outer ring) has two windings with their axes at 90° to each other. AC is supplied at a constant voltage to one of these windings, giving an alternating magnetic field of constant amplitude. An aluminium cup in the annular gap between the stator poles and a soft iron core, is rotated at the speed to be measured. An alternating current, at right angles to both the magnetic field and the direction of motion, is generated in the metal of the cup, circulating as shown, with amplitude proportional to the speed of rotation. This eddy current has an associated magnetic field, and this generates AC in the other winding. Thus the output signal appears as the amplitude modulation of a 'carrier wave' voltage which has the same frequency as the AC supplied to the transducer. The frequency response of the transducer is therefore limited to the range from zero to about one-fifth of the frequency supplied. The supply frequency is usually in the range 50 to 500 Hz. Reversing the direction of rotation reverses the eddy currents, causing a 180° phase shift in the output AC.

## Exercises on chapter 5

1 A velocity measurement system, which takes a reading of velocity every 0.1 seconds, is used on an object which is subjected to a *constant* acceleration from 0 to 4.5 m/s in 0.6 seconds. Because the signal is corrupted by electronic 'noise', the system records the following values of velocity:

| Time (seconds) | 0 | 0.10 | 0.20 | 0.30 | 0.40 | 0.50 | 0.60 |
|---|---|---|---|---|---|---|---|
| Velocity (m/s) | 0 | 0.95 | 0.90 | 2.30 | 3.35 | 3.60 | 4.70 |

a) Calculate the true acceleration and the true distance travelled in 0.6 seconds.
b) On a graph of velocity (vertical) against time (horizontal), plot the true velocity and the recorded velocity; the recorded velocity graph to be made up of straight lines joining the plotted points.
c) On a graph of acceleration (vertical) against time, plot the true acceleration and the accelerations obtained from the gradients of the straight lines which make up the recorded velocity graph.
d) On a graph of displacement (vertical) against time, plot the true displacement and the displacement obtained from the recorded velocity graph; these displacements to be obtained by integrating (i.e. adding up) the areas of 0.1 second strips underneath the respective velocity graphs.
e) By reference to the acceleration and the displacement graphs, compare and comment on the effects of differentiation and of integration on the accuracy of results obtained by those two processes.

2 A pitot tube installed in a water pipe shows a pressure difference of 6.25 kPa between the pitot and static connections. Calculate the velocity of water flowing past the pitot tube. (Density of water is 1000 kg/m$^3$)

3 a) Calculate the pressure difference between the pitot and static connections of an aircraft pressure head if the aircraft is flying at 400 knots near sea level, where the density of the air is 1.226 kg/m$^3$. (1 knot = 0.514773 m/s)
b) What speed would be indicated by the airspeed indicator if the aircraft was flying at the same speed at 30 000 ft, where the density of the air is 0.458 kg/ m$^3$, assuming that it had indicated the true speed (400 knots) in part a)?

4 a) Explain the correct procedure for using a stroboscope to measure an unknown rotational speed of a shaft.
b) Why is it possible to get the wrong answer if the stroboscope is used to observe a mark on the side of the shaft instead of on the end?

5 When a stroboscope was used to observe a radial line on the end of a shaft rotating at an unknown speed, a single stationary image was obtained at 5940, 4750, 3960 and 3390 flashes/min. There were no single stationary images at any speeds between those values. Determine the speed of the shaft.

6 Explain, with the aid of a diagram, the toothed wheel and proximity pickup system for measuring rotational speed. Briefly describe an alternative type of proximity pickup for use with the toothed wheel.

7 Explain the Hall effect, illustrating your answer with a sketch. State three alternative uses for a Hall-effect transducer. What are the advantages of using a Hall-effect transducer as a switch, in place of a mechanical switch?

8 a) Explain briefly how the following rotational speed transducers work:
   i) DC tachogenerator
   ii) AC tachogenerator without external excitation
   iii) AC drag-cup tachogenerator.
  b) Compare their relative advantages and disadvantages.

9 Compare the following speed measuring devices:
  a) stroboscope
  b) toothed wheel and proximity pickup
  c) eddy current drag-cup tachometer
  d) DC tachogenerator
  e) AC drag-cup tachogenerator.

  by answering the following questions for each device:
   i) Does it require a human being to operate it?
   ii) In its most usual form, does it have an output signal for use at a distance from the transducer?
   iii) In its most usual form, does it apply a significant torque load to the rotating body?
   iv) Can it be used to measure transient speed variations occurring in a fraction of a second?
   v) Is its output in the form of a digital signal?
   vi) Is it suitable for use in an automatic control system?
   vii) Does it require some form of external power supply?
   viii) Might a phase-sensitive detector form part of the measurement system?
   ix) Might a synchronous generator and synchronous motor form part of the measurement system?

Give your answers in the form of a table with rows labelled a) to e) and columns labelled i) to ix), inserting the letters Y for 'Yes' or N for 'No' in each space.

# 6
# Acceleration and vibration transducers

## Seismic pickups

The first seismic pickups were used to measure the vibrations resulting from earthquakes, hence the name: the Greek word for earthquake is *seismos*.

In instrumentation, seismic pickups are used to measure the motion of the surfaces to which they are fixed. They are sensitive to motion along one axis only, so if the motion is three-dimensional, three seismic pickups are needed to determine the components of the motion along three mutually perpendicular axes. The principal features of a seismic pickup are shown diagrammatically in Figure 6.1. The essential component is the seismic mass. This is a body of metal, suspended from a *resilient* support. This is a support whose deflection is proportional to the force applied to it. The inertia of the seismic mass causes it to lag behind the motion of the casing when the casing is accelerated, causing a deflection in the support. This deflection forms the input to a transducer, which produces a proportional output signal. In Figure 6.1 the transducer is represented by a potentiometer, but any suitable type of transducer may be used; for example Figure 1.4 shows a seismic pickup in the form of an accelerometer in which the transducer is an unbonded strain gauge bridge.

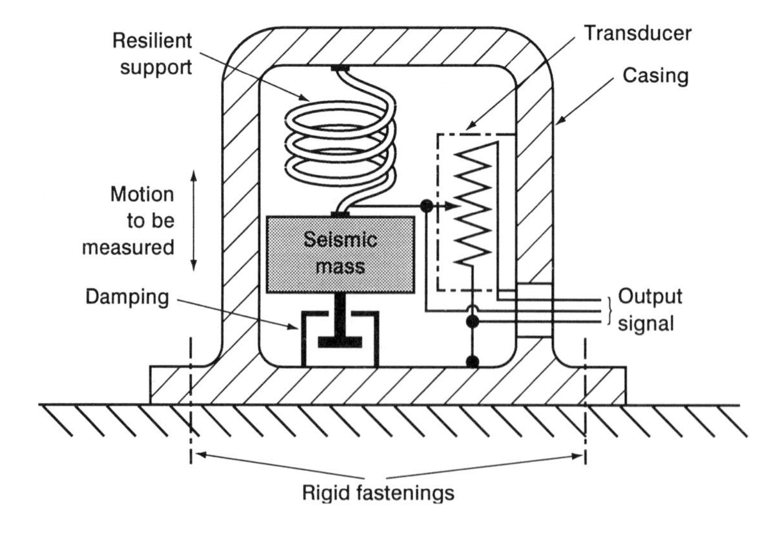


**Figure 6.1** The essential features of a seismic pickup

The 'dashpot' in Figure 6.1 represents the damping which will inevitably be present. This damping may consist only of the hysteresis of the support material, or it may be increased by filling the casing with a silicone fluid of suitable viscosity for example.

By choosing suitable values for the mass, the stiffness of the support and the damping, and by using an appropriate transducer, the same basic arrangement of seismic pickup can be designed as a displacement pickup, a velocity pickup or an acceleration pickup (*accelerometer*). The seismic pickup is essentially a damped spring-mass system, and will have a natural frequency of vibration given by:

$$\omega_n = \sqrt{\frac{\lambda}{m}}$$

where

$\omega_n$ is the natural angular frequency (rad/s)
$\lambda$ (lambda) is the spring stiffness (N/m)
$m$ is the mass (kg).

## Displacement pickups

This type of pickup is used to measure the displacement of a vibrating body when there is no fixed reference point available, for example in determining the movement of the chassis of a vehicle. We therefore want the seismic mass to behave (as far as possible) as though it was fixed in space. This can be arranged by using a relatively large seismic mass and a relatively 'floppy' resilient support. This gives a low value of $\omega_n$ to the spring-mass system. Figure 6.2 shows the frequency response of such a pickup with various values of damping ratio $\varsigma$ (zeta).

$$\varsigma = \frac{\text{Actual damping}}{\text{Critical damping}}$$

Critical damping is the value of damping which, if the mass is displaced from its equilibrium position and released, allows it to return in the shortest possible time without overshooting. If the actual damping is greater than critical ($\varsigma > 1$) the mass returns more slowly, again without overshooting. If the actual damping is less than critical ($\varsigma < 1$) the mass returns more quickly, but overshoots and oscillates about the equilibrium position with a decaying oscillation.

Figure 6.2(a) shows that for frequencies of vibration well above $\omega_n$ the displacement of the seismic mass relative to the casing is practically equal to the displacement applied to the casing, while Figure 6.2(b) shows that those displacements will be nearly 180° out of phase with each other. This means that as the casing moves in one direction, the seismic mass moves in the opposite direction relative to it – it virtually stands still. Heavier damping reduces the phase lag somewhat from 180° but this is not usually important since we are usually more interested in the amplitude of a displacement than its phase angle. It can be shown mathematically that $\varsigma = 0.707$ gives the least variation of displacement ratio for values of $(\omega/\omega_n) > 1$ (see Figure 6.2(a)). At this value of $\varsigma$ we can bring $(\omega/\omega_n)$ down to about 1.75 before the error in displacement measurement exceeds 5%, so displacement pickups are often designed to have a damping ratio of about 0.7.

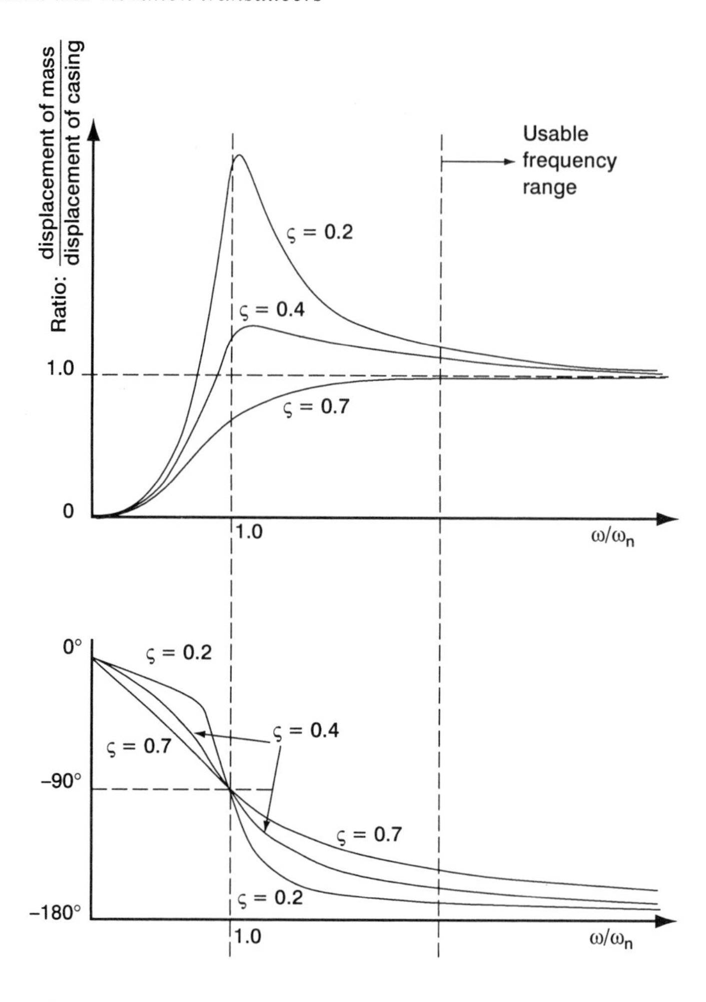


**Figure 6.2** Frequency response of a seismic displacement pickup; (a) amplitude, (b) phase shift

## Velocity pickups

A signal proportional to velocity may be obtained from a vibration by:

1 differentiating the signal from a displacement pickup by passing it through a differentiating circuit
2 integrating the signal from an accelerometer by passing it through an integrating circuit
3 using a seismic velocity pickup. This is similar in principle to Figure 6.1, but with a velocity transducer in place of the displacement transducer.

Integrating from an accelerometer gives much more accurate results than differentiating from a displacement pickup, because differentiation amplifies any errors in the signal, whereas integration diminishes them (for an example see Question 1 of the Exercises in chapter 5). However, a velocity pickup gives a velocity signal directly, and this can be passed through an integrating circuit to give a displacement signal as well if required.

A velocity pickup is designed like a displacement pickup, to have a low value of $\omega_n$ and to operate at angular frequencies well above $\omega_n$, so that the motion of the seismic mass is virtually the same as that of the casing but (almost) opposite in phase. The transducer is usually a coil of wire carried by the seismic mass. The coil is suspended in a radial magnetic field so that a voltage proportional to velocity is generated in the coil when it is vibrated axially. Figure 6.3 shows the construction of a typical velocity pickup. The seismic mass consists mainly of a central rod with its associated nuts, washers and coil former. The rod connects together two flexible diaphragms, whose stiffnesses add to form the 'spring'. The coil former suspends the coil in a narrow annular slot in a cylindrical magnet, the magnetic field acting radially across the slot. The coil former may be made of metal, so that eddy currents are generated in it to provide eddy current damping.

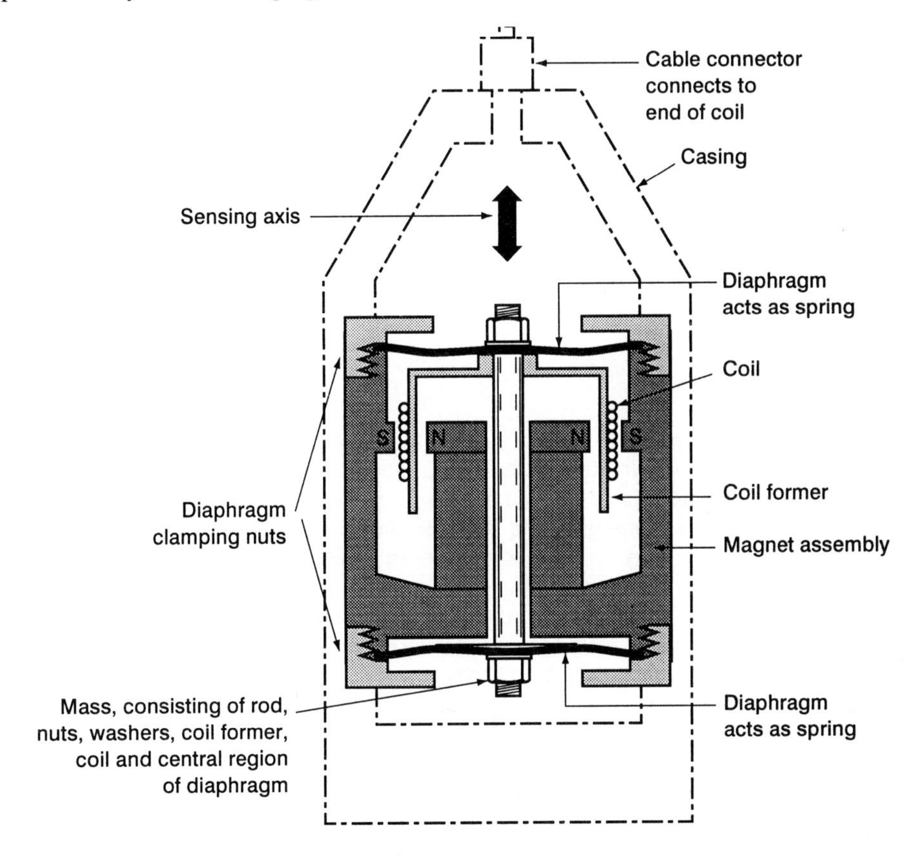


**Figure 6.3** Section through a vibration velocity pickup

## Acceleration pickups (accelerometers)

We have seen how, by designing the pickup system of Figure 6.1 to have a low value of $\omega_n$, we can use it as a displacement pickup or a velocity pickup for angular frequencies well above $\omega_n$. To design it as an acceleration pickup we must go to the opposite extreme.

Figure 6.2(a) shows that at angular frequencies well *below* $\omega_n$ the displacement of the seismic mass relative to the casing tends to zero. Therefore at these much lower frequencies the seismic mass must be accelerating with the same acceleration as the casing. To give it these accelerations, corresponding forces must be applied by the spring because:

force = mass × acceleration.

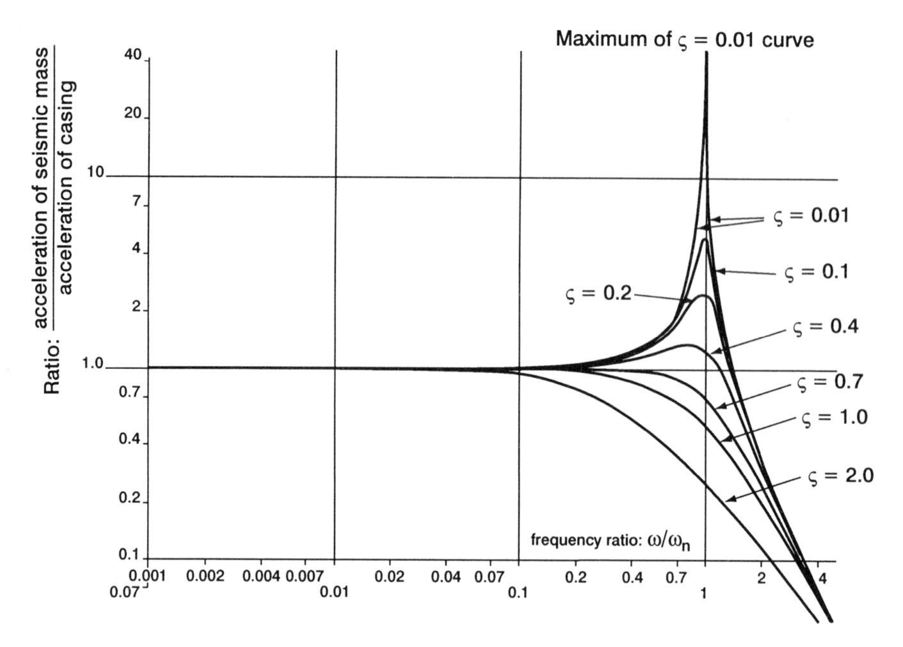


**Figure 6.4** The frequency response curves of seismic accelerometers with various values of damping ratio $\varsigma$

Therefore we can use the spring as a transducer, to tell us the force applied to the mass, its acceleration, and hence the acceleration of the casing. The graphs of Figure 6.4 show the ratio (acceleration of seismic mass)/(acceleration of casing) plotted against $\omega/\omega_n$ on logarithmic scales, for various values of damping ratio $\varsigma$. Because the horizontal scale is logarithmic, the left-hand end of the curves may be extrapolated to an infinitely small value of $\omega/\omega_n$, the acceleration ratio remaining constant at 1.0.

The curves indicate that provided the damping ratio does not exceed 1.0, a seismic accelerometer will give accurate readings of acceleration for frequencies of vibration from zero up to about 0.2 of its undamped natural frequency. For heavier damping than this, the upper frequency limit will be somewhat less. Most accelerometers, however, use a piezoelectric crystal as a combined 'spring' and transducer, and the damping ratio of a crystal is almost zero – in fact the frequency response of a piezoelectric accelerometer can be assumed to be that shown by the curve for $\varsigma = 0.01$. The ideal damping ratio would be $\varsigma = 0.7$ as this would allow accurate measurement up to about 0.5 of the undamped natural frequency. Clearly, to obtain the widest range of usable frequency response we want an accelerometer with the highest possible value of undamped natural frequency, and referring to the formula

$$\omega_n = \sqrt{\frac{\lambda}{m}}$$

we can see that this is given by a spring-mass system with a high value of spring stiffness, $\lambda$, and a low value of mass, $m$. For this reason a piezoelectric crystal is usually employed as the connection between the seismic mass and the casing, because it has a very high modulus of elasticity and so a very high spring stiffness. However it has the disadvantage that very low frequencies of vibration give time for the charge on the crystal to start to leak away, so there is a low frequency limit (usually about 5 Hz) below which the output of a piezoelectric accelerometer is unreliable. For even lower frequencies, and for the measurement of slowly varying or steady accelerations, some other form of spring and transducer must be used.

Suitable types are discussed in the section on low frequency accelerometers later in this chapter.

## Piezoelectric accelerometers

Tension cannot be applied to a crystal without using some kind of adhesive to make the tensile connection, and such a connection would be unreliable, so in the simplest form of piezoelectric accelerometer the crystal is kept permanently compressed by the seismic mass. Thus the effect of accelerations in alternate directions is an alternating increase and decrease in the compressive force on the crystal. The compressive preload is applied by screwing the seismic mass down on to the crystal to a given torque. The electrical connections to the crystal are made by metallizing (depositing a thin film of metal on) the end faces. This type of construction gives an accelerometer which is rugged, but because the casing is part of the 'spring' in the spring-mass system, it may be subject to spurious inputs. These include temperature change (causing expansion or contraction of the casing), acoustic noise, base bending (distorting the casing), cross-axis motion, and magnetic fields.

One of the difficulties of measuring accelerations on a light, flexible structure is that the mass of the accelerometer may alter the frequency and amplitude of the vibration we are trying to measure. For such applications we need an accelerometer of the smallest and lightest possible type – a *piezoelectric shear* type. In this type of accelerometer a ring of piezoelectric material is bonded to a central pillar. The seismic mass is a metal ring bonded to the outside of the piezoelectric material. Acceleration in the direction of the sensing axis causes shear stresses in the piezoelectric material, which is arranged so that corresponding values of electrical charge appear between the central pillar and the seismic ring. Because the spring-mass system in this type of accelerometer consists only of the seismic ring and the piezoelectric material it is not sensitive to stresses caused by distortion of the casing or the base. Units can therefore be made very much lighter – one design has an outside diameter of 4 mm and a total mass of 0.14 g. Decreasing the seismic mass in this way has the advantage of increasing the undamped natural frequency, and hence of increasing the frequency range of the accelerometer, but it has the disadvantage of seriously reducing sensitivity, because the electrical output comes from the work done by the seismic mass. However, a reduced sensitivity may be acceptable if the accelerometer is only being used to find the natural frequency of vibration of a structure by finding the frequency of excitation at which the output of the accelerometer is a maximum.

## Types of low-frequency accelerometer

*Strain gauges* may be used as the transducer system; the strain gauges can be of either the unbonded type, as in Figure 1.4, or the bonded type. Bonded strain gauges are usually applied to a thin flexible beam or cantilever which acts as the spring supporting the seismic mass. They are connected into a bridge circuit and positioned so as to have maximum sensitivity to accelerations along the sensing axis, minimum sensitivity to transverse accelerations, and to cancel out temperature effects. Damping is usually by means of silicone oil, to give a damping ratio of 0.6 to 0.8.

Other forms of transducer which may be used in accelerometers are the potentiometer, the linear variable differential transformer, and the differential capacitor.

## Servo accelerometers

Servo accelerometers (another name for them is *null-balance accelerometers*) are used in preference to other types of accelerometer where greater accuracy is required. In the usual form of servo accelerometer the seismic mass is attached to the casing by material which has been thinned down, by machining, to make it flexible enough to act as a hinge. The seismic mass is maintained in an almost constant position relative to the casing by an automatic control system. This controls the position of the seismic mass by adjusting the current through an electromagnet which consists of a pair of coils attached to the seismic mass, and annular permanent magnets fixed to the casing. The current through the coils is also passed through a sensing resistance, $R$, so that an output voltage proportional to the acceleration is obtained from the voltage drop across $R$. The position of the seismic mass relative to the casing is sensed by an inductive or capacitive displacement transducer, the output of which is amplified and applied to the electromagnet to provide the restoring force.

Servo accelerometers have the following advantages over other types of accelerometer:

1. Because the mechanical spring is replaced by an electrical 'spring', linearity is improved and hysteresis eliminated.
2. Damping can be built into the characteristics of the electrical circuit and can therefore be made less sensitive to temperature change.
3. By introducing an offset current from an external source through the electromagnet coils, the servo accelerometer can be used as an acceleration controller.
4. Similarly, by means of offset currents, the static and dynamic performance of the accelerometer can be checked out before the start of expensive tests on a vehicle.

## The calibration of accelerometers

Accelerometers for the measurement of steady or slowly varying accelerations may be calibrated up to an acceleration of $\pm 1\ g$ (the standard value of $g$ is 9.80665 $m/s^2$) by using the earth's gravitational attraction. The accelerometer is mounted on a tilting table from which the angle $\theta$ between the sensing axis and the vertical can be measured. At $\theta = 0$ the force of gravity on the seismic mass is the same as the inertia force due to an acceleration of 9.8 $m/s^2$. At any other angle of $\theta$ the corresponding acceleration is $9.8 \cos \theta\ m/s^2$. For accurate calibration the true value of $g$ at the location where the calibration is taking place should be used. The standard value, given above, is approximately correct for temperate latitudes, but $g$ varies from 9.832 $m/s^2$ at the poles to 9.780 $m/s^2$ at the equator.

Some steady-state accelerometers have provision for applying known forces to the seismic mass along the sensing axis, by means of weights, so that if the value of the seismic mass is known, the accelerometer can be calibrated for accelerations greater than $g$ by applying the equivalent of the inertia force. If the construction of the accelerometer does not permit this it may be mounted on a turntable so that its sensing axis is radial; the turntable is then run at known angular velocities of $\omega$ rad/s, so that known centripetal accelerations of $\omega^2 r$ $m/s^2$ are applied, where $r$ is the radius in metres to the centre of the seismic mass.

Piezoelectric accelerometers cannot usually be calibrated by means of static loadings because their charge leaks away, although if the piezoelectric material is quartz the time constant of the leakage may be several days due to its high electrical insulation. It is usual, however, to calibrate piezoelectric accelerometers by shaking them with simple harmonic motion along the sensing axis, by means of an electro-mechanical exciter. For a primary calibration the amplitude of the motion is measured by means of an interferometer, using a laser as the light source and a phototransistor to convert the interference fringes into electrical pulses. By this means both the amplitude, $x$, and the angular frequency, $\omega$, of the motion may be accurately measured; the amplitude of the acceleration is then $\omega^2 x$.

For a secondary calibration, the accelerometer to be calibrated is mounted 'back-to-back' with one which has already been calibrated to act as a transfer standard, and the same simple harmonic motion is applied by the exciter to both. The acceleration applied to the accelerometer to be calibrated is then read from the one which has been previously calibrated.

## Exercises on chapter 6

1 Show by means of a simple labelled diagram the essential features of a seismic pickup. List the three main types of seismic pickup.

2 For a damped spring-mass system, sketch curves showing how the ratio (displacement of mass)/(displacement of support) varies with the exciting frequency, for various values of damping ratio. Hence show the usable frequency range of a seismic displacement pickup in relation to the undamped natural frequency of the spring-mass combination.

3 State three ways in which a signal proportional to the velocity of a vibration may be obtained, listing them in order of preference (assume that the necessary transducers are available) and giving your reasons for placing them in that order.

4 Draw a labelled diagram showing a section through a velocity pickup and explain how the signal is obtained. Is its usable frequency range above or below the undamped natural frequency of the spring-mass combination?

5 a) Describe how the seismic mass and crystal are arranged in:
   i) an accelerometer in which piezoelectric material is in compression,
   ii) an accelerometer in which piezoelectric material is in shear.
  b) State the main advantage and disadvantage of each type relative to the other.
  c) State an approximate value of the damping ratio of a piezoelectric accelerometer.
  d) Can a piezoelectric accelerometer be used to measure accelerations which only change very slowly – if not, why not?

6 For a damped spring-mass system sketch the shape of graphs with logarithmic scales along both axes, which show how the ratio (acceleration of mass)/(acceleration of support) varies with the exciting frequency for various values of damping ratio between $\varsigma = 0.01$ and $\varsigma = 1.0$. Indicate the undamped natural frequency, and the usable frequency range of an accelerometer.

7 Describe two alternative ways in which damping may be applied to the seismic mass of an accelerometer. State the ideal value for the damping ratio of such an instrument.

8 a) Explain the principle of a servo accelerometer.
  b) State three advantages which a servo accelerometer has compared with other types of accelerometer.

9 How would the following calibrations usually be carried out?
  a) The calibration of a strain-gauged accelerometer over the range $+1\,g$ to $-1\,g$ only.
  b) A primary calibration of a piezoelectric accelerometer (brief description only).
  c) A secondary calibration of a piezoelectric accelerometer.

10 Which of the accelerometer types i) to iv), listed below, would you use for which of the following applications a) to d)? Give reasons for your answers. Assume that constraints such as cost, availability and size are not applicable.
   i) piezoelectric
   ii) potentiometric
   iii) strain-gauged
   iv) servo.
  a) To measure accelerations with frequencies not greater than 1 Hz. Weight must be kept to a minimum.
  b) To measure accelerations with frequencies not greater than 1 Hz, where considerable electrical interference is present.
  c) To measure accelerations with the maximum possible accuracy.
  d) To measure accelerations with frequencies not less than 100 Hz. Weight must be kept to a minimum.

# 7
# Flow measurement

## The venturi meter

This is a restriction in a pipe, in the form of a tapering section which leads to a narrow throat, as shown in Figure 7.1. Beyond the throat, the cross-section tapers more gradually back to the original pipe diameter. There is a reduction in pressure at the throat because the velocity of the liquid is greater in the smaller cross-section; except for a small loss of pressure due to fluid friction this pressure drop is regained as the cross-section enlarges again to the original pipe diameter. Pressure tappings at the entry to the venturi and at the throat are led to a differential pressure transducer, from which a signal proportional to the pressure difference may be obtained.

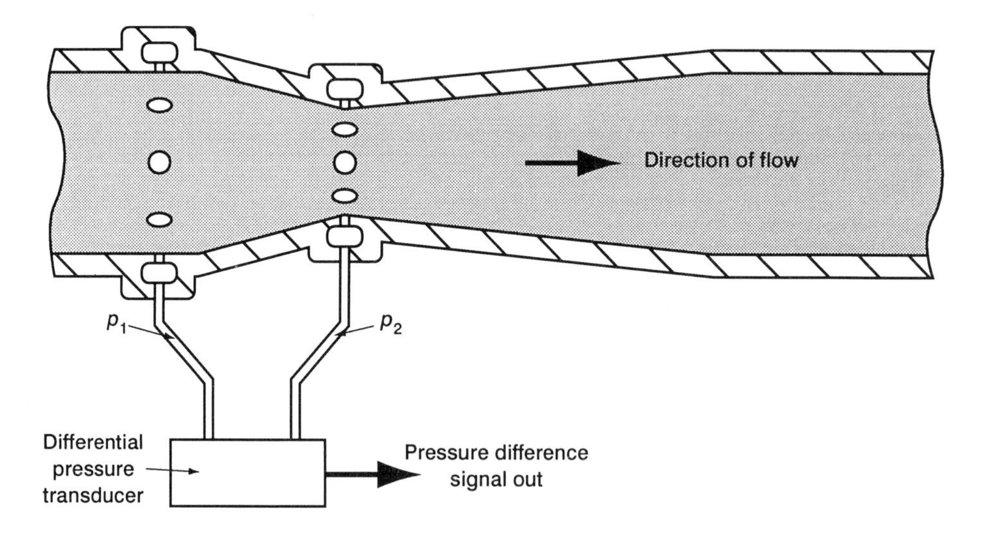


**Figure 7.1** Section through a venturi meter

The pressure difference in a venturi meter is governed by Bernoulli's law; that in any continuous body of liquid the sum of potential energy, pressure energy and kinetic energy is constant at all points. In a horizontal pipe potential energy is constant and thus cancels out from any equation. So equating the sum of pressure energy and kinetic energy at the upstream tapping (subscript 1) and at the throat (subscript 2) for unit mass (1 kg) of liquid we get the equation that appears on the next page.

$$\frac{p_1}{\rho} + \frac{v_1^2}{2} = \frac{p_2}{\rho} + \frac{v_2^2}{2}$$

where

$p$ is pressure (N/m$^2$)
$v$ is velocity (m/s)
$\rho$ is the density of the liquid (kg/m$^3$).

Likewise the volume flow rate

$$\dot{Q} = a_1 v_1 = a_2 v_2 \ \frac{\text{m}^3}{\text{s}}$$

where

$a$ is the cross-sectional area of flow (m$^2$).

Furthermore:

$$\therefore \dot{Q} = \frac{a_1 a_2}{\sqrt{(a_1^2 - a_2^2)}} \sqrt{\frac{2}{\rho}(p_1 - p_2)} \ \text{m}^3/\text{s}$$

This can be simplified:

$$\dot{Q} = a_2 \sqrt{\frac{2(p_1 - p_2)}{\rho(1 - \beta^4)}}$$

where

$$\beta \text{ is the ratio} : \frac{\text{throat diameter}}{\text{pipe diameter}}$$

Again this is a theoretical flow rate, and to obtain the actual flow rate it must be multiplied by a coefficient of discharge, $C_d$, which is found by calibration and is usually in the range 0.97 to 0.99.

The above equation shows that for a given venturi meter and a given liquid the flow rate depends only on the square root of the pressure difference. Differential pressure transducers tend to become less accurate at pressure differences less than about one-tenth of the upper limit of their range, so the accuracy of the combination of venturi and transducer can only be relied on down to about one-third of the maximum allowable flow rate (because $\sqrt{0.1} = 0.316$).

Another possible source of error at low flow rates is the decrease in the coefficient of discharge as Reynolds number decreases.

## Reynolds number

Osborne Reynolds found that the flow of a liquid through a round pipe could be either streamlined (laminar) or turbulent, and that Reynolds number (symbol: Re) determined which of these two types of flow would occur in any particular case. Reynolds number may be calculated as:

$$\frac{vd\rho}{\eta} \text{ or as } \frac{vd}{\nu}$$

depending on whether the dynamic or the kinematic viscosity is given in the data (both formulae give the same number, of course).

In the formulae on the previous page:

$v$ is the velocity of the liquid (m/s)
$d$ is the inside diameter of the pipe (m)
$\rho$ is the density of the liquid (kg/m$^3$)
$\eta$ (eta) is the dynamic viscosity of the liquid (Ns/m$^2$)
$\nu$ (nu) is the kinematic viscosity of the liquid (m$^2$/s).

- **Dynamic viscosity** is also known as coefficient of viscosity. It is usually quoted in the non-SI unit the centipoise (abbreviation cP). 1 cP = $10^{-3}$ Ns/m$^2$. In *basic,* SI units of Ns/m$^2$ can be expressed as kg/m s.
- **Kinematic viscosity** is dynamic viscosity divided by density. It is usually quoted in centistokes (abbreviation: cSt). 1 cSt = $10^{-6}$ m$^2$/s.
- **Reynolds number** is dimensionless; if we put the units into the above fractions along with the numbers, the units cancel out completely. So Reynolds number has exactly the same numerical value whether we are working in SI, Imperial, or any other *consistent* system of units.

If Reynolds number is less than 2000, the flow in a pipe is *laminar* – that is, the liquid moves in the form of infinitely thin sleeves which slide smoothly, one inside another, so that the liquid at the centre of the pipe is moving fastest while the liquid in contact with the wall of the pipe is stationary. At Reynolds numbers greater than 2000, vortices form in the liquid and the flow is said to be *turbulent.* The turbulence increases as Reynolds number increases, and complete turbulence is reached at a Reynolds number of about $2 \times 10^5$. The velocity of the liquid is then virtually constant at any radius from the axis of the pipe, except for a thin boundary layer 'attached' to the wall of the pipe. The Reynolds number of the flow in the throat of a venturi is always greater than that in the pipe because, for a given flow rate of a given liquid, Reynolds number is inversely proportional to diameter, because:

$$\text{Re} \propto v \times d \propto \frac{1}{a} \times d \propto \frac{1}{d^2} \times d \propto \frac{1}{d}$$

So we need only consider the Reynolds number of the flow in the pipe itself. The coefficient of discharge of a venturi meter is only constant for Reynolds numbers greater than $2 \times 10^5$. Figure 7.2, which applies to a typical venturi in a 100 mm diameter pipe, shows how it decreases at lower Reynolds numbers. Because, for a given venturi meter, the coefficient of discharge depends only on the Reynolds number, it may be found by calibration using one liquid (often water) and applied to any other liquid through the same meter at the same Reynolds number.

*Example 7.1*
A pipe of 100 mm internal diameter carries water at 20°C, which has a density of 998 kg/m$^3$ and a dynamic viscosity of 1.002 cP. Calculate the velocity:

a) below which the flow will be completely laminar
b) above which the flow will be completely turbulent.

*Solution*

$$\text{a)}\quad Re = \frac{vd\rho}{\eta} \therefore 2000 = \frac{v \times 0.1 \times 998}{1.002 \times 10^{-3}} \quad \frac{\cancel{m}}{\cancel{s}} \times \cancel{m} \times \frac{\cancel{kg}}{\cancel{m}^3} \times \frac{\cancel{m}\,\cancel{s}}{\cancel{kg}}$$

$$\therefore v = \frac{2000 \times 1.002 \times 10^{-3}}{0.1 \times 998} = 0.0201 \text{ m/s}$$

$$\text{b)}\quad v = \frac{0.0201 \times 2 \times 10^5}{2000} = 2.01 \text{ m/s}$$

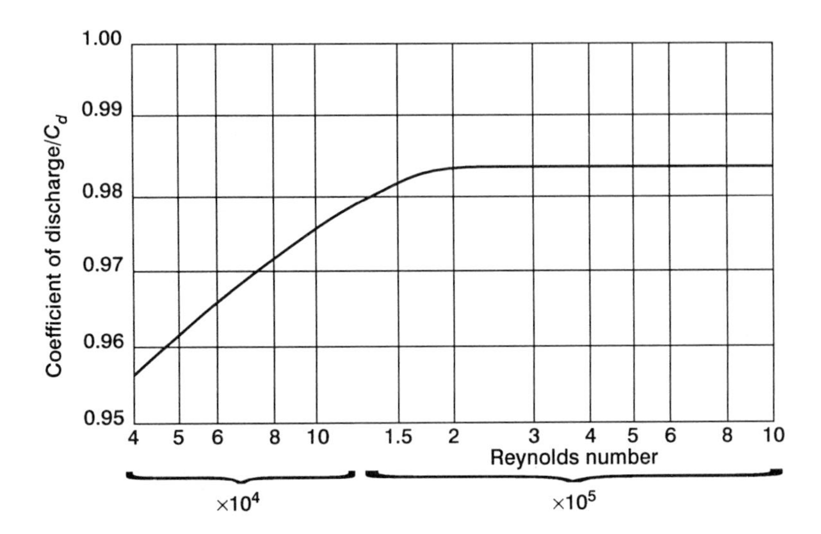


**Figure 7.2** Coefficient of discharge of a typical venturi meter in a 100 mm diameter pipe, plotted on a logarithmic scale of Reynolds number

The above example shows that in a pipe of reasonable size, water at room temperature would have to be moving quite slowly for the flow to be laminar; usually flow velocities exceed the 2 m/s obtained in the answer to b) above.

*Example 7.2*
The throat and full bore diameters of a horizontal venturi meter are 30 mm and 60 mm respectively.

a) Calculate the coefficient of discharge of the venturi meter if the pressure at the full bore section is 47.7 kN/m$^2$ above that at the throat when water is being pumped through it at the rate of 0.420 m$^3$/min. Take the density of water as 1000 kg/m$^3$.
b) If a liquid with a relative density of 0.75 and a dynamic viscosity of 0.00157 Ns/m$^2$ is pumped through the same venturi meter, calculate the flow rate which would give the same Reynolds number, and hence the same coefficient of discharge. Take the dynamic viscosity of water as 0.001 Ns/m$^2$.

*Solution*

a) $\dot{Q} = C_d a_2 \sqrt{\dfrac{2(p_1 - p_2)}{\rho(1-\beta^4)}}$

$$a_2 = \frac{\pi}{4} \times 0.030^2 = 0.000707 \text{ m}^2; \ \beta = \frac{30}{60} = 0.5$$

$$\therefore \frac{0.420}{60} = C_d \times 0.000707 \times \sqrt{\frac{2 \times 47700}{1000(1-0.5^4)}} \quad \therefore 0.007 = 0.00713 C_d$$

$$\therefore C_d = 0.982$$

b) $a_1 = \frac{a_2}{\beta_2} = \frac{0.000707}{0.25} = 0.00283 \text{ m}^2$

$$v_1 = \frac{\dot{Q}}{a_1} = \frac{0.007}{0.00283} = 2.47 \text{ m/s}$$

$\therefore$ in part a)

$$\text{Re} = \frac{vd\rho}{\eta} = \frac{2.47 \times 0.060 \times 1000}{0.001} = 148200$$

For the other liquid:

$$148200 = \frac{v \times 0.060 \times 0.75 \times 1000}{0.00157}$$

$$\therefore v = \frac{148200 \times 0.00157}{0.06 \times 750} = 5.17 \text{ m/s}$$

Flow rate:

$$a_1 v_1 = 0.00283 \times 5.17 = 0.01463 \text{ m}^3/\text{s}$$

or

$$0.01463 \times 60 = 0.878 \text{ m}^3/\text{min}.$$

## The orifice plate

This is another way of making an obstruction in a pipe so that the flow rate can be measured from the pressure difference it causes. Shown in section in Figure 7.3, it is simply a flat plate with a sharp-edged hole in it, through which the fluid has to pass. This constricts the moving fluid into a more or less streamlined volume, as though it were passing through a venturi. The shaded region on either side, adjacent to the plate, is stationary fluid which serves to transmit the pressures from the moving body of fluid to the pressure transducer.

Everything which has been written about the flow through a venturi meter in the preceding pages applies also to the flow through an orifice plate. The formula for calculating the flow rate is identical with that given in the preceding section.

The difference is that the coefficient of discharge is much less than that of a venturi meter; it is about 0.63 for completely turbulent flow, *increasing* somewhat for less turbulent flow, so the coefficient of discharge depends on the Reynolds number of the flow in the pipe, just as it does for a venturi meter. There are two main advantages to an orifice plate.

1 It gives a larger pressure difference for a given flow rate than a venturi meter does.
2 It is cheap and very compact – it is often possible to insert it into an existing pipeline as a ready-made and calibrated ring containing orifice plate and pressure tappings, which can be sandwiched between existing pipe flanges.

The main disadvantage of an orifice plate compared with a venturi meter is that while some of the immediate pressure drop at the downstream side of the orifice plate is regained, the recovery is much less than would be obtained from a venturi, so there is a greater power loss in pumping through an orifice plate than through a venturi. Its lower initial cost may be cancelled out by higher running costs. Also, liquids containing solids in suspension (slurries) tend to silt up in the 'dead' region on either side of the plate, which alters the calibration; this may be overcome by lowering the position of the orifice in the plate, so that the silt is carried

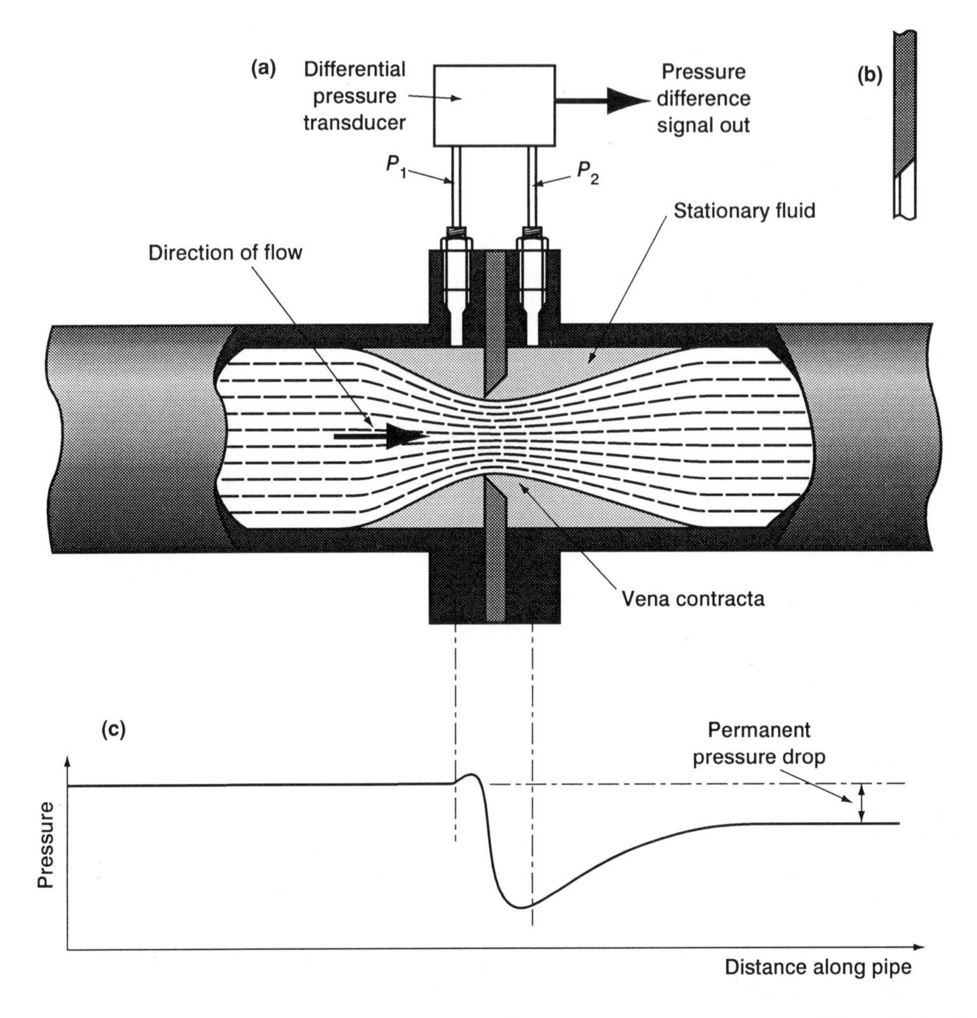


**Figure 7.3** The orifice plate flowmeter (a) section showing orifice and pressure tappings (b) detail of edge of orifice (c) pressure distribution adjacent to the orifice

through. Rounding-off of the sharp edge of the orifice will also alter the calibration. Such wear may be caused by long use or by abrasive particles in the fluid.

## Mechanical flowmeters

In meters of this type the fluid is made to do work on some kind of machine, the quantity of fluid passing through the machine being proportional to the number of oscillations or rotations of the mechanism, as indicated by a counter. The domestic gas meter is an example of this type of meter; the gas inflates alternately each chamber of a pair of bellows, being diverted to the other chamber by a valve mechanism as one chamber becomes completely full.

There are many variations of this principle; almost any sort of 'pump' mechanism which can be driven by forcing fluid through it is a suitable basis for a flowmeter of this type. It is basically a volume measurement, but rate of flow can be given by the same mechanism with a tachometer in place of a counter.

## The turbine flowmeter

In this type of mechanical flow meter the fluid flows past a rotor with skewed blades, spinning the rotor at a speed proportional to the flow rate. Figure 7.4 shows the construction of such a meter. To ensure consistent readings, the flow upstream of the rotor is straightened by stationary radial vanes, which also act as spacers to centralise the rotor bearings. The rotor may drive a mechanical counter through reduction gearing or it may generate a digital signal by means of a magnetic transducer similar to that described on p.72, one cycle of AC being generated each time a rotor blade passes the pick-off coil – the blades must, of course, contain magnetic material.

Because of friction in the bearings, the actual speed of the rotor is always slightly less than the speed at which it should theoretically be rotating (based on the pitch of the helix traced out by the blades), so such meters are classified as *inferential*. This error becomes more serious at lower flow rates, and there is obviously a minimum flow rate below which the rotor does not rotate at all. The error is usually less than 2% provided that the flow rate is greater than about 7% of the rated maximum. To keep the accuracy constant fluid must be clean so that the bearings run freely.

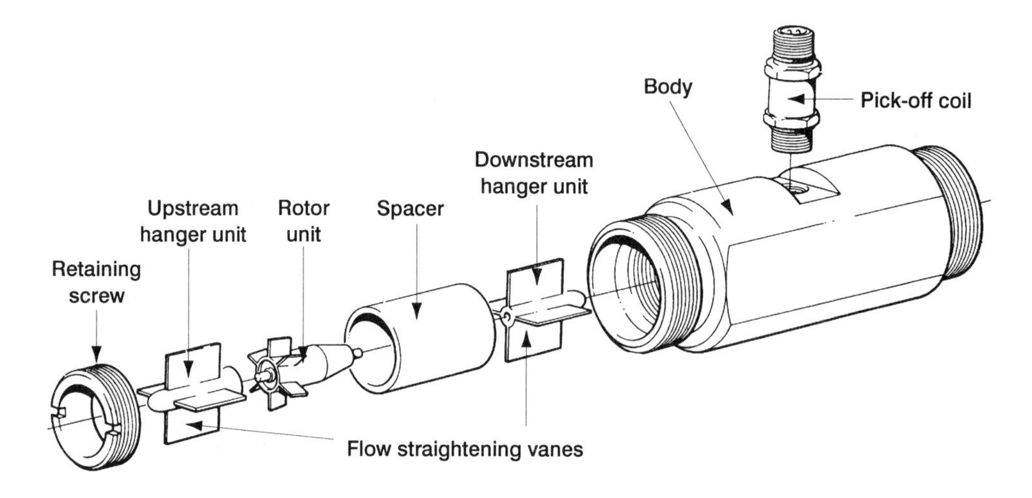


**Figure 7.4** Turbine meter unit (Figure courtesy of Kent Instruments)

## The rotameter

This variable area flowmeter, illustrated in Figure 7.5, consists of a 'float' in a tapered tube. The tube, which is usually of high-strength glass, is arranged with its axis vertical so that the fluid enters the narrow end of the tube and rises to exit at the wide end. The float does not actually float but appears to do so, as it rises or falls to a level where it is in equilibrium under the action of its weight (acting downwards) and the drag of the fluid plus the buoyancy force acting upwards. The fluid rises past the float through the annular space between the float and the wall of the tube and this restriction creates a pressure drop between bottom and top of the float. The upward drag of the fluid is mainly due to this pressure drop acting on the cross-sectional area of the float. If the flow rate increases, the pressure drop increases, causing the float to rise until the increased annular space between float and wall of tube reduces the pressure drop to its equilibrium value again.

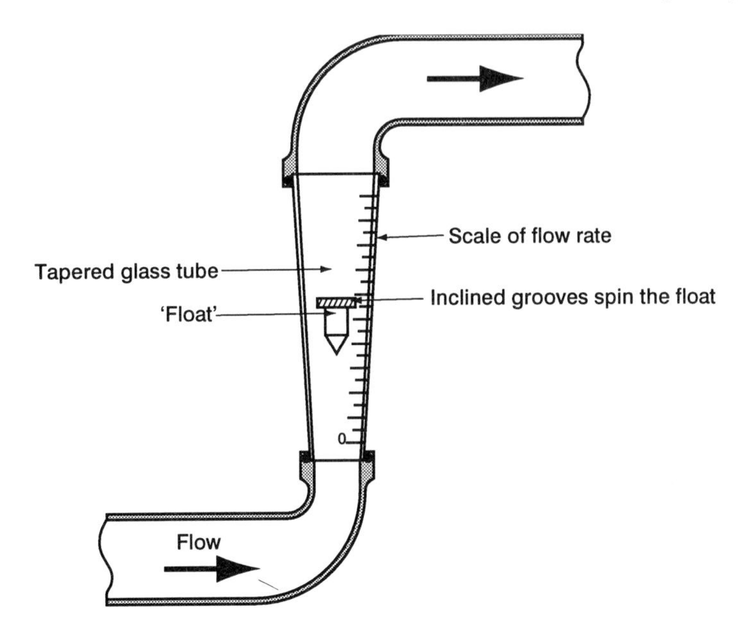


**Figure 7.5** The Rotameter

To take a reading of flow rate from a rotameter the top of the float is sighted against a scale engraved on the glass tube. To withstand higher pressures, the tapered tube may be made of metal, and the float position detected magnetically through the wall of the tube.

The float is usually cylindrical, with a pointed bottom end, sharp edges to create turbulence, and helical grooves around a rim at the top so that the fluid spins it, to stabilise its axis gyroscopically. Alternatively the float may be spherical, or its axis may be stabilised by arranging it to slide up and down a vertical guide wire coaxial with the tube.

Rotameters usually have an accuracy of about ±2% of full scale and a repeatability of about 0.25% of the reading. They have an accurate range of about 10:1 – much better than flowmeters in which the reading depends on the square root of a pressure drop.

The calibration of a rotameter applies to only one particular density of fluid. Rotameters may be used to measure the flow rates of gases or liquids. They are limited to fairly small rates of flow.

## The electromagnetic flowmeter

The principle of the electromagnetic flowmeter is shown in Figure 7.6; it follows from Fleming's right-hand rule for generators. If a conductor is moving through a magnetic field, the forefinger, second finger and thumb of the right hand, placed at right angles to each other represent respectively the directions of flux, EMF and motion. In Figure 7.6 the conductor is the liquid, the motion is the flow of that liquid through the pipe, the magnetic field is at right angles to that flow, and the electrodes then pick off the voltage generated at right angles to both the field and the flow. The voltage obtained is directly proportional to the rate of flow of the liquid.

In practical electromagnetic flowmeters, the magnetic field is produced by coils immediately above and below the pipe. The short length of pipe which forms the flowmeter is of non-magnetic material, so that the field can pass through it, and if not an electrical insulator

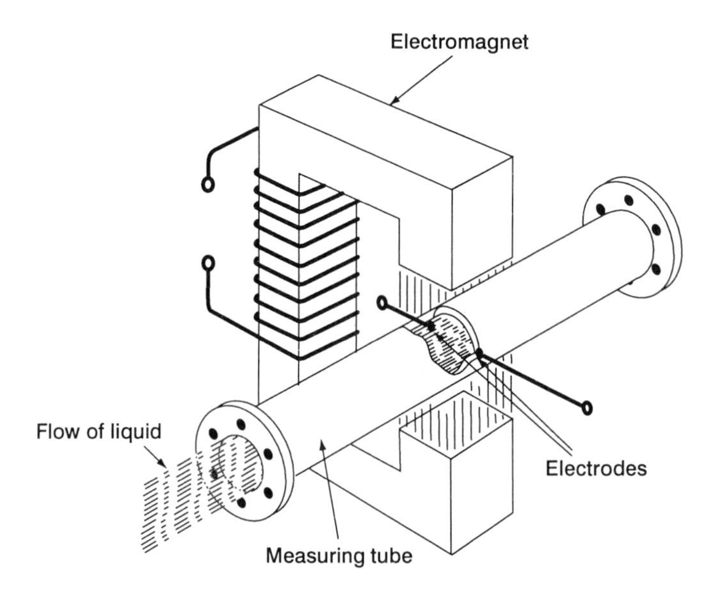


**Figure 7.6** Principle of the electromagnetic flowmeter

itself, it is lined with insulating material so that it does not short-circuit the output voltage. Because the flowing liquid must conduct electricity, petroleum products cannot be measured by this type of flowmeter. Any liquid which separates into ions has sufficient conductivity, however, so solutions of acids, alkalis and even water can be measured by this means, provided that the water is not completely pure. A minimum conductivity of about 1 mS/m is necessary in the liquid to be measured.

With most liquids a constant magnetic field would cause polarization of the electrodes – that is the formation, by electrolytic action, of an insulating layer of neutral molecules on one or both of the electrodes, causing a falling-off of the output voltage. There would also be a magnetohydrodynamic effect (a difference in the velocity of the liquid between one side of the pipe and the other) which would introduce errors in the readings. To eliminate these errors the electromagnets are supplied either with AC at 50 or 60 Hz or with *interrupted* DC.
There are several advantages to electromagnetic flowmeters.

1 There is no obstruction in the pipe, therefore the flowmeter causes no pumping losses – this can make a considerable saving in running costs.
2 The calibration is unaffected by changes in viscosity or by disturbances in the density or flow of the liquid, provided that the velocity distribution is symmetrical about the vertical centre line of the pipe. Thus this type of flowmeter can measure the flow rates of viscous slurries and non-Newtonian liquids such as blood or non-drip paint.
3 There is a wide linear range of measurement.
4 They are able to measure reverse flows.
5 With suitable materials for the pipe lining and the electrodes, the flow rates of corrosive liquids and liquids carrying abrasive solids in suspension can be measured.

## The hot-wire anemometer

A greatly enlarged view of a velocity measuring head is shown in Figure 7.7. It consists of a fine tungsten wire stretched between the tips of a streamlined forked support. Typical data for

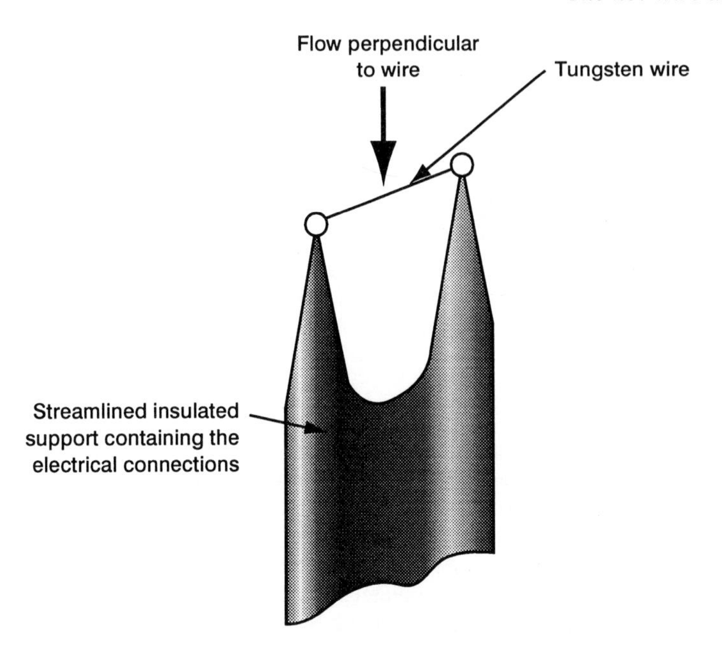


**Figure 7.7** Hot-wire anemometer

the wire would be diameter 0.008 mm, length 1 mm, resistance 1 Ω. It measures velocity from the effect, on the wire's electrical resistance, of the cooling caused by the flow of the fluid past the wire. It is mainly used to measure the flow velocities of gases but it can also be applied to the measurement of liquid velocities. Because of its small thermal capacity it is particularly useful for the measurement of rapid fluctuations in velocity.

For a given temperature difference between the wire and the fluid the heat transfer from the wire to the fluid is determined by the following equation:

Heat transfer rate (watts) = $A + B\sqrt{v}$

where

$v$ is the velocity of flow while
$A$ and $B$ are constants.

Equating heat power output to electrical power input gives

$$I^2R = A + B\sqrt{v}$$

where

$I$ is the current through the wire and
$R$ is its resistance.

The hot wire anemometer is used with a bridge circuit. The best method is to keep the resistance of the hot wire (and hence its temperature) constant, by adjusting the bridge power supply voltage to keep the bridge balanced. (Higher rates of flow have a greater cooling effect on the wire and the bridge power supply voltage must be increased to put more current through the wire to keep its temperature constant.) The temperature output signal comes from the voltage drop across a series resistor in the power supply lead. Flow velocity fluctuations with frequencies up to 50 kHz can be measured if the bridge is automatically balanced by an electronic circuit.

There are two practical difficulties which could occur with a hot-wire anemometer:

1 The hot wire may vibrate in high flow velocities, causing it to fatigue and break
2 Unless the fluid is very clean the hot-wire may become coated with dirt, which will alter the calibration, or it may even be broken by impact with large dirt particles.

To overcome the hot-wire's liability to fracture and to enable this type of anemometer to operate in high-temperature fluid flows, the hot wire may be replaced by a thin film of platinum deposited on glass. The film may be deposited on the 'cutting edge' of a wedge of glass facing into the flow, or for very high temperature applications it may be deposited on the outside of a glass tube through which cooling water is passed.

## The vortex flowmeter

When fluid flows around a suitably shaped obstacle, such as a cylinder, vortices (eddies) peel off from alternate sides of the obstacle and are carried downstream. These are called *von Karman vortices*, after their discoverer. If the flow has a Reynolds number above about 10 000, the distance between successive vortices is constant for a given cross-section of obstacle, thus the number of vortices passing a given point in a given time interval is a measure of the velocity of the fluid. Liquids, gases and steam may be metered by this method.

The vortices may be detected by a variety of means. Since their creation causes alternating side forces on the obstacle, it may be strain gauged or fitted with piezoelectric transducers to count them, or it may detect them as local velocity changes picked up on the surface of the obstacle by hot-film anemometry. Downstream of the obstacle they may be detected by ultrasonic vibrations. Figure 7.8 shows how this is done. The diagram shows a longitudinal section through a vortex flowmeter. The obstacle is shown in the form of a cylinder, though some manufacturers prefer a cross-section with sharp corners as being more effective in creating vortices. The ultrasonic vibrations are emitted in a narrow beam by a piezoelectric crystal (the transmitter), which is vibrated at a frequency of a few megahertz by an oscillator applying an AC voltage at that frequency. The receiver is a similar piezoelectric crystal which converts the received vibrations back into an AC voltage. Vortices passing through the beam interrupt it to some extent. The output of the receiver is an AC voltage in the form of a carrier

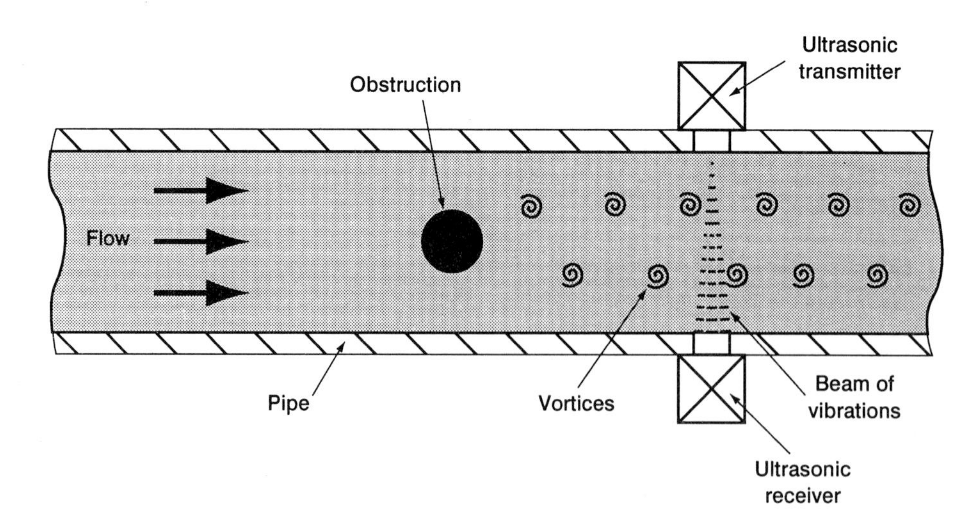


**Figure 7.8** Flow measurement by vortex shedding

wave at the transmitter frequency, modulated by the passage of successive vortices. If the carrier wave is filtered out by a low-pass filter and the resulting low-frequency signal applied to a frequency-to-voltage converter, the output will be a DC voltage signal directly proportional to the velocity of flow.

Because of this direct proportionality, the maximum to minimum flow rate ratio is quite large. A typical installation might have a maximum to minimum ratio of 20:1, corresponding to a rate of vortex creation ranging from about 200 per second to about 10 per second.

## Doppler flowmeters

The Doppler effect has already been explained in chapter 5. It can be used to measure flow velocity in pipes either by a beam of ultrasonic vibrations or by a beam of laser light projected through a transparent section of pipe. The ultrasonic method is mainly used for the measurement of liquids, though it can be applied to gases. The laser method is equally applicable to liquids and gases.

A Doppler flowmeter cannot measure a flow of absolutely pure fluid; it can only operate on fluids carrying particles or air bubbles in suspension, because the Doppler effect acts on reflections from the particles or bubbles. If an ultrasonic Doppler flowmeter is to measure flow in a 200 mm diameter pipe, for example, particle sizes of about 100 μm to 40 μm are a minimum requirement (they can be at the lower end of the range if there is a greater concentration of them in the fluid). A laser Doppler flowmeter, on the other hand, can operate with particles so small that it can measure the flow of almost any transparent fluid of 'ordinary' purity.

### Ultrasonic Doppler flowmeters

Figure 7.9 shows the principle of a 'clamp-on' ultrasonic Doppler flowmeter. It consists of a piezoelectric transmitter and a piezoelectric receiver, similar to those forming the ultrasonic sensing device in the vortex flowmeter described in the preceding section of this chapter, and working at a similar frequency.

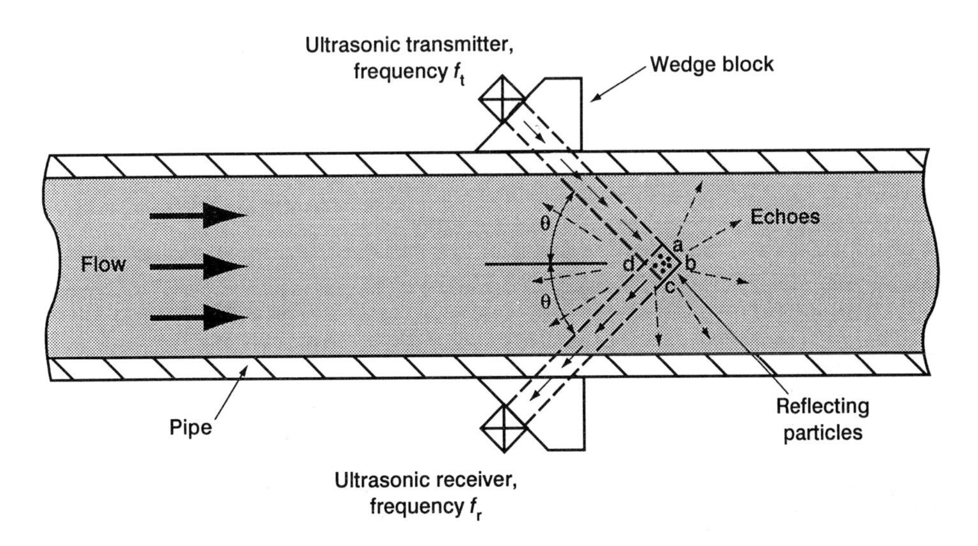


**Figure 7.9** Doppler ultrasonic flowmeter with separate receiver and transmitter

The transmitter applies a narrow beam of vibrations to a wedge block; they pass through the wedge block and the wall of the pipe, and echoes are reflected in all directions from the particles that lie in their path through the fluid. The receiver, mounted on another wedge block, can pick up echoes only from particles which are in the narrow cylindrical volume corresponding to that through which the transmitter's vibrations are propagated. Thus the only echoes picked up by the receiver are those in the volume indicated by abcd in Figure 7.9. The received echoes are very weak and have to be amplified. Their frequency has been decreased by the Doppler effect. In theory the decrease is given by:

$$f_t - f_r = \frac{2f_t \cos\theta}{c} \times v$$

where

$f_t$, $f_r$ and $\theta$ are shown in Figure 7.9
$c$ is the speed of sound in the fluid
$v$ is the velocity of the particles.

In practice the above equation is found to give results which are only approximately correct because of factors which have not been taken into account in its derivation, and the relationship between frequency shift and velocity should be obtained by calibration.

There are two main methods by which the difference between the transmitted and received frequencies may be obtained:

1. The receiver output is mixed with a portion of the oscillator output to produce a beat frequency equal to the frequency difference; this is extracted by a low-pass filter.
2. An up-down counter is used to count-up the cycles of transmitter frequency over a period of from 5 to 20 seconds; the received frequency is then counted-down over the same time interval. The remainder is a number proportional to the frequency difference. This method gives slower response to flow changes but has an averaging effect, cancelling out false signals due to random noise.

## Laser Doppler flowmeters

A laser enables light to be used for Doppler velocity measurements because it produces monochromatic (single frequency) light at frequencies which are precisely known and very stable. Figure 7.10 shows the principle of a laser Doppler flowmeter. The beam of light from the laser enters an optical system which splits it into an intense main beam and a much weaker reference beam. Both beams are focused on to the point in the fluid at which they cross. The particles in the fluid reflect the light of the main beam in all directions. A very small proportion of this reflected light will be in the direction of a photodetector which is on the receiving end of the reference beam, and so the photodetector will produce a beat frequency which is the difference between the laser light frequency and the Doppler-shifted frequency of the reflections from the moving particles. The beat frequency is applied to an up-down counter, operating as described for the ultrasonic Doppler flowmeter in the preceding section, its output being a number proportional to the velocity of the fluid at the point in the cross-section on which the beams were focused.

## The pitot tube

This has already been described and illustrated in chapter 5, as a development of the airspeed measurement transducer for aircraft.

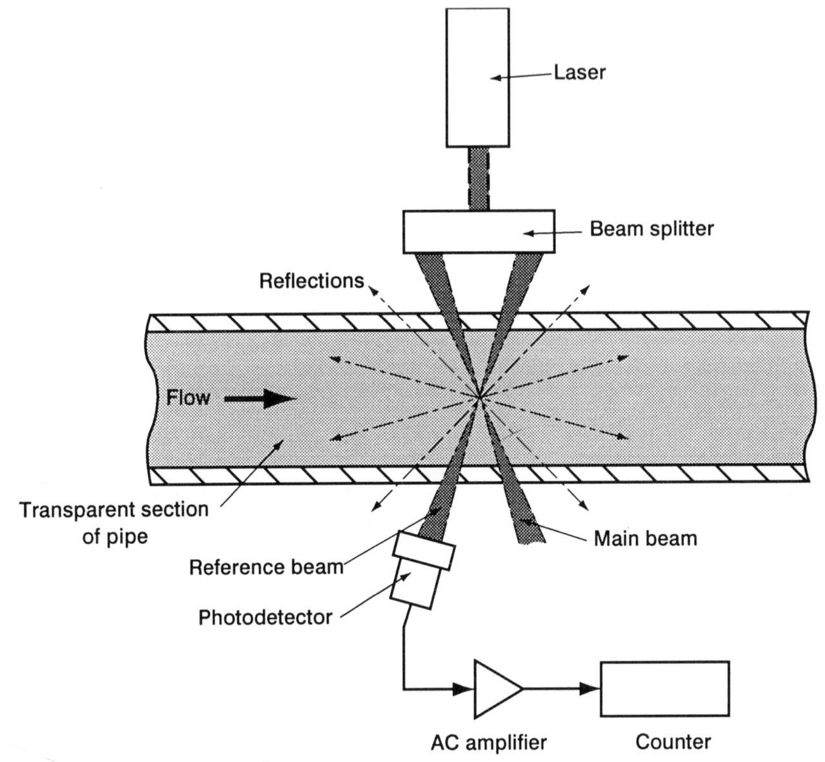


**Figure7.10** A laser Doppler flowmeter

## The calibration of flowmeters

Fluid flow is one of least satisfactory industrial measurements because the calibration of a flowmeter can be affected by variations in the density and viscosity of the fluid as its temperature and pressure change and by flow disturbances due to elbows, tees or valves immediately upstream or downstream of the flowmeter. Accurate calibration is important because the cost control of plant operation depends largely on flow measurement.

**The primary calibration** of a flowmeter for a **liquid** is done by setting a steady rate of flow and measuring the time taken to collect a given weight or volume of the outflow in a tank or reservoir. By repeating this process for a number of different flow rates covering the useful range of the meter a calibration curve is obtained.

The calibration of a flowmeter for a **gas** can be done in a similar way, by measuring the time taken to collect a given volume of the gas. The gas is collected over water or some other sealing liquid in a *gasometer*; a counterbalanced inverted cylindrical container which rises as the gas is accumulated. Figure 7.11 illustrates the principle. The temperature and pressure of the gas must be measured so that the calibration can be corrected for variations in gas density. This will require the gauge pressure to be converted to absolute, so a reading of ambient barometric pressure must also be taken.

Where such equipment is not available the principle of dynamical similarity may be used, the flowmeter being calibrated with a liquid and the calibration applied to gas flows with the same values of Reynolds number. Velocity probes (anemometers) may be calibrated by moving the probe at constant velocity through the stationary fluid, either in a straight line or at the end of a rotating arm.

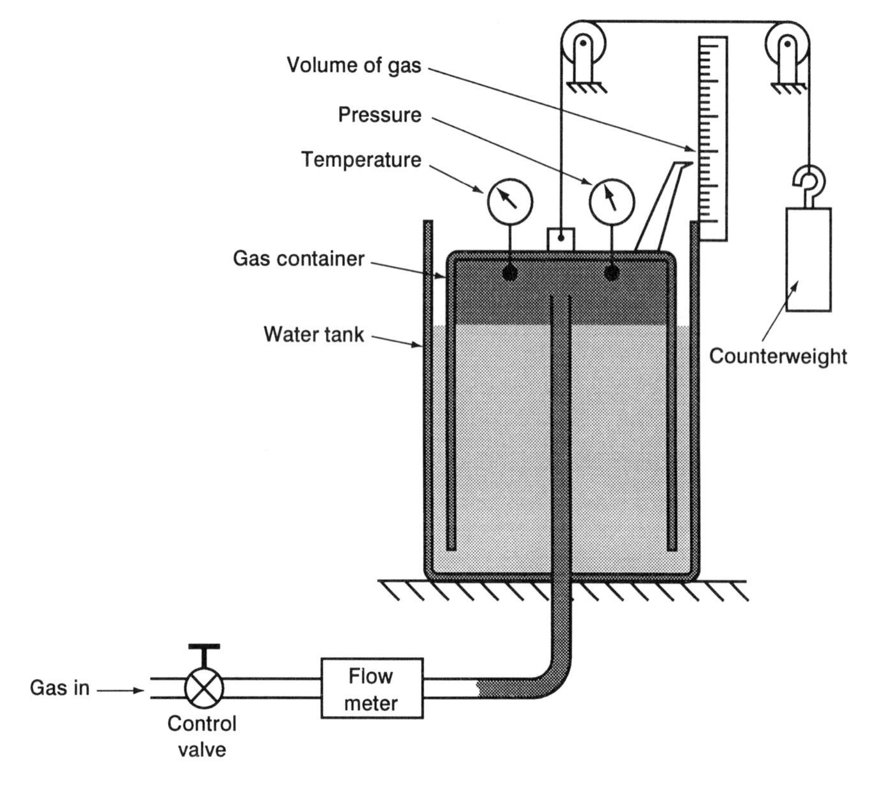


**Figure 7.11** Calibration of a gas flowmeter

**Secondary calibration** of a flowmeter may be carried out by connecting it in series with another flowmeter which has already been calibrated, and using the flow rates measured by the calibrated flowmeter to calibrate the uncalibrated one.

## Exercises on chapter 7

1 A venturi meter has an inlet of 150 mm and a throat of 100 mm diameter. Its coefficient of discharge is 0.985.
   a) Calculate the volume flow rate of water through the pipe if the pressure difference between inlet and throat is 34.5 kN/m$^2$. Take the density of water as 1000 kg/ m$^3$.
   b) Calculate the Reynolds number of the flow at inlet if the kinematic viscosity of the water is 1.004 cSt.
   c) Calculate the volume flow rate of a liquid with a dynamic viscosity of 1.31 cP and a density of 1050 kg/m$^3$ which would have the same Reynolds number.
   d) Calculate the pressure difference between inlet and throat produced by the flow rate of the liquid calculated in c).

2 a) What is meant by 'laminar flow' and 'turbulent flow'?
   b) State the values of Reynolds number:
      i) below which the flow is completely laminar
      ii) above which the flow is completely turbulent.
   c) Convert into SI units: i) 0.282 cP, ii) 0.294 cSt.
   d) If the values given in part c) refer to the same liquid and temperature, what is its density?

e) Calculate the Reynolds number if this liquid is flowing at the rate of 14.4 litres per second in a pipe of 75 mm diameter.

3 a) Sketch a section through i) a venturi meter, and ii) an orifice plate installation in a pipeline.
b) List the advantages and disadvantages of using an orifice plate instead of a venturi meter to measure the flow rate of a liquid which may contain solids in suspension.

4 Describe, with a simple sketch, the principle of a turbine flowmeter. Suggest two possible reasons why it might under-read the flow rate.

5 Explain the principle of the rotameter type of variable-area flowmeter. Why does it need to be recalibrated if it is used with a fluid of a different density from the one it was originally calibrated with?

6 a) Explain, with the aid of a simple sketch, the principle of an electromagnetic flowmeter.
b) State four advantages and one possible disadvantage of an electromagnetic flowmeter, compared with a venturi meter.
c) Why is a constant magnetic field not normally used?
d) What are the two alternatives to a constant field?

7 a) Sketch a hot-wire anemometer transducer.
b) Explain how the transducer is operated in the constant temperature mode.

8 a) What are the two main faults which may occur with a hot-wire anemometer?
b) What development of the hot-wire principle enables it to withstand high-temperature gas flows?

9 Explain, with the aid of a diagram, the principle of the vortex flowmeter. How is the passage of the vortices sensed and converted into a signal proportional to velocity? Suggest two alternative ways in which the formation of vortices may be detected.

10 Explain how the Doppler effect is used in the measurement of fluid velocity taking as examples
a) measurement of the velocity of a liquid by ultrasonic means
b) measurement of the velocity of a gas by means of a laser.
In each case illustrate your answer with a diagram.

11 Briefly explain how the primary calibration of a flowmeter could be carried out,
a) for liquid flows
b) for gas flows.

12 Considering the following flow measurement devices:
a) venturi meter
b) orifice plate
c) turbine flowmeter
d) rotameter
e) electromagnetic flowmeter
f) hot-wire anemometer
g) vortex flowmeter
h) Doppler flowmeter
i) pitot tube
list any which:
i) are unsuitable for fluids carrying particles
ii) cause little or no obstruction to flow
iii) can only measure fluids which carry particles
iv) have a linear relationship between flow rate and output, over a wide range of flow rates
v) have a fragile transducer
vi) cannot measure fluids which are electrical insulators
vii) can measure reverse flows without change of calibration
viii) have a calibration which changes if the density of the fluid changes.

# 8
# Electronics for instrumentation

## The decibel

The power gain of an electrical component such as an amplifier is the ratio (output power)/(input power). In many cases this fraction may have a value of several millions and it is more convenient to express it in logarithmic form. The logarithmic unit of power gain is the **bel**, (symbol B) calculated as

$$\text{Power gain (bels)} = \log_{10}\left(\frac{P_2}{P_1}\right)$$

where $P_2$ and $P_1$ are the output and input power respectively. The bel is rather too large a unit for most purposes, so we actually work in tenths of a bel, decibels (abbreviation: dB). Thus:

$$\text{Power gain (dB)} = 10 \times \log_{10}\left(\frac{P_2}{P_1}\right)$$

Usually, however, we want to relate the input and output *voltages* in terms of decibels.

$$P = V \times I \text{ and } I = \frac{V}{R} \text{ by Ohm's Law.}$$

Thus

$$P = \frac{V^2}{R} \text{ (i.e. } P \text{ is proportional to } V^2.)$$

$$\therefore \text{Power gain (dB)} = 10 \times \log_{10}\left(\frac{V_2^2}{V_1^2}\right)$$

i.e.

$$\therefore \textbf{Power gain (dB)} = \mathbf{20 \times \log_{10}\left(\frac{V_2}{V_1}\right)}$$

The decibel scale, being logarithmic has three great advantages over a numerical scale of gain:

1 Very large variations of gain can easily be plotted on a graph.
2 Graphs which are curved when plotted as numerical gain values become straight lines when plotted as dB.
3 When a system consists of a number of devices in series (e.g. amplifiers, filters, etc.) the overall output/input voltage ratio is obtained by *multiplying* together their individual

voltage ratios. The same result can also be obtained by *adding* together their corresponding dB values of gain (adding logs to multiply). This enables us to add the ordinates of individual frequency-response graphs in dB to get an overall frequency-response graph in dB for the system.

*Example 8.1*
A signal processing unit consists of a low-pass filter, a pre-amplifier and a main amplifier. At a particular input frequency their output/input voltage ratios are 0.75, 63 and 2250, respectively. Calculate the overall voltage ratio of the system at that frequency:

i) by multiplication
ii) by adding dB values.

*Solution*
i) Overall voltage ratio = $0.75 \times 63 \times 2250$ = **106 300**

ii)

$$\begin{aligned}\text{Filter gain} &= 20\log_{10} 0.75 = -2.50\text{ dB}\\ \text{Pre-amp gain} &= 20\log_{10} 63 = 35.99\text{ dB}\\ \text{Main amplifier gain} &= 20\log_{10} 2250 = 67.04\text{ dB}\\ \text{Total} &= 100.53\text{ dB}\end{aligned}$$

$$100.53 = 20\log_{10}\left(\frac{V_2}{V_1}\right)$$

$$\log_{10}\left(\frac{V_2}{V_1}\right) = 5.0265$$

$$\left(\frac{V_2}{V_1}\right) = 10^{5.0265} = \mathbf{106\,300}$$

The decibel scale takes a little getting used to. Its main features are:

0 dB is a voltage ratio of 1 (i.e. $V_2 = V_1$)
3 dB is a voltage ratio of 1.41 (i.e. $\sqrt{2}$)
6 dB is a voltage ratio of 2
20 dB is a voltage ratio of 10

(Each additional 20 dB multiplies the voltage ratio by 10.)

Negative values of dB give voltage ratios which are the reciprocals of the ratios given by the positive dB values. For example, –3 dB is a voltage ratio of

$$\frac{1}{\sqrt{2}} = 0.71$$

that is, the output voltage is 0.71 of the input voltage. Figure 8.1 shows graphically the relationship between decibels and voltage ratio.

Although essentially a ratio between two signals, decibels can also be used to express single, absolute measures of magnitude. A reference magnitude is set as 0 dB, then other values of dB on this scale indicate the multiplier of the reference magnitude. Some multimeters have a scale graduated in dB in this way for ease of making frequency response and signal-to-noise measurements. Examples of this type of scale are:

1) 0 dBV refers to 1 V RMS; thus 20 dBV is 10 V.
2) 0 dBm refers to 1 mW into 600 Ω (i.e. 0.775 V RMS);
thus –6 dBm is $0.5 \times 0.775 = 0.387$ V RMS.

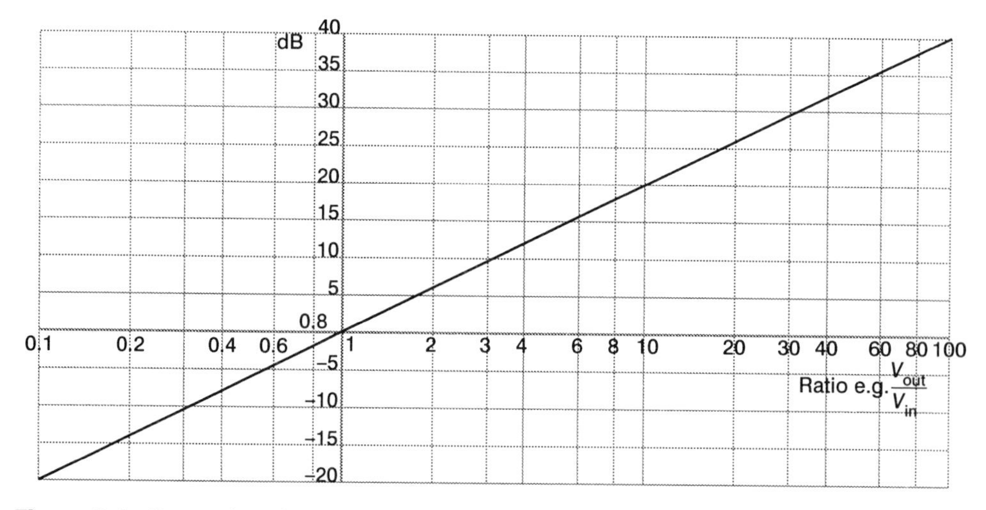


**Figure 8.1** Conversion of voltage ratio to decibels

Similarly, a component such as a transducer, which has its output and input measured in completely different units can still have its gain expressed in dB. For example, a pressure gauge has an input measured in pascals (N/m$^2$), while its output is the angle turned through by the pointer. Its gain, therefore, has units of degree/Pa. In such a case the dB value of gain is calculated as:

$$20\log_{10}\frac{\text{(gain at a particular frequency)}}{\text{(gain at the flat part of the response curve)}}$$

Thus the gain for the flat part of the frequency response curve is taken as 0 dB. Figure 8.2 (on the next page) shows some examples of gain calculated in this way.

*Example 8.2*
What amplitude will a centre-zero moving-coil meter show when an AC voltage of true amplitude 3 V is applied to it, if the frequency of the AC is such that the instrument's response is 3 dB down?

*Solution*
The amplitude of the pointer's motion diminishes as the frequency of the input AC is increased, because of damping and inertia of the pointer/moving-coil assembly. Because the scale is graduated in volts instead of degrees of rotation we can work directly in volts:

$$-3 = 20\log_{10}\left(\frac{V_2}{V_1}\right)$$

$$\therefore \log_{10}\left(\frac{V_2}{V_1}\right) = -0.15$$

$$\left(\frac{V_2}{V_1}\right) = 10^{-0.15}$$

$$= 0.708$$

$\therefore$ the meter indicates an amplitude of $0.708 \times 3\text{V} = 2.12$ V

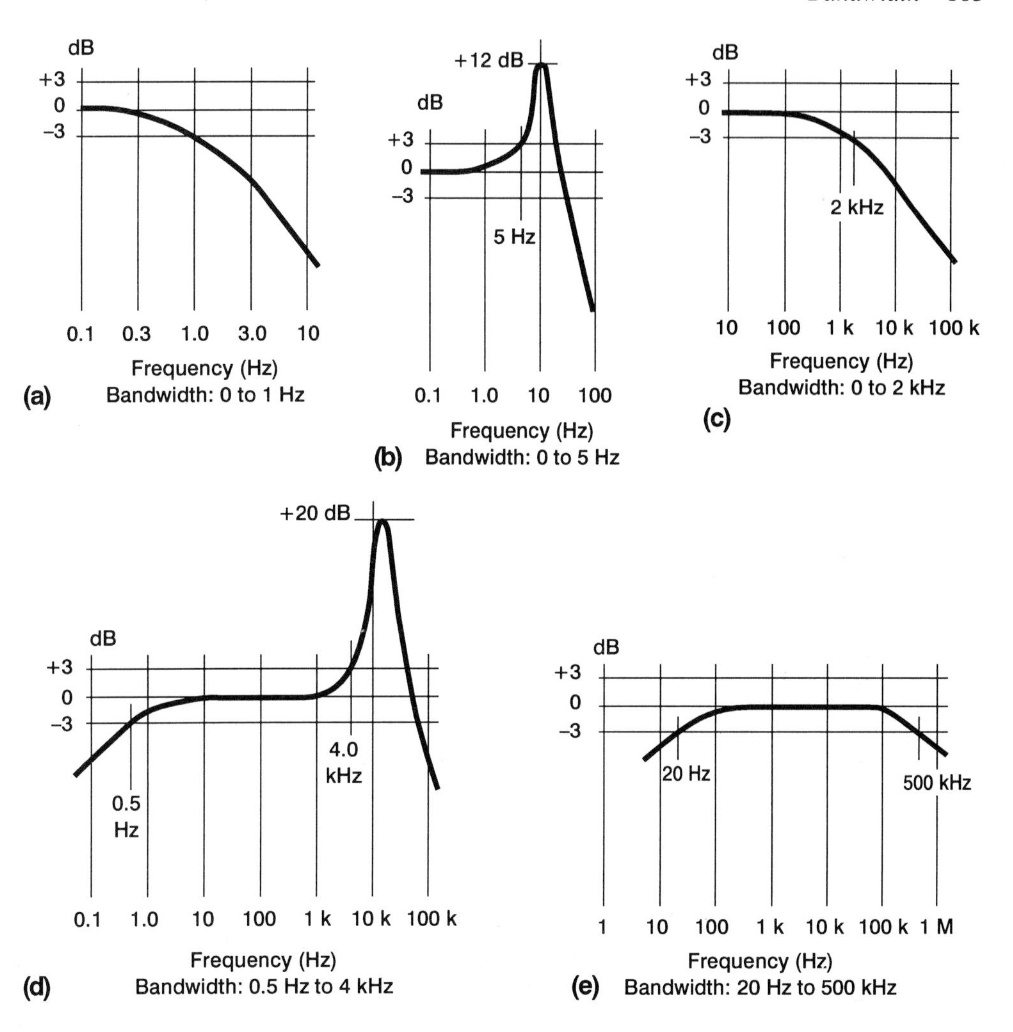


**Figure 8.2** Frequency response curves and bandwidths of typical instrumentation system components: (a) moving-coil meter, (b) pressure gauge, (c) UV recorder galvanometer, (d) piezoelectric accelerometer, (e) magnetic tape recorder

## Bandwidth

The bandwidth of an amplifier or any other component of an instrumentation system is the range of frequencies over which the gain is constant to within ± 3 dB. Figure 8.2 shows the frequency response curves of some typical transducers and display devices and indicates how their bandwidths are determined.

*Example 8.3*

A pressure gauge has the frequency response curve shown in Figure 8.2(b). A sinusoidal pressure variation of ± 20 kPa at a frequency of 0.1 Hz causes the pointer to move through ± 10°. Calculate:

i) the gain of the instrument at the given frequency
ii) the angular movement of the pointer if a pressure variation with the same amplitude is applied at a frequency of 10 Hz

*Solution*

i) $$\text{gain} = \frac{\text{output}}{\text{input}} = \frac{10}{20} = 0.5\frac{\text{degree}}{\text{kPa}}$$

ii) From Figure 8.2(b), the gain of the instrument is 12 dB greater at 10 Hz than it is at 0.1 Hz.

$$\therefore 12 = 20\log_{10}\left(\frac{\text{gain at 10 Hz}}{0.5}\right)$$

$$\therefore \text{gain at 10 Hz} = 0.5\times 10^{0.6} = 1.99\frac{\text{degree}}{\text{kPa}}$$

angular movement of pointer = ± 20 × 1.99 = ±39.8°

Or, more simply:

From Figure 8.1, 12 dB corresponds to a ratio of 4/1
∴ angular movement of pointer = ± 10° × 4/1 = ± 40°

## Amplifiers

In instrumentation, the word 'amplifier' usually means an electronic circuit which increases the voltage of a signal. It can be thought of as a 'black box', as in Figure 8.3(b), which has an input impedance, $Z_{in}$ and an output impedance, $Z_{out}$, and which requires a DC power supply.

Impedance ($Z$) and resistance ($R$) have the same units: ohms. *Resistance* implies that voltage and current are in phase; *impedance* implies that there may be a phase difference between them. Usually, when dealing with instrumentation amplifiers in which input and output impedances are almost entirely resistive we can think of impedance as meaning the same thing as resistance.

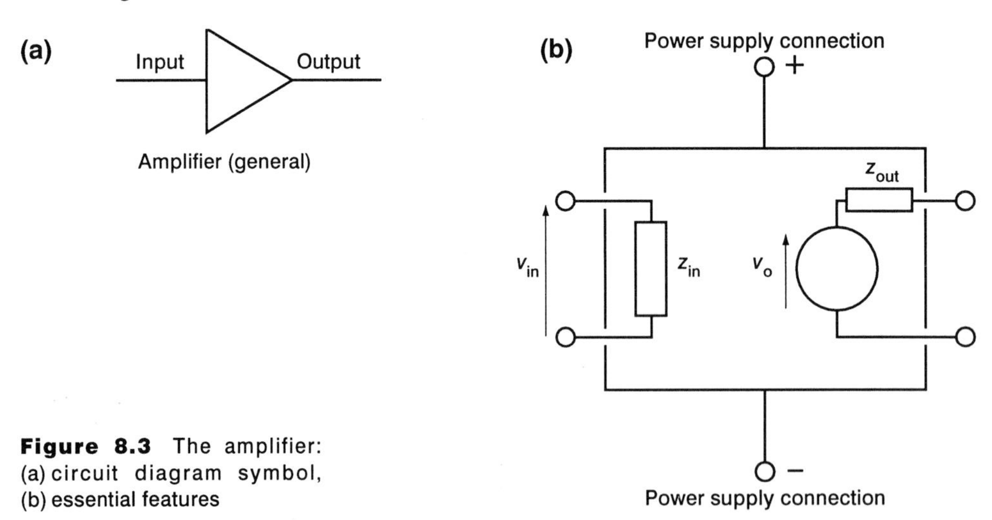


**Figure 8.3** The amplifier: (a) circuit diagram symbol, (b) essential features

The output of the amplifier comes from the power supply, and the purpose of the input signal is to control the transfer of power from the power supply to the output, rather like a hand turning a tap on and off. An amplifier cannot, therefore, deliver an output voltage greater than its power supply voltage – in fact its maximum output voltage is always slightly less than the power supply voltage. If an AC voltage input is big enough have an output amplitude which would be greater than this limit, the output waveform gets 'clipped' to the limit, becoming something like a square wave.

Some amplifiers give little or no increase of *voltage* between output and input (their voltage gain is about 0 dB), but can keep that gain constant while giving a much greater output *current* than would be obtainable from the previous component in the system. In small-signal form, as used in integrated circuits, they are called *buffer amplifiers*, or just *buffers*. Their main features are:

1 that the gain is approximately unity
2 that the amplifier is non-phase-inverting
3 that the input impedance is high
4 that the output impedance is low.

In large-signal form they are called *power amplifiers*; for example the final stage of amplification in any audio system driving a moving-coil loudspeaker is a power amplifier.

Most amplifiers are *AC coupled*, the input connection and the coupling between internal stages of amplification being through either capacitors or transformers. This sets a low-frequency limit to their bandwidth.

*DC amplifiers* have direct connection of input and direct connection between internal stages of amplification, so their frequency range extends down to 0 Hz (i.e. to unidirectional signals). AC amplifiers should be used where possible as the output of DC amplifiers is inclined to drift due to temperature changes and the ageing of components (see the following section on Noise).

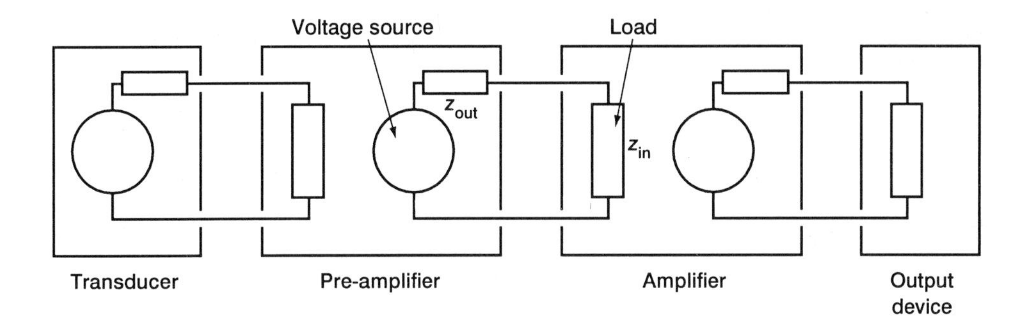


**Figure 8.4** An instrumentation system as a chain of voltage sources and loads

An amplifier may receive its input from a pre-amplifier or from a transducer, and its output may go to a recording or display device. An instrumentation system may consist of a chain of modules as shown in Figure 8.4, in which the circles represent sources of voltage, the input impedance of one module acting as a load on the output of the preceding module. It can be shown that maximum power is transferred to the load when the source and load impedances are equal, that is, when $Z_{out} = Z_{in}$ in Figure 8.4. In instrumentation, however, when the signal is a voltage, what we need to transfer is not maximum power but maximum voltage. This suggests that $Z_{in}$ should be infinitely greater than $Z_{out}$, but in that case no power would be transferred. The usual practice therefore is to make the input impedance of one module 10 or 20 times as big as the output impedance of the preceding module, so that the voltage across the input terminals is a fairly high proportion of the source voltage.

## Transformers

Provided that the signal is in the form of AC and not DC, the impedance ratio between the output of one module and the input of the following one can, if necessary, be modified by coupling them through a transformer. The impedance ratio of the two windings is equal to the square of the turns ratio. Thus if an impedance is connected to the secondary winding of a transformer which has $n_1$ turns of wire in the primary winding and $n_2$ turns in the secondary, the effective impedance, seen at the terminals of the primary of the transformer, is

$$Z' = \left(\frac{n_1}{n_2}\right)^2 \times Z$$

*Example 8.4*
Calculate the effective input impedance of an amplifier when its input, which has an impedance of 1.5 kΩ is fed from the secondary winding of a transformer which has 200 turns of wire in the primary and 40 in the secondary.

*Solution*
From the given equation:

$$\text{effective impedance} = \left(\frac{200}{40}\right)^2 \times 1.5\ \text{k}\Omega = 37.5\ \text{k}\Omega$$

## Integrated circuits

Operational amplifiers, logic gates, counters, etc. are usually in the form of integrated circuits, as shown in Figure 8.5. In this type of device complex circuits consisting of transistors, resistances and capacitances are reduced to microscopic size and by a process of masking and the diffusion of the appropriate impurities, are formed on a very small rectangle of silicon, the 'chip'. As an example, the type 741 integrated circuit operational amplifier contains 20 transistors, 11 resistances and one capacitance on a chip with dimensions of only one or two millimetres per side.

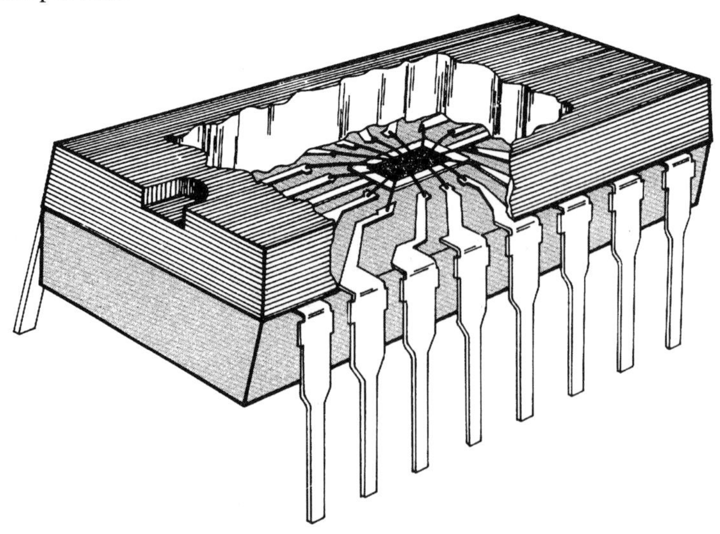

**Figure 8.5** Enlarged view of an integrated circuit showing connections to chip

In integrated circuit manufacture a pattern of hundreds of identical copies of the circuit is projected on to a disc of silicon which, after the masking and diffusion processes are complete, is cut up into the individual chips. This enables integrated circuits to be produced very cheaply. Integrated circuits can be soldered directly into printed circuit boards, but for prototype work they should be mounted in sockets so that substitutions can be made when checking for faults.

## The operational amplifier

This is an extremely useful device which has largely replaced the discrete transistor in many applications. An operational amplifier (op-amp) is a high-gain amplifier with differential inputs. Its circuit diagram symbol is shown in Figure 8.6. It has two input connections; one, marked '–' on the symbol, is the *inverting input*, the other, marked '+', is the *non-inverting input.*

Inverting input
Non-inverting input
Output

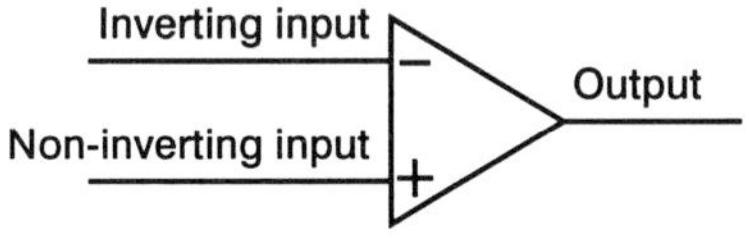

**Figure 8.6** Symbol for an operational amplifier

The operational amplifier is unaffected by voltages which are applied equally to both inputs, but amplifies the *difference* in voltage between the two inputs. If the positive of the voltage difference is applied to the non-inverting input, the output voltage becomes more positive; if applied to the inverting input the output becomes less positive (or more negative).

The open-loop gain of an op-amp is very high; depending on the type, 100 dB (= $10^5$) is quite common but it can be as high as $10^7$. Because of this, unless there is negative feedback to limit the gain, the output voltage is always at either the positive or negative limit of the range determined by the power supply voltage – it is virtually impossible to apply a small enough input voltage difference to give an output voltage anywhere between the two limits. Thus an op-amp can be used as a *zero detector*. As the input voltage changes polarity in passing through zero, the output voltage changes from maximum positive (logic 1) to maximum negative (logic 0) or vice versa.

### Inverting and non-inverting amplifiers

For negative feedback the output must be connected to the inverting input in some way. Figure 8.7 on the next page shows how the connection can be made to produce an amplifier which would have a reduced gain but is stable. Because the open loop gain of the op-amp is very high, $v_a$, the voltage between the two inputs, must be very small (a matter of a few microvolts) if $v_{out}$ is to be within the limits imposed by the power supply voltage. So point P, the junction between $R_1$ and $R_f$ is virtually earthed. Also, because the input impedance of the op-amp is very high, the current $i_a$ going into the op-amp itself is virtually zero.

Thus

$$i_f \approx i_i$$

$$\therefore \frac{v_{out}}{R_f} \approx -\frac{v_{in}}{R_1}$$

$$\therefore \text{voltage gain} \frac{v_{out}}{v_{in}} \approx -\frac{R_f}{R_1}$$

Due to the virtual earth at P, the input impedance is $R_1$. The minus sign indicates that the output and input are of opposite polarity at every instant; i.e. this is an *inverting amplifier*.

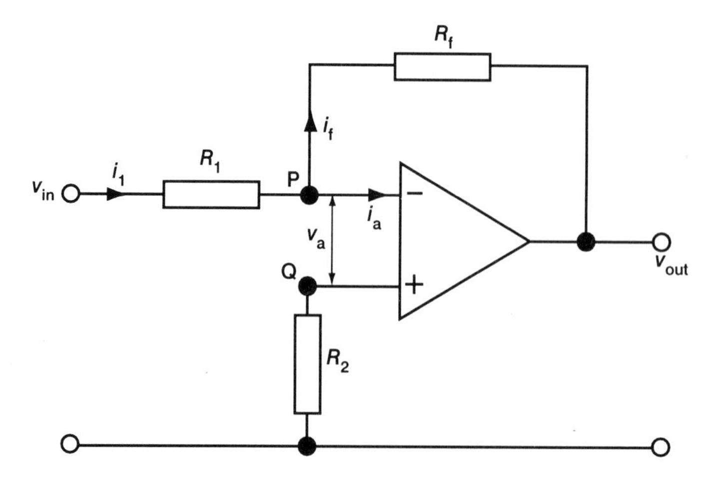


**Figure 8.7** An inverting amplifier

Figure 8.8 shows the op-amp circuit for a *non-inverting amplifier*. In this case the voltage gain is:

$$\frac{v_{out}}{v_{in}} \approx \frac{R_1 + R_f}{R_1}$$

and the input impedance is $R_2$.

In both of these circuits $R_2$ is normally equal to $R_1$ and is introduced to reduce the effects of offset voltage and thermal drift; its value is not critical and in some versions of Figure 8.7, the op-amp's non-inverting input is directly connected, as in Figure 8.10.

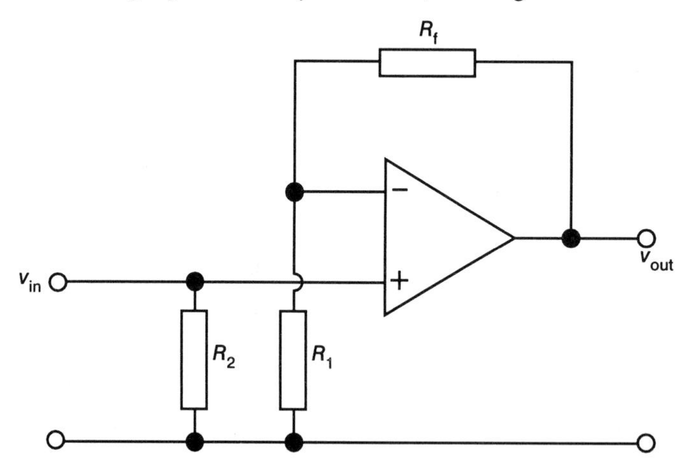


**Figure 8.8** A non-inverting amplifier

## Balanced-input amplifier

Figure 8.9 shows a *balanced-input amplifier*. This configuration is used when the connections to the transducer pass through a 'noisy' environment in which both conductors pick up equal noise voltages. These are blocked by the amplifier while the voltage difference (the signal) is amplified with a gain of $R_f/R_1$. The input impedance is $R_1 + R_2$, because the voltage difference between the two points P and Q is virtually zero, due to the op-amp's very high open loop gain. In this circuit it is essential that $R_2 = R_1$ and $R_3 = R_f$ for balance.

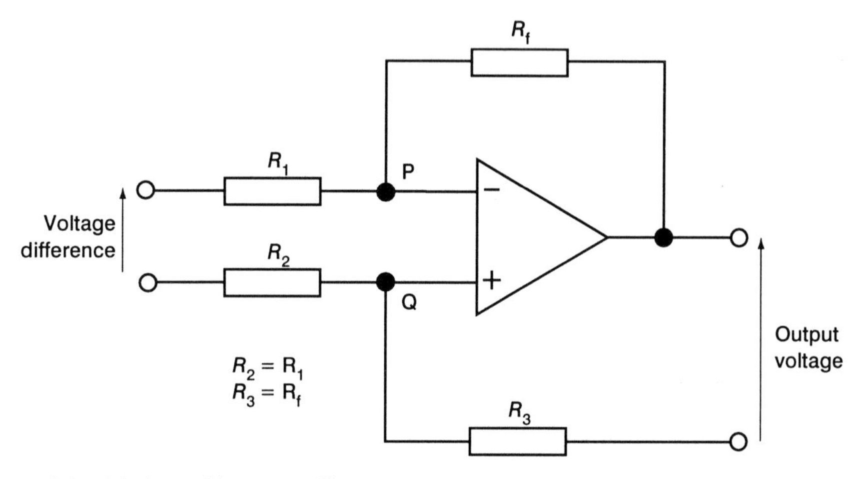


**Figure 8.9** A balanced-input amplifier

## Inverting summing amplifier

Figure 8.10 shows the op-amp as an *inverting summing amplifier* in which a number of input signals can be added together and their sum amplified. The ratio of the feedback resistance to the individual input resistance determines the contribution of that input to the total result.

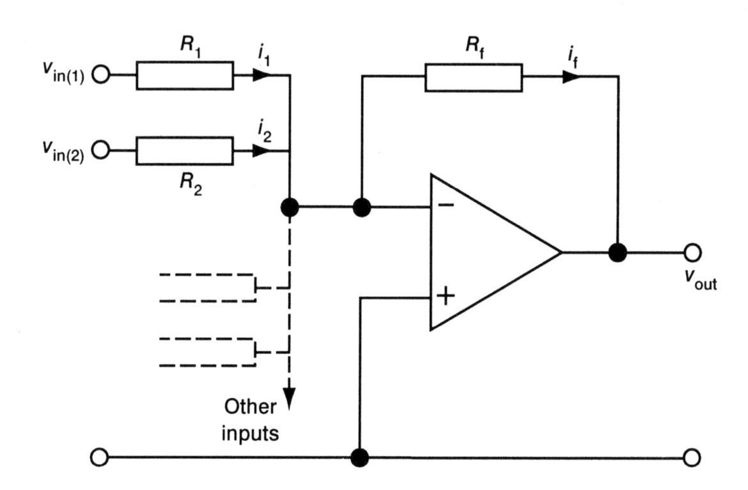


**Figure 8.10** An inverting summing amplifier

Figure 8.10 uses the same reasoning as in the simple inverting amplifier shown in Figure 8.7:

$$i_1 + i_2 + \cdots = i_f$$

$$\frac{v_{in(1)}}{R_1} + \frac{v_{in(2)}}{R_2} + \cdots = -\frac{v_{out}}{R_f}$$

$$v_{out} = -\left(\frac{R_f}{R_1} v_{in(1)} + \frac{R_f}{R_2} v_{in(2)} + \cdots\right)$$

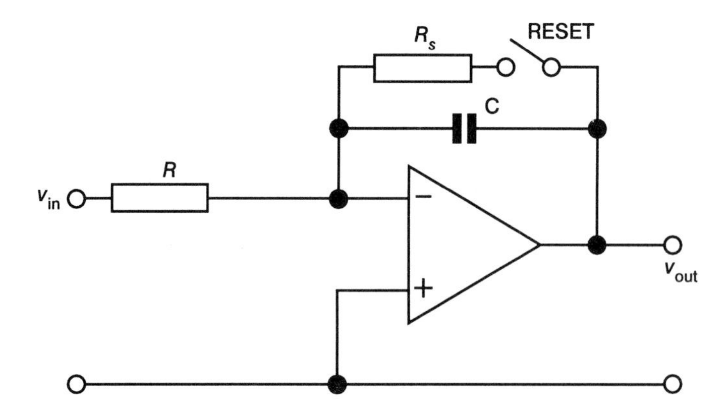


**Figure 8.11** The op-amp as an integrator

## The op-amp as an integrator

Figure 8.11 shows the op-amp as an integrator. The reset switch is closed for an instant to discharge the capacitor through $R_s$, to start the integration. Then

$$v_{out} = -\frac{1}{RC}\int v_{in}dt$$

if $v_{in}$ is constant, this gives the voltage-time graph shown in Figure 8.12. The capacitor must be a low-leakage type, not an electrolytic, otherwise the dotted curve is obtained. In practice the switch would be a solid-state device.

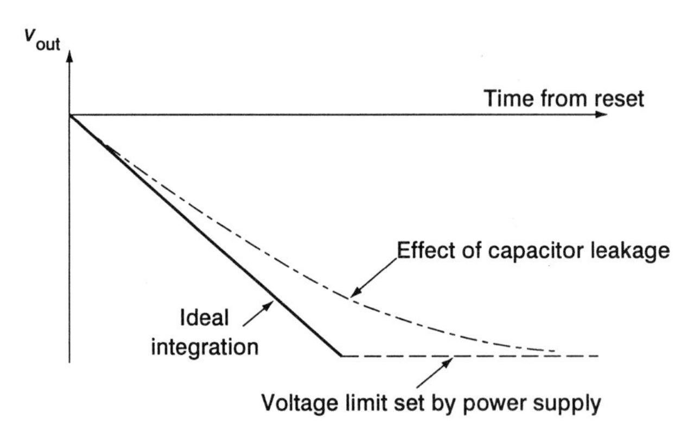


**Figure 8.12** Graph of output voltage of integrator

# Noise

The output from an electrical transducer is usually in the form of a voltage. Unfortunately, between the transducer and the display or recording device, noise voltages are liable to appear and be amplified together with the wanted signal. If we define 'noise' as any unwanted alteration of the signal, it can appear in the following forms:

1 drift
2 internally generated random noise
3 noise picked up from external sources.

**Drift**

Drift of DC signal voltages may be due to changes in the gain of DC amplifiers, caused by variations in power supply voltage or by temperature change within the amplifier. Drift may also be caused by thermocouple effects when temperature changes occur where two different metals meet in a circuit. The remedy is to use stabilised power supplies and to keep temperatures constant as far as possible by the use of heat sinks and temperature-controlled ventilation.

A much slower *permanent drift* effect may be caused by the ageing of components within an amplifier.

Drift may be cancelled out by temporarily short-circuiting the input to the amplifier and subtracting any resulting output voltage from the system's output when it is working normally. This may be programed as an automatic routine. Alternatively, since AC amplifiers are much less subject to drift, the DC output of a transducer may be converted to AC by means of a *chopper*. In its simplest form, the chopper is a transistor either in series with the amplifier input, as an amplifier, or in parallel with it, as a short-circuit. The base of the transistor is supplied with a square-wave voltage from an oscillator so that the transistor regularly interrupts the DC, chopping it into a kind of AC at the frequency of the oscillator, with amplitude proportional to the DC value.

**Internally generated random noise**

Internally generated random noise has two main sources: thermal noise and shot noise. *Thermal noise* comes from electron movement due to the thermal vibration of the atoms of a conductor. It is a *white noise* (that is, a noise made up of all frequencies). Its RMS voltage is proportional to the square root of the absolute temperature of the conductor. *Shot noise* is also a white noise. It comes from random fluctuations in the passage of charge carriers through semiconductor material.

Usually the signal-to-noise ratio is so high that we are unaware of these two types of noise, but their effects can be seen as a 'snowstorm' on a television screen if the aerial is disconnected from the receiver, and in areas of poor reception the resulting grainy effect can spoil the picture. In that case, as we cannot do anything about the noise, the remedy is to increase the signal strength by fitting the first stage of amplification to the aerial itself in the form of a mast-head preamplifier. This amplifies the signal before it has lost too much of its strength in the downlead to the TV set. In instrumentation, a similar situation would be a transducer connected to the rest of the system by a long cable in which signal strength may be lost and external noise picked up. In such a case, to improve the signal-to-noise ratio, a preamplifier should be inserted as close to the transducer as possible.

**External noise**

Noise may be picked up from the surroundings by capacitive pick-up (from electrostatic fields) or inductive pick-up (from electromagnetic fields), or it may be introduced into equipment through its power supply leads. Often, the signal-to-noise ratio can be greatly improved just by positioning instrumentation cables and equipment as far away as possible from sources of noise such as 50 Hz power cables, electric motors, generators or switchgear.

## Screening techniques

Capacitive and inductive pick-up may be reduced by *screening*. This means enclosing cables, amplifiers, etc., in metal which is earthed (grounded) so that the voltages picked up by the screening are short-circuited to earth. In this context, 'earth' means a connection to the soil or to the structure of a vehicle, which is assumed to be at zero voltage under all conditions. To screen against capacitive pick-up, the metal forming the screen need only be a good conductor

– copper or aluminium. To screen against inductive pick-up, the screening must be of iron or other magnetic material, and it must be thick enough to be effective – two or three millimetres thickness is needed for screening out 50 Hz magnetic fields.

If one conductor of the pair carrying the signal can be earthed it can act as the screen against capacitive pick-up by being woven as a tube of copper wires surrounding insulation through which the other conductor passes – this is the principle of *co-axial cable*, as used for TV aerials. Where both conductors must be insulated from earth, screened cable must be used. This consists of insulated wires which are twisted together in pairs and enclosed in a tube of copper braid, or wrapped in aluminised polyester tape, which acts as the screen. The screen greatly reduces capacitive pick-up by capturing charges which would otherwise reach the signal conductors, but it must be earthed or the charges will still reach the conductors through screen-to-conductor capacitance, as shown in Figure 8.13. The twisting together of the conductors reduces inductive pick-up by ensuring that signal-carrying pairs will pick up identical inductive noise voltages. These can be cancelled out by connecting the output of the cable to an op-amp type of amplifier.

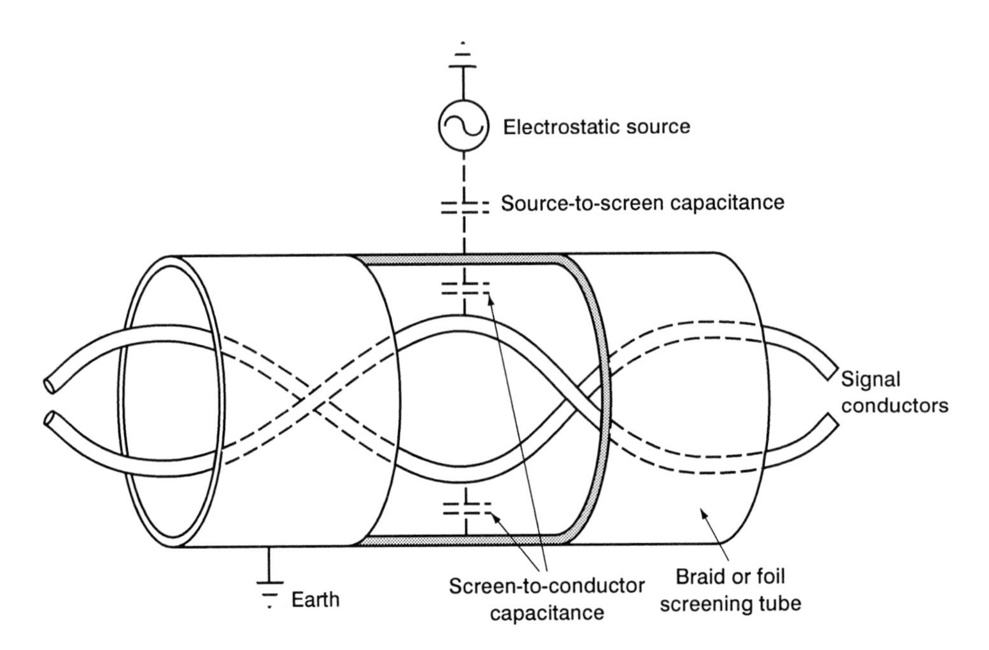


**Figure 8.13** Electrostatic screening of conductors

*Earth loops* can create large noise voltages. They occur when earth connections are made at more than one point, as in Figure 8.14 in which separate earth connections are made to the screening at either end of a cable. This arrangement can pick up noise in two ways:

1 From current flow. If a current happens to be flowing through the material to which the earthing points are connected (as could happen in a motor vehicle with single-conductor electrical systems, using earth return) some of the current will flow through the screening, which, if the current is noisy, will transmit the noise by capacitance to the signal wires.
2 By induction. The complete earth loop can act as a loop of wire in a magnetic field, having noise currents induced in it by the alternating magnetic field of a noise source. The noise is transferred to the signal wires by capacitive pick-up.

Another earth loop exists if one of the signal conductors is also earthed at each end by being connected to the earthed casing of the transducer at one end and the earthed casing of the amplifier at the other end, as in Figure 8.14(a). Figure 8.14(b) shows how these earth loops should be broken. The screening and the signal conductor are earthed at the transducer end only and the input to the differential amplifier is allowed to 'float'. Note that although there may be no direct connection between the transducer and earth it may in effect be earthed by capacitance.

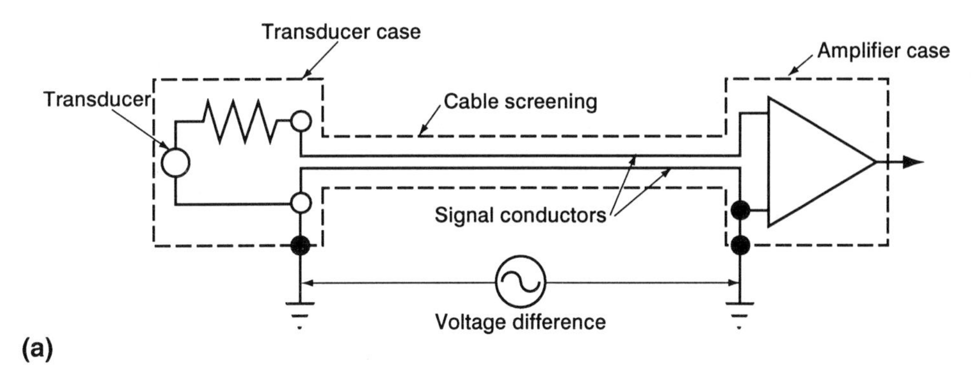


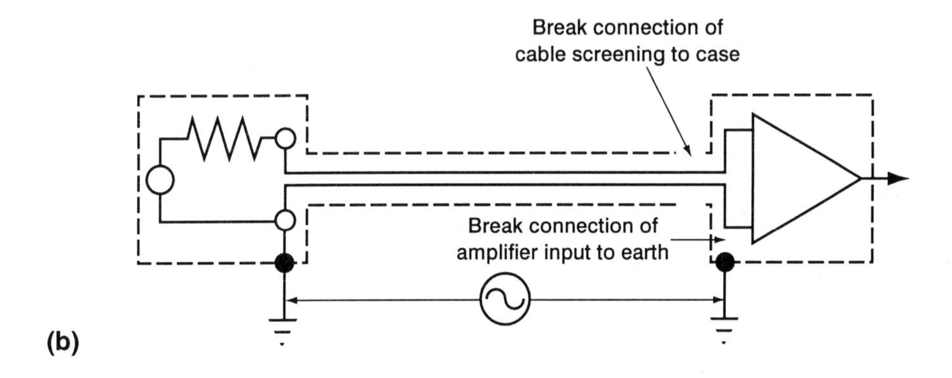


**Figure 8.14** (a) Typical earth loop (b) breaking the loop

## Suppressors

These are used to suppress electrical noise caused by sparking at the brushes of electric motors or by sparking at the contacts of relays which are switching reactive loads. The noise may be transferred by electromagnetic radiation or along the power supply cable if the noise source and instrumentation equipment share the same power supply. Suppression, which is best applied as close to the noise source as possible, can take the form of capacitors, chokes or varistors.

### Capacitors

Capacitors are used as reservoirs to smooth out voltage variations and to conduct high frequency transients to earth. Figure 8.15 shows a circuit of the type used to suppress electrical noise from small electric motors.

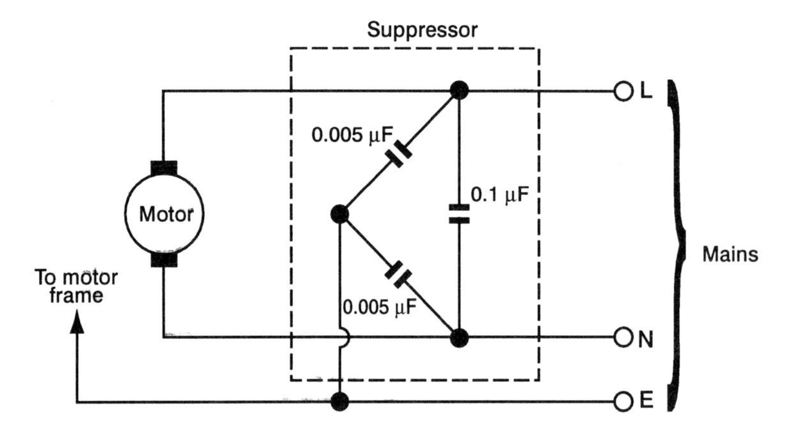


**Figure 8.15** Suppressor-capacitor for a small electric motor

### Chokes

Chokes are simply inductances which smooth out rapid variations of current, so they are put in series with power supply leads to act as barriers to noise.

### Varistors

Varistors are connected across power supply leads to short-circuit transient high-voltage 'spikes'. They are semiconductors which have a high resistance at voltages up to a stated maximum; voltages in excess of that maximum however, cause the resistance to drop to a low value until the excess energy has been dissipated.

### Filters

Filters are used to exclude unwanted bandwidths from a signal. The filter may be needed to exclude high frequencies because they are carrying only noise, in which case a *low-pass filter* is required. Or it may be needed to exclude low frequencies for the same reason; in this case a *high-pass filter* is used. If the wanted signal is contained in a narrow frequency band within a frequency range which contains unwanted signals and/or noise, a *band-pass filter* is employed. Conversely, if the unwanted signal or noise is concentrated in a narrow frequency band within the signal frequency range, a *band-stop* or *notch filter* is used. Figure 8.16 shows typical frequency response curves for these four types of filter. The diagrams are sketches of *Bode plots*: graphs plotted on log-linear graph paper with frequencies on the logarithmic scale and the gain of the filter in decibels (normally zero or a loss) on the linear scale. This use of logarithms on both axes reduces the low-pass and high-pass curves to a horizontal and an inclined straight line in each case, with a rounded portion where the straight lines meet.

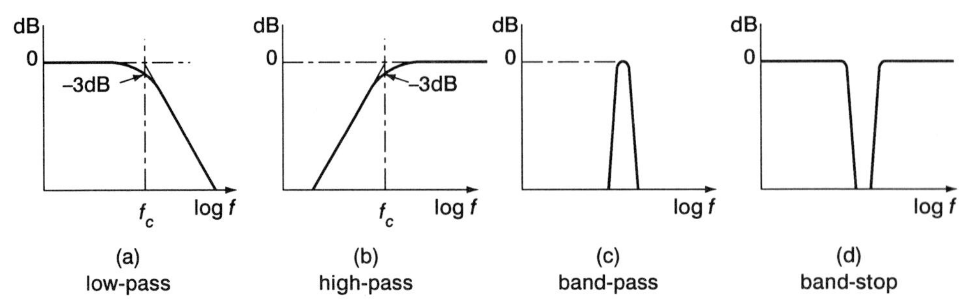


**Figure 8.16** Typical frequency response curves of the main types of filter (a) low-pass, (b) high-pass, (c) band-pass, (d) band-stop

Extending the straight lines to where they intersect defines the *cut-off frequency*, $f_c$ (more accurately called the *corner frequency*). This frequency intersects the curve at –3 dB (the gain is 3 dB down at this point). The slope of the inclined line depends on the *order* of the filter (that is, the order of the polynomial in its transfer function). For a first-order filter the slope is 20 dB per decade (for every ten-fold increase in frequency, the gain changes by 20 dB). For a second-order filter the slope is 40 dB per decade and for an $n^{th}$ order $20n$ dB per decade. Thus sharpness of cut-off depends on the order of the filter. Higher orders of filtering are obtained by 'cascading' filters putting them in series so that their orders add together.

- Filters may be *passive, active or digital.*
- A **passive** filter uses resistance, capacitance and/or inductance to attenuate the unwanted frequencies.
- An **active** filter uses an op-amp with resistors and capacitors only, to perform the same function. Active filters are better for low frequency, small signal work and can have gains greater than 0 dB, but they need a power supply. Passive filters tend to be simpler and are more suitable for frequencies above the audio range, where active filters are limited by the bandwidth of the op-amp.
- **Digital** filters are used with digital signals (see chapter 10). They are not applicable to this section, which is concerned only with analogue signals.

**Low-pass filters**

Low-pass filters can attenuate high frequencies either by 'short-circuiting' them through a capacitor, or by 'choking' them through the impedance of an inductor. Figure 8.17 shows low-pass first-order passive filters of both types, together with a circuit for a low-pass second-order active filter.

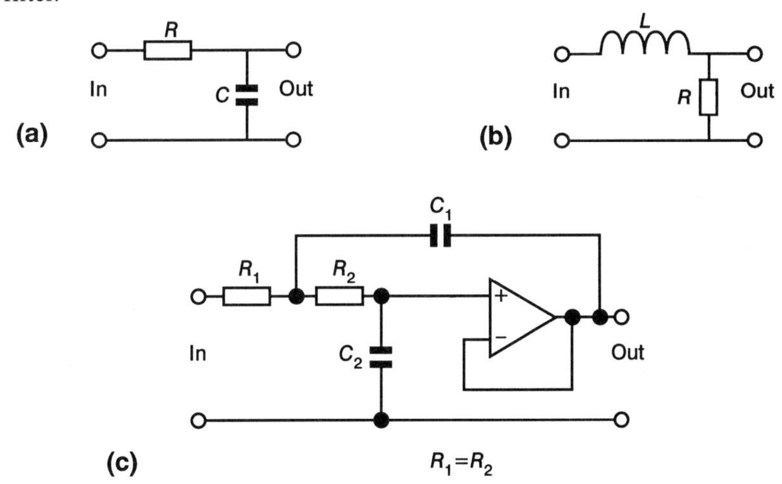


**Figure 8.17** Low-pass filter circuits: (a) and (b) are first-order passive alternatives, (c) second-order active

**High-pass filters**

As Figure 8.18 shows, the circuits of passive high-pass filters are similar to those of the low-pass filters, but the inductor and capacitor change places. Thus the inductor 'short-circuits' the low frequencies but cannot short-circuit the high frequencies and the capacitor conducts the high frequencies but attenuates the low frequencies. In the case of the active filter, resistors are replaced by capacitors, and capacitors by resistors.

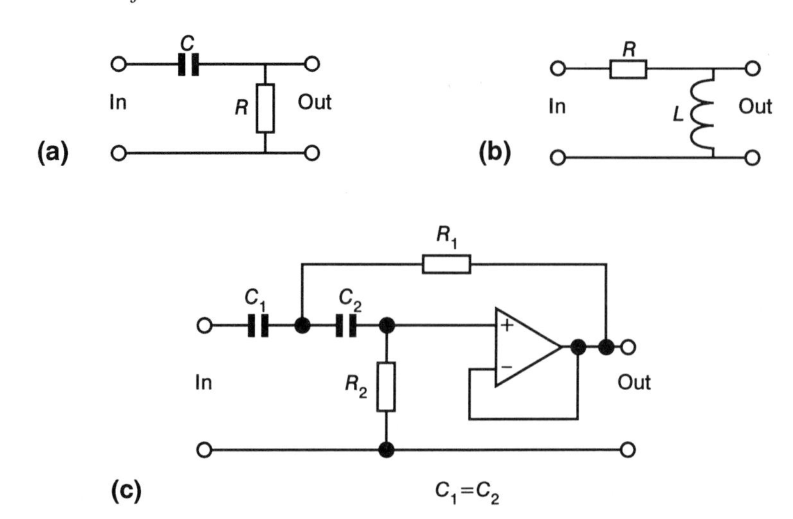


**Figure 8.18** High-pass filter circuits: (a) and (b) are first-order passive alternatives, (c) second-order active

Simple low-pass and high-pass filters of the type shown in Figures 8.17(a) and 8.18(a) are easily constructed. In both cases, the corner frequency is given by

$$f_C = \frac{1}{2\pi CR}$$

where $C$ and $R$ are the capacitance in farads and the resistance in ohms, respectively.

*Example 9.1*

A simple low-pass filter is required to reduce the amplitude of 50 Hz mains hum in a DC signal. Choose values of $C$ and $R$ in Figure 8.17(a) to give a corner frequency of about 5 Hz and estimate the attenuation at 50 Hz.

*Solution*

Take

$$C = 10\ \mu\text{F}$$

$$f_C = \frac{1}{2\pi CR}$$

$$\therefore R = \frac{1}{2\pi C f_C}$$

Then

$$R = \frac{1}{2\pi \times 10 \times 10^{-6} \times 5} = 3180\ \Omega$$

As this may be rather a high resistance to put in series with the signal, we will take $C = 100\ \mu\text{F}$, which gives $R = 318\ \Omega$ .The nearest standard value resistor is 330 Ω, which gives a corner frequency of

$$\frac{1}{2\pi \times 100 \times 10^{-6} \times 330} = 4.82\ \text{Hz}$$

A first-order low-pass filter such as that in Figure 8.17(a) gives an attenuation of 20 dB per decade at frequencies above the corner frequency, and 50 Hz is 1 decade above 5 Hz.

From Figure 8.1:

$$-20 \text{ dB Gives } \frac{v_{in}}{v_{out}} = 0.1$$

Therefore as the actual corner frequency is slightly less than 5 Hz, the amplitude of the 50 Hz ripple should be less than a tenth of its original value. A precise value for the attenuation could be obtained from a Bode plot, but such accuracy could be spurious in view of the normal tolerance of ± 20% on capacitor values.

**Band-pass filters**

These may be made by passing the signal through a high-pass filter and a low-pass filter in series so that the frequency range passed by the filter is the difference between the corner frequencies of the two filters. The corner frequency of the low-pass filter must of course be higher than that of the high-pass filter, otherwise we should have an 'all-stop' filter – which would not be very useful!

The high-pass filter should usually precede the low-pass filter, and if they are passive filters they should have a buffer amplifier between them because it is usual for a filter to receive the signal from a low-impedance source and pass it on to a high impedance load. The buffer amplifier can be an op-amp with 100% negative feedback, giving a gain of 0 dB. The active filters of Figures 8.17(c) and 8.18(c) are already of this form and do not need a buffer between them. So the block diagram of a band-pass filter is as shown in Figure 8.19.

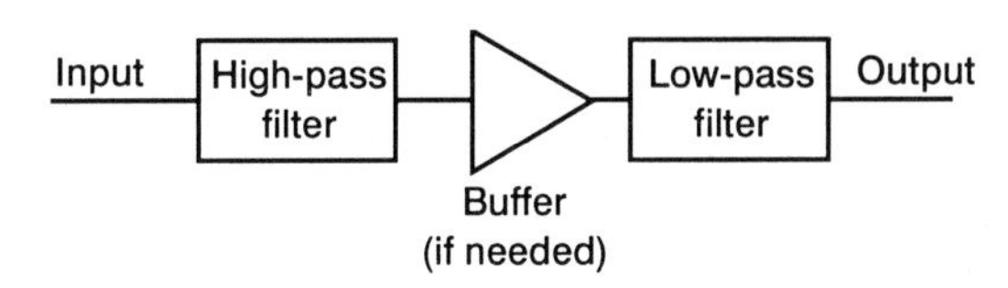


**Figure 8.19** Band-pass filter

**Band-stop filters**

These are made from a low-pass filter and a high-pass filter in parallel, as shown in Figure 8.20, so that the low-pass filter blocks the signal at the low frequency end of the stop band, while via the alternative path the high-pass filter conducts it at frequencies above the stop band's high-frequency limit.

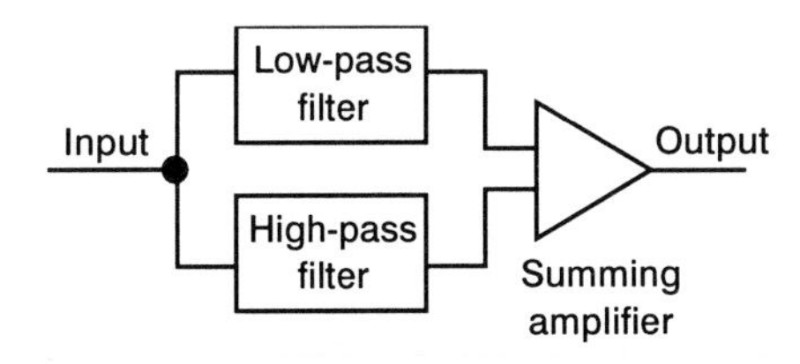


**Figure 8.20** Band-stop filter

**Notch filters**

Notch filters are similar to band-stop filters but are designed to remove one specific frequency from the signal rather than a frequency band. However, frequencies close to the specific frequency are attenuated by the filter to some extent.

**Phase effects**

Capacitors and inductors are energy storage devices and so they cause phase shifts in the signal. The first-order low-pass filter of Figure 8.17(a) introduces a *phase lag* which is negligible at frequencies up to about $0.1f_C$, increases to 45° at $f_C$ and approaches 90° at frequencies above about $10f_C$. The first-order high-pass filter of Figure 8.18(a) introduces a similar *phase lead* over the same frequency range. (Figures 8.17(a) and 8.18(a) are an integrating circuit and a differentiating circuit, respectively.) With continuous signals a phase shift is not of great significance but if the signal consists of pulses or sudden changes of level, phase effects can cause considerable distortion and phase correction circuits (all-pass filters) may have to be used.

## Modulation

Modulation is a method of transmitting a signal by causing it to modify a high-frequency sinewave known as the *carrier wave*. It is mainly used for transmitting signals by radio, but it has applications in instrumentation as a way of reducing the signal-to-noise ratio when signals have to be sent along 'noisy' cables. At the receiving end, the signal is recovered from the carrier by *demodulation*. The two main methods of modulation are *amplitude modulation* and *frequency modulation*. *Pulse code modulation*, in which the signal is digital instead of analogue, is dealt with in Chapter 10.

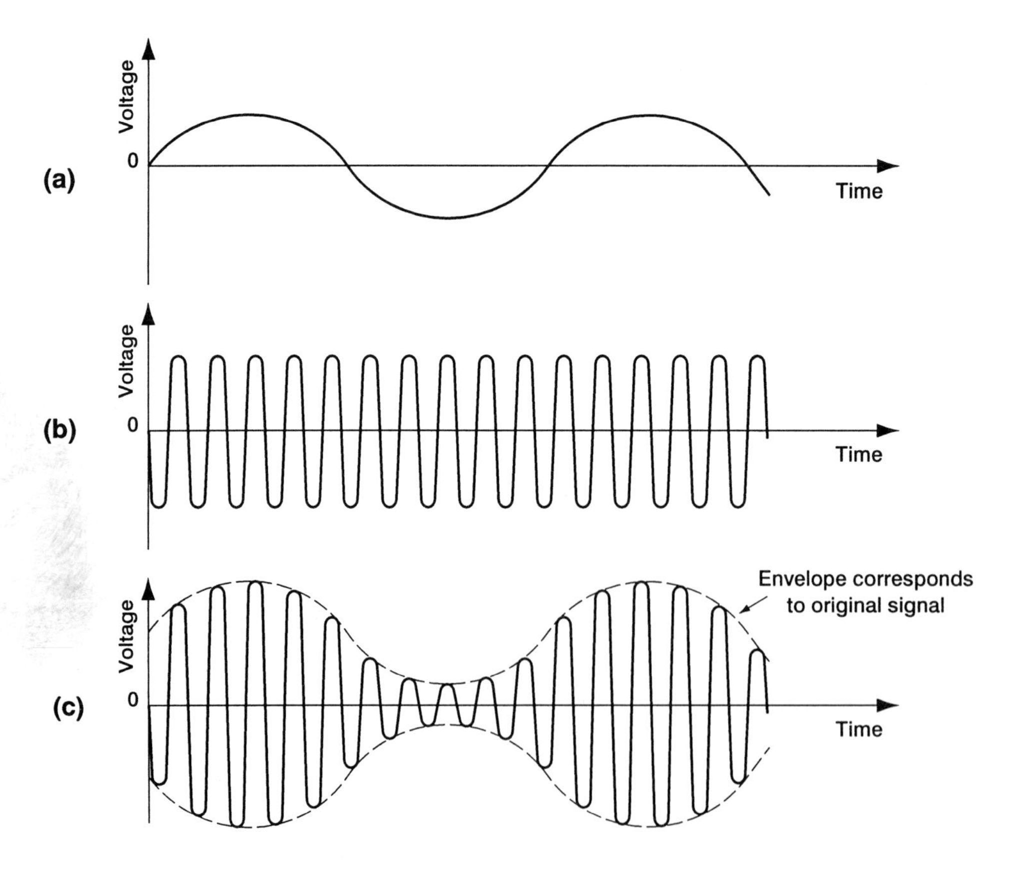


**Figure 8.21** Amplitude modulation: (a) signal, (b) carrier wave, (c) amplitude-modulated carrier

## Amplitude modulation (AM)

This was the original method of modulation, invented as a way of sending audio frequencies by radio. It is still used for all 'long wave', 'medium wave' and 'short wave' broadcasting. Figure 8.21 shows how a signal (shown for simplicity as a sinewave) modulates the amplitude of the carrier wave so that the upper and lower profiles of the carrier wave *envelope* correspond to the original signal. To recover the signal, the carrier wave is passed through a diode (the *detector*) so that either the positive or the negative half-cycles of the carrier wave are blocked. A low-pass filter then smooths out the remaining half-cycles, so that only the original signal remains.

There is a limit to the modulation which can be imposed on the carrier: the upper and lower profiles of the envelope must not overlap; if they do, a completely different shape of envelope results and the original form of the signal is lost.

## Frequency modulation (FM)

In this type of modulation, the amplitude remains constant but the signal voltage changes the *frequency* of the carrier wave as shown in Figure 8.22. This frequency-modulated carrier wave is generated by a voltage-controlled oscillator, which increases or decreases its frequency as the signal voltage varies.

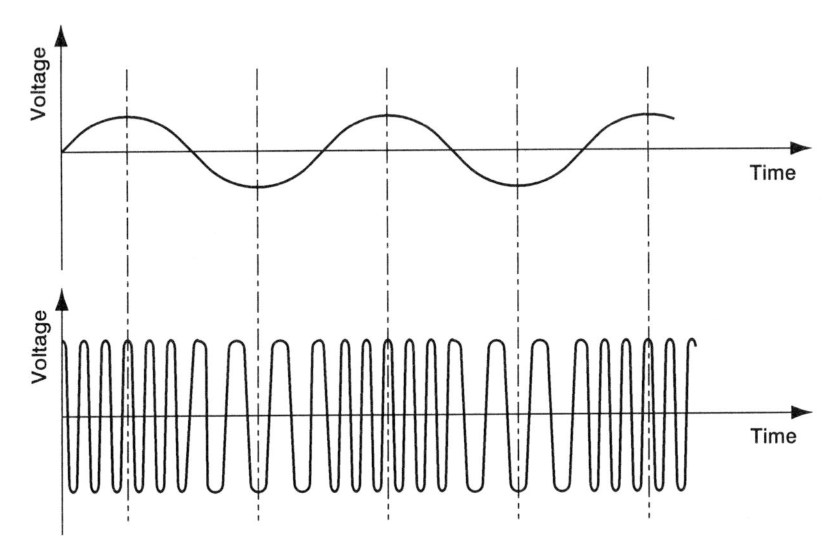


**Figure 8.22** Frequency modulation: (a) signal, (b) frequency-modulated carrier wave

The usual method of demodulating the FM carrier is by a *phase-locked loop*, which is shown in block diagram form in Figure 8.23. If the voltage-controlled oscillator is generating a frequency sufficiently close to that of the incoming FM signal, the phase detector recognises a difference in phase and produces a corresponding error signal voltage. After low-pass filtering and amplification, this is fed back to make the voltage-controlled oscillator adjust its frequency to that of the incoming frequency – the loop is then said to be 'in lock'. As the FM carrier wave's frequency continually varies, the corresponding variation in the control voltage to keep the loop 'in lock' has the same form as the original signal before it was modulated on to the carrier wave, and thus is the demodulated output.

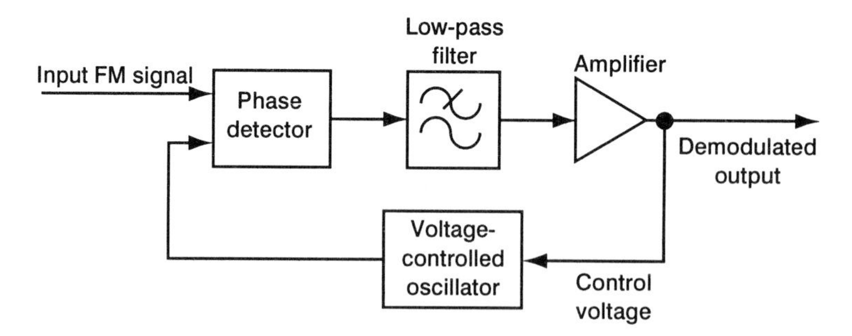


**Figure 8.23** Demodulation of an FM signal by phase-locked loop

The cut-off frequency of the low-pass filter should correspond to the maximum frequency in the original signal, to reduce interference effects and noise at frequencies above the cut-off.

The main advantage of frequency modulation is that it eliminates the type of noise which causes amplitude variations in the carrier wave. Thus it is used in instrumentation tape recorders to eliminate noise caused by microscopic variations in the magnetic coating of the tape. The main disadvantage of FM is that the carrier frequency should be at least five times the maximum frequency in the signal, so the maximum signal frequency when using FM in a tape recorder can only be about a fifth of the maximum which could be recorded directly.

## Exercises on chapter 8

1 Calculate the gain in dB between an input voltage of 2.0 V and an output voltage of a) 8.8 V, b) 0.04 V.

2 A signal processing system consists of a preamplifier, an amplifier and a filter, in cascade. At a given frequency, the preamplifier and the amplifier have gains of 12 dB and 31 dB respectively, while the input and output voltages of the filter are 7.0 and 0.35 respectively. Calculate the overall gain of the system at that frequency.

3 a) What voltage is i) 0 dBV, ii) 25 dBV, iii) –12 dBV?
 b) What power is i) –3 dBm, ii) 6 dBm, iii) 30 dBm?
  (Note: use power ratio, not voltage ratio, for dBm.)
 c) What are the corresponding voltages if the power is dissipated in 600 Ω?

4 The following readings were obtained when a galvanometer of an ultra-violet recorder was connected to a variable-frequency AC source delivering a sinewave with a constant amplitude of 0.5 V:

| Frequency (Hz) | 20 | 60 | 200 | 600 | 1000 | 1400 | 2000 | 2600 | 3200 | 4000 |
|---|---|---|---|---|---|---|---|---|---|---|
| Galvanometer trace amplitude (mm) | 63 | 63 | 63 | 65 | 62.5 | 58 | 49 | 29 | 22.5 | 15 |

a) For each frequency:
 i) calculate the amplitude ratio as:

$$\frac{\text{output amplitude}}{\text{input amplitude}}\left(\frac{\text{mm}}{\text{V}}\right)$$

 ii) calculate the gain in decibels as:

$$20\log_{10}\left(\frac{\text{amplitude ratio at that frequency}}{\text{amplitude ratio where constant}}\right)$$

b) Plot the dB values against frequency, on log-linear graph paper, with frequency on the logarithmic scale. Hence determine the upper limit of the bandwidth of this galvanometer.

5 a) Draw simple circuit diagrams to show how an operational amplifier may be used as:
   i) an inverting amplifier
   ii) a summing amplifier
   iii) an integrator.
   
   b) In the inverting amplifier circuit the input and feedback resistors are 10 kΩ and 470 kΩ respectively. Calculate the gain of the amplifier in dB.

6 a) Explain how noise may be
   i) generated within an instrumentation system
   ii) picked up from an external source.
   
   b) Give brief explanations of the various ways in which noise pick-up from an external source may be reduced.

7 a) Explain how
   i) a suppressor
   ii) a filter can reduce noise in a signal.
   
   b) Sketch a frequency response curve for each of the four main types of filter.

8 A band-pass filter is to be made by combining a low-pass and a high-pass filter, each filter consisting of a simple arrangement of a resistor and a capacitor. The pass band cut-off frequencies are to be 300 and 3400 Hz.
   a) Draw:
   i) the circuit diagram of each filter (state which is which)
   ii) the block diagram of the band-pass filter.
   
   b) For each filter choose a standard value of capacitor, calculate the resistance to give the required cut-off frequency, and add the relevant values to the circuit diagrams of part a) i).

9 Explain *amplitude modulation* and *frequency modulation*. Your explanation should include sketched voltage/time graphs of a sinusoidal signal, an unmodulated carrier wave and the carrier wave as modulated by each of the two methods.

   What is the main advantage and the main disadvantage of each of the two types of modulation?

10 Draw a block diagram of a phase-locked loop and explain how it demodulates an FM carrier wave.

# 9
# Digital instrumentation principles

## Why digital?

Almost all the transducers we have considered so far have had an *analogue* (also called *analog*) output; that is, the output is in a different form to the input, but its magnitude is proportional to the magnitude of the input. For example, the output of a strain gauge bridge is a *voltage* proportional to *strain*. Why should we go to the trouble and expense of turning an analogue signal into a digital one in which the output quantity is converted into a series of voltage pulses representing its numerical equivalent? The main reason is precision. Once a quantity has been expressed as a number it cannot change its value like an analogue signal may do while it is being processed (if for example the gain of the amplifier following the strain gauge bridge should alter). Also, if the signal is expressed in binary digits the processing can be done numerically in a microprocessor or computer.

## Binary digit voltage levels

In digital systems, the voltage of the power supply to the integrated circuit chips ($V_{cc}$) is usually +5 V DC. The inputs to and outputs from the chips are the binary digits 1 or 0, '1' being indicated by any voltage from $V_{cc}$ down to +2 V, and '0' by any voltage from 0 V up to +0.8 V. (Somewhat different voltages may apply to some families of integrated circuit chips but the principle is the same.) These two states may be referred to as 'high' and 'low' instead of '1' and '0'.

## Data signals and control signals

In a digital system data signals are numerical values in binary digit form. Control signals, also in logic 1 and logic 0 form, control the devices which produce or process this data. Examples include:

- start
- ready (indicates that a device has completed its task)
- inhibit (makes a device wait until the system is ready to receive its output)
- stop.

Data signals and control signals are generally kept separate from each other, each having its own set of conductors (called a *bus*) connecting the devices of the system together.

## Series/parallel data transfer

Data in the form of a binary number consists of a sequence of binary digits (*bits*). The bits are grouped together into *words* (also called *bytes*). A word or byte may contain 8, 12, 16 or more bits.

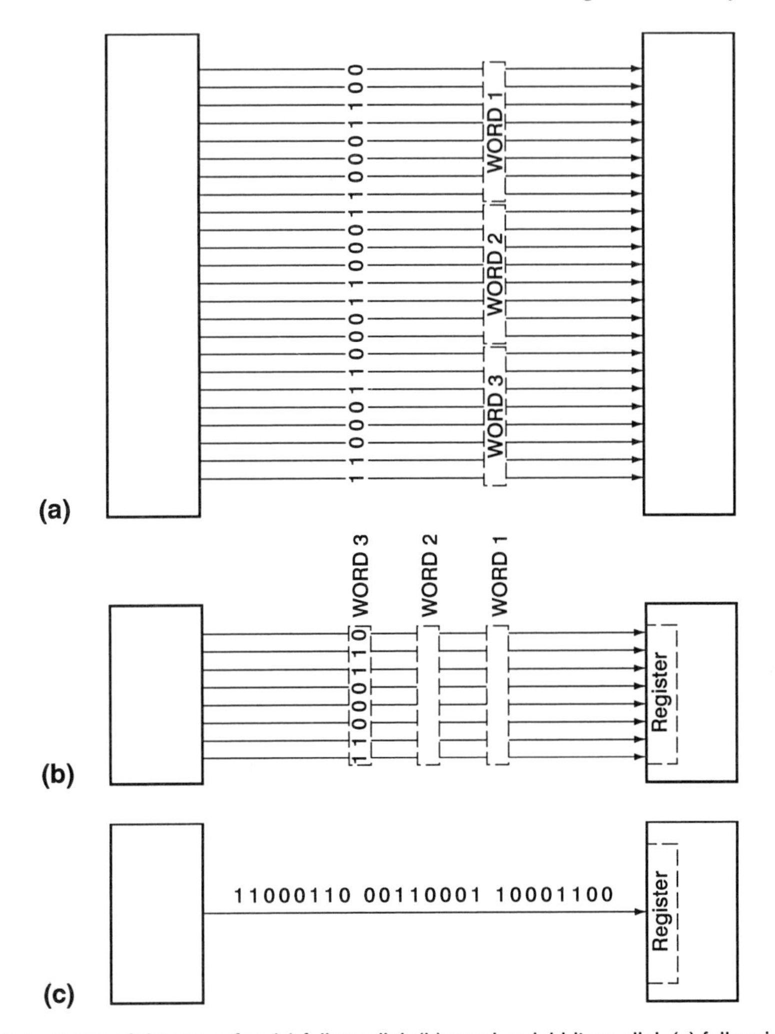


**Figure 9.1** Types of data transfer: (a) full parallel, (b) word serial-bit parallel, (c) full serial

The three methods of transferring data are illustrated in Figure 9.1.

- **Full parallel** requires a separate conductor for each bit but all the bits are transferred instantly.
- **Word serial-bit parallel** needs only as many conductors as there are bits in a word, but this method requires a temporary memory (a *register*) at the receiving device, to re-assemble the words into the original data value. Data transfer obviously takes longer by this method.
- **Full serial** means that the bits in a data value are transferred one at a time (though they may be still divided up into words). Clearly this is the slowest method of data transfer but it only requires one signal line, which is a great advantage when data has to be transferred by telephone line or radio link. Again, a register is necessary to re-assemble the binary number at the receiving end.

## Integrated circuit families

The two main families of integrated circuit chips are TTL and CMOS. The same types of devices (logic gates, counters, etc.) are available in both families. However, the two families are continually being developed and improved; at present the main developments are low-power TTL (LSTTL), high speed CMOS (HC) and advanced high speed CMOS (AC). Some *typical* characteristics of the various families are summarised below:

| | TTL | LSTTL | CMOS | HC | AC |
|---|---|---|---|---|---|
| Propagation delay time (ns) | 10 | 10 | 105 | 10 | 5 |
| Maximum clock frequency (MHz) | 35 | 40 | 12 | 40 | 120 |
| Fan out | 40 | 20 | 4 | 10 | 60 |
| Power dissipated per gate (mW) when switching at 100 kHz | 10 | 2 | 0.1 | 0.17 | 0.17 |

To make sense of this table, we shall have to define the terms:

- TTL (*Transistor-Transistor Logic*) – in this type of integrated circuit, logic gates, etc. are formed from networks of transistors diffused into the surface of the silicon chip.
- CMOS (*Complementary Metal-Oxide Semiconductor*) – this type of integrated circuit has MOS transistors in place of the more usual bipolar transistors of the TTL chips. MOS transistors have much higher impedance (resistance), and therefore much reduced current consumption, but the higher impedance causes them to be slower at switching, except in the case of the HC and AC versions. The CMOS configuration is more compact than TTL, enabling CMOS circuit chips to be made smaller; thus they are more suitable for large-scale integration.
- CMOS integrated circuits may become faulty due to the static electricity voltages generated on a person's body by friction with nylon clothing or floor coverings. To prevent this both the operative's body and the workbench must be well earthed when they are handled.
- Propagation delay time – this is the interval between a digital change of input and the corresponding change of output. The maximum number of such changes which can be completed in a second is called the *maximum clock frequency*.
- Fan out – this is the maximum number of integrated circuit inputs which can be connected to one output without overloading it.
- Power dissipation per gate – this is the power required to operate a logic gate. Except for battery operated devices, the power requirement is negligible; its importance is in the heating effect it causes at the junctions of semiconductors in large scale integrated circuits.

## Logic gates

A logic gate is an integrated circuit which, when values of binary 0 or binary 1 are applied to its inputs, gives an output of 0 or 1 in accordance with the rule governing the behaviour of the particular type of gate. Logic gates are mainly used as solid-state switches. They are thousands of times faster than mechanical switches and, unlike them, cannot suffer from contact bounce. The table on the next page shows both the American and the IEC (International Electrotechnical Commission) standard symbols for the most important types, together with the rule which determines the output of each type.

The symbols in the table are drawn with their inputs on the left of the symbol and their outputs on the right:

- AND, NAND, OR, NOR and EXCLUSIVE OR gates are not limited to two inputs; they can have any number of inputs. An INVERTER however has only one input.
- NAND and NOR gate outputs are the opposite of the outputs of AND and OR; this is indicated by their initial letter N (for negate) and by the small circle on their symbols at the output, which indicates that the output is inverted.

| **Type of gate** | **American symbol** | **IEC symbol** | **Binary value of output** |
|---|---|---|---|
| AND | | & | 1 if all inputs are 1, otherwise 0 |
| NAND | | & | 0 if all inputs are 1, otherwise 1 |
| OR | | ≥1 | 0 if all inputs are 0, otherwise 1 |
| NOR | | ≥1 | 1 if all inputs are 0, otherwise 0 |
| EXCLUSIVE OR | | =1 | 1 if only one input is 1, otherwise 0 |
| INVERTER | | 1 | 1 if input is 0, 0 if input is 1 |

*Example 9.1*
Draw up a truth table for a three-input NOR gate.

*Solution*

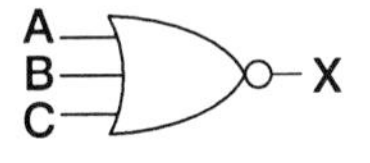


| Inputs | | | Output |
|---|---|---|---|
| A | B | C | X |
| 0 | 0 | 0 | 1 |
| 0 | 0 | 1 | 0 |
| 0 | 1 | 0 | 0 |
| 0 | 1 | 1 | 0 |
| 1 | 0 | 0 | 0 |
| 1 | 0 | 1 | 0 |
| 1 | 1 | 0 | 0 |
| 1 | 1 | 1 | 0 |

## Schmitt trigger

Logic gates are available with Schmitt trigger circuits incorporated in their inputs. A Schmitt trigger circuit is useful where the transition between binary 0 and binary 1 voltage levels is liable to be slow or affected by electrical noise. As shown in Figure 9.2, its output is triggered 'high' by a higher input voltage than that which triggers it 'low'. Between the two triggering actions the output voltage remains constant, so a Schmitt trigger 'cleans up' a sloping input pulse, turning it into a square pulse, and the hysteresis effect gives some immunity from noise. A Schmitt trigger circuit is indicated by the symbol within the logic gate symbol.

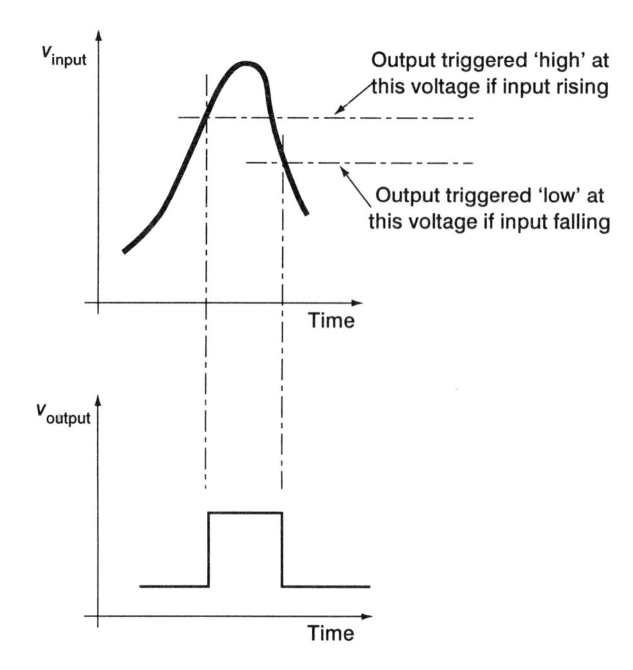


**Figure 9.2** Relationship between input and output of a Schmitt trigger circuit

## Oscillators

Logic circuits are usually operated by a transition from binary 0 to binary 1 ('low' to 'high') on one of the inputs. The circuit is said to be 'clocked' by the input pulse. To ensure that a steady stream of clock pulses is always available, an oscillator is necessary. For accurate timing of operations a crystal-controlled oscillator is used. The crystal is a wafer of quartz. Quartz is a very stable piezoelectric material (see chapter 4) and the frequency of the output waveform is determined by the natural frequency of mechanical vibration of the quartz wafer, which depends on its physical dimensions and thus is virtually constant. Where accuracy of timing is essential, the frequency of oscillation can be altered slightly by means of a trimming capacitor. This may be done first to adjust the natural frequency of the crystal, because crystals cannot be cut with sufficient accuracy, and later to compensate for the ageing of the crystal.

The frequency of oscillation will vary if the voltage of the supply to the oscillator varies or if the temperature of the crystal varies. The effect of temperature variation may be reduced by a temperature compensation circuit. Even better frequency stability is obtained by enclosing the crystal in a temperature-controlled 'oven' which is kept at about 70°C – well above the highest possible ambient temperature.

Crystal-controlled oscillators usually operate at frequencies between 1 MHz and 20 MHz. Where the oscillator is only needed for setting a time interval and great accuracy is unnecessary, a simple circuit based on an op-amp or a NAND gate may be used. These operate by the charging and discharging of a capacitor, and oscillate at much lower frequencies.

## Binary counters

A binary counter is an integrated circuit which receives a clock input from the oscillator, and has various outputs, each of which gives a square-wave at some fraction of the clock input frequency. Binary counters divide the clock frequency by powers of two. For example, output frequencies of $1/2^1$, $1/2^2$, $1/2^3$, etc. of the clock frequency are available from the appropriate pins, which are usually denoted $Q_1$, $Q_2$, $Q_3$, etc., respectively, on the pin-out diagrams.

The outputs are all kept 'low' as long as a 'high' voltage is maintained on a *reset* input pin. Switching the reset input to 'low' starts the various outputs alternating at their respective frequencies. If the outputs are read simultaneously in the correct order (that is, $Q_n$, ... $Q_2$, $Q_1$) they form a binary number which is the total number of pulses received at the clock input while the reset input is 'low'. Figure 9.3 illustrates the principle of this method of counting.

An up/down counter can count up a total or count down from it depending on whether a 'high' or a 'low' voltage is applied to its up/down control pin.

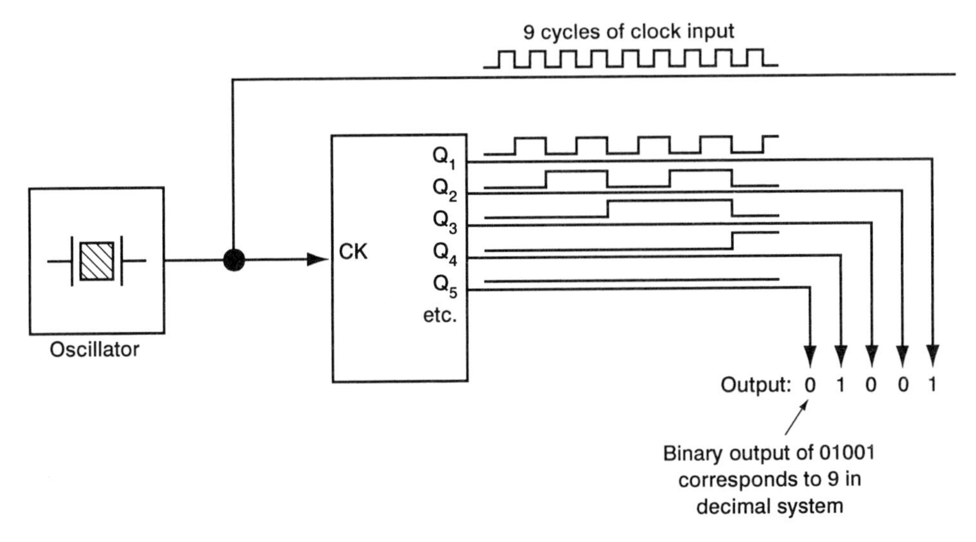


**Figure 9.3** Example showing how a binary counter counts clock input pulses

## BCD codes

An instrument which has a digital display will show the measured quantity in decimal form; for example, a digital voltmeter might show a reading of 153 volts. Each of those three digits normally has its own independent display device, with its own four-binary-digit input in accordance with the usual decimal-to-binary truth table as shown below:

| Displayed numeral | Binary inputs | | | |
|---|---|---|---|---|
| | MSB | | | LSB |
| 0 | 0 | 0 | 0 | 0 |
| 1 | 0 | 0 | 0 | 1 |
| 2 | 0 | 0 | 1 | 0 |
| 3 | 0 | 0 | 1 | 1 |
| 4 | 0 | 1 | 0 | 0 |
| 5 | 0 | 1 | 0 | 1 |
| 6 | 0 | 1 | 1 | 0 |
| 7 | 0 | 1 | 1 | 1 |
| 8 | 1 | 0 | 0 | 0 |
| 9 | 1 | 0 | 0 | 1 |

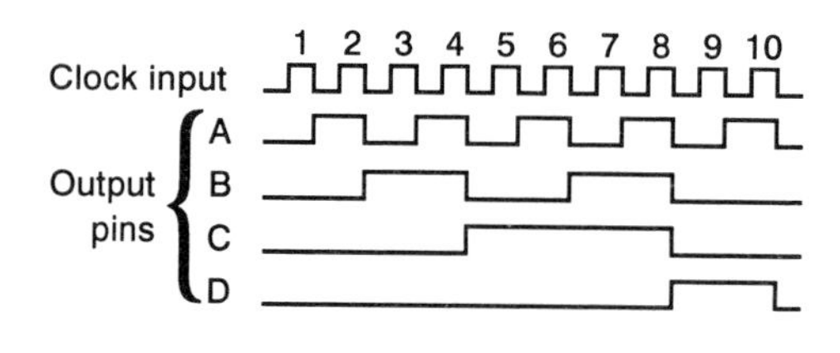


| Cycles of clock input | Output pins | | | |
|---|---|---|---|---|
| | D | C | B | A |
| 0 | 0 | 0 | 0 | 0 |
| 1 | 0 | 0 | 0 | 1 |
| 2 | 0 | 0 | 1 | 0 |
| 3 | 0 | 0 | 1 | 1 |
| 4 | 0 | 1 | 0 | 0 |
| 5 | 0 | 1 | 0 | 1 |
| 6 | 0 | 1 | 1 | 0 |
| 7 | 0 | 1 | 1 | 1 |
| 8 | 1 | 0 | 0 | 0 |
| 9 | 1 | 0 | 0 | 1 |
| 10=0 | 0 | 0 | 0 | 0 |

**Figure 9.4** Output voltage/time graphs and corresponding truth table of a decade counter

Thus the decimal quantity 153, which in *pure* binary form would be 10011001 becomes 0001 0101 0011. This method of representing the digits of a decimal number by groups of four binary digits is called *binary-coded decimal* or BCD for short. There are several different ways of coding groups of four binary digits to represent decimal numbers. Each different code has a different truth table from that shown above and has been devised for a particular purpose. For example the *XS-3 Gray code* changes only one binary digit at a time when the decimal number changes by one and is commonly used on binary-coded discs, to avoid ambiguity, as already mentioned on p.49. The code shown in the truth table above is called the NBCD code (the *natural* BCD code) to distinguish it from the other BCD codes.

## Decade counters

A decade counter is an integrated circuit which counts cycles of clock input and gives the total as a decimal digit from 0 to 9 in NBCD form on four output pins. Figure 9.4 shows 10 cycles of clock input and the corresponding 'high' and 'low' voltages which would appear at output pins A to D as the count progressed from 0 to 9 and back to 0 again. To count a number greater than 9, decade counters are *cascaded*. The switching of output D from 'high' to 'low' sends a 'carry' pulse to the clock input of the next decade counter in the cascade, which is counting tens. Its 'carry' pulse, in turn, operates the clock input of the next decade counter, which counts hundreds, and so on, as shown in Figure 9.5. Decade counters can, of course, be cascaded within a single integrated circuit.

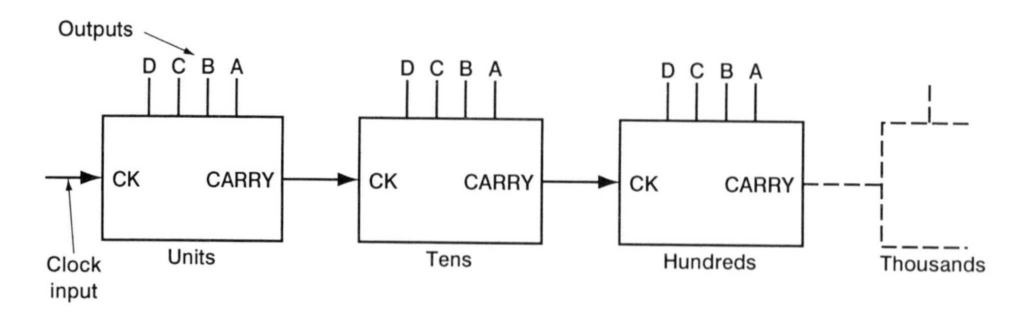


**Figure 9.5** Decade counters in cascade

## Decimal digit display

The usual method of displaying decimal digits is by splitting them up into combinations of seven standard segments, marked a to g in Figure 9.6 and switching on those segments necessary to display the particular digit required. The segments may be of light-emitting diode or liquid crystal type.

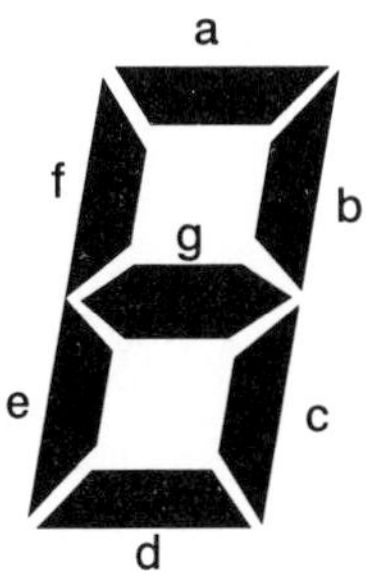

**Figure 9.6** A 7-segment digit

The output of a decade counter is on four pins, A to D, while the input to a seven-segment display has seven pins, a to g, so a special logic circuit, a *4-to-7 decoder*, with truth table shown in Figure 9.7, must be interposed between the counter and the digit display.

| Decimal | D | C | B | A | a | b | c | d | e | f | g | Display |
|---|---|---|---|---|---|---|---|---|---|---|---|---|
| 0 | 0 | 0 | 0 | 0 | 1 | 1 | 1 | 1 | 1 | 1 | 0 | 0 |
| 1 | 0 | 0 | 0 | 1 | 0 | 1 | 1 | 0 | 0 | 0 | 0 | 1 |
| 2 | 0 | 0 | 1 | 0 | 1 | 1 | 0 | 1 | 1 | 0 | 1 | 2 |
| 3 | 0 | 0 | 1 | 1 | 1 | 1 | 1 | 1 | 0 | 0 | 1 | 3 |
| 4 | 0 | 1 | 0 | 0 | 0 | 1 | 1 | 0 | 0 | 1 | 1 | 4 |
| 5 | 0 | 1 | 0 | 1 | 1 | 0 | 1 | 1 | 0 | 1 | 1 | 5 |
| 6 | 0 | 1 | 1 | 0 | 1 | 0 | 1 | 1 | 1 | 1 | 1 | 6 |
| 7 | 0 | 1 | 1 | 1 | 1 | 1 | 1 | 0 | 0 | 0 | 0 | 7 |
| 8 | 1 | 0 | 0 | 0 | 1 | 1 | 1 | 1 | 1 | 1 | 1 | 8 |
| 9 | 1 | 0 | 0 | 1 | 1 | 1 | 1 | 1 | 0 | 1 | 1 | 9 |

**Figure 9.7** Truth table of an NBCD to 7-segment decoder

## Scalers

A binary counter can also be used as a scaler to scale down an oscillator frequency to a more suitable value, a square wave being taken from one of the output pins of the counter at a frequency which is a fraction of the original clock input frequency. This much lower frequency may have a period as long as several minutes, which can be used to make a time delay, the rise from 'low' to 'high' at the end of the first half-cycle, switching on some device such as the alarm bell in a burglar alarm system.

## Counter-timers

Figure 9.8 shows how the same principle is used in the time-base of a counter-timer, the oscillator frequency of 10 MHz being divided by 10 at each output of a decade scaler to give

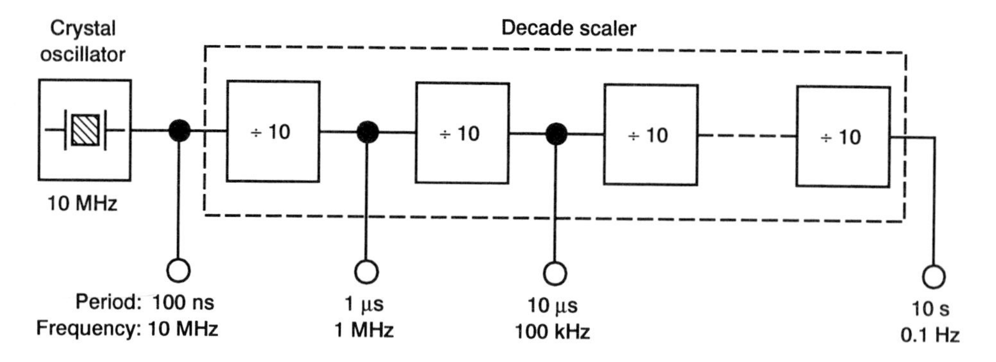


**Figure 9.8** Timebase of a counter-timer

timing intervals from 100 ns to 10 s or frequencies from 10 MHz to 0.1 Hz. The oscillator is crystal-controlled so that its frequency is as accurate as possible. A counter-timer consists of:

1. the timebase
2. a decade counter with a decimal display of seven or more digits
3. a logic gate to switch the input to the counter on and off
4. input amplifiers with Schmitt trigger circuits which can cut out noise and 'shape' the input signal
5. gate control circuits.

Figure 9.9 shows how these components are connected together for the basic processes of measuring frequency and period.

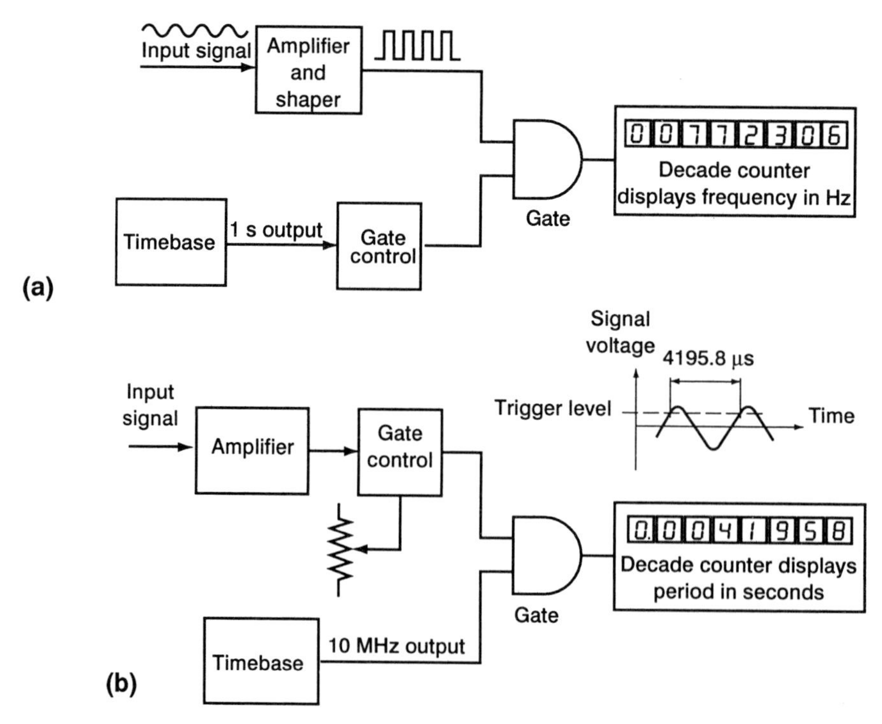


**Figure 9.9** Examples of use of counter-timer: (a) frequency measurement, (b) period measurement

- **Frequency** is measured by admitting the input signal to the decade counter for an exact period of time, set by the timebase. The count displayed at the end of that period is then the input frequency.
- **Period** is measured by admitting the 10 MHz output of the timebase to the decade counter from the instant when the signal voltage first exceeds a trigger voltage level until the instant when it again *rises* through that level (thus the gate control circuit must be able to distinguish between positive and negative slope). The instrument's range switch adjusts the position of a decimal point in the displayed count of oscillator cycles to show the result in seconds, milliseconds or microseconds.
- Time intervals, pulse widths, and the phase angle between two identical waveforms can also be measured by this type of instrument if it has dual inputs.

Sources of error are:

1. a random error of ± 1 in the frequency count, because the decade counter cannot show fractions less than 1
2. time-base error due to variations in oscillator frequency caused by (i) temperature change, (ii) supply voltage change, (iii) ageing of the crystal and (iv) short term instability
3. trigger error due to noise, which causes early or late crossing of the trigger level
4. channel mismatch error when two inputs are in use, with rise times and propagation delay times which differ.

Because of the possible random counting error of ± 1 noted above, frequency measurement is more accurate for high-frequency signals than for low-frequency. The frequency of a low-frequency signal is more accurately found by measuring its period and taking the reciprocal. The break-even point between the two methods is determined by the square root of the oscillator frequency. For a 10 MHz oscillator this is 3160 Hz. Thus frequency measurement is the more accurate above 3160 Hz, period measurement below 3160 Hz.

A development of the counter-timer is the computing counter. This actually contains two counters: one counts cycles of input, the other simultaneously counts cycles of the 10 MHz clock. The instrument can be set to measure over any length of time; the start of measuring is triggered by a set voltage level on the input signal, the end is triggered at the same point on the input signal cycle immediately after the set time period has expired. A microprocessor then divides one count by the other to obtain either frequency or period, and displays the result. Figure 9.10 shows a block diagram of the instrument. Counting only complete cycles of input eliminates the frequency error of ± 1 cycle mentioned earlier; in its place is a ± 1 clock pulse error (± 100 ns) which is normally very much smaller.

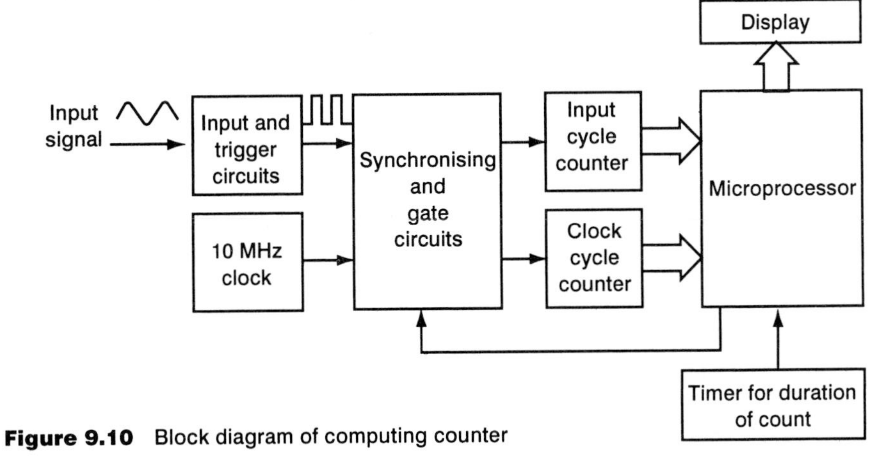


**Figure 9.10** Block diagram of computing counter

## Multiplexing

This means using a single channel for conveying several independent signals. In telecommunications, the multiplexing is usually done by *frequency-division*; each signal occupies a narrow frequency band, the frequency bands being 'stacked' one above the other in frequency and modulated onto a single high-frequency carrier wave. At the receiving end the signals are separated by band-pass filters into their separate channels. For example, in a radio link in a telephone system, a speech channel occupies the range 300 to 3400 Hz. The frequency band to carry it is 4 kHz wide and the bands are modulated on to the carrier wave at frequencies 60 to 64 kHz, 64 to 68 kHz, 68 to 72 kHz and so on.

The alternative to frequency-division is *time-division*, in which each signal in turn is connected to the communication channel for a short period of time. This is normally used where the signals are in the form of binary digits. An example of time-division multiplexing is seen in the usual arrangement of multiplexed connections to the 7-segment digits of an instrument with a digital display. Figure 9.11 is a diagram of a typical circuit for a four-digit display of the light emitting diode type. The BCD value of the four-digit number to be displayed is stored in some form of temporary memory, either RAM (random access memory) or latches (logic circuits which store the sequence of bits and connect them to the next stage on receipt of a 'write' pulse). The clock, which can be a simple oscillator, is connected to a ÷ 4 counter. The output of this counter, 00, 01, 10, or 11, tells the memory to connect the BCD value of either the first, second, third or fourth digit to the 4-to-7 decoder, and this decoder sets the appropriate segments for this digit 'high' on all the display digits. The common cathode pin of each display digit

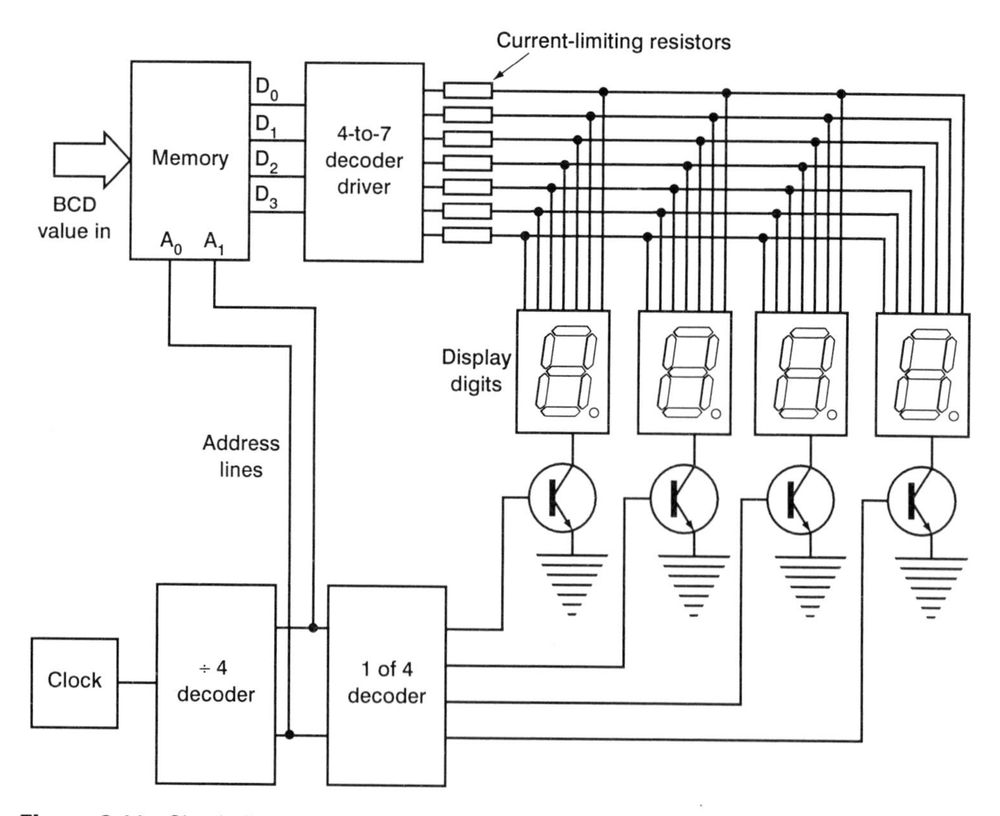


**Figure 9.11** Circuit diagram of a multiplexed display

is connected to ground via a transistor, which acts as a switch. The 1-of-4 decoder, in response to the output of the ÷ 4 counter, switches on the first transistor so that although all four display digits have the same segments 'high', only the first digit lights up. After one clock period the counter is incremented, the bits for the second digit are fed to the 4-to-7 decoder and the 1-of-4 decoder switches off the first display digit and switches on the second one. And so on. The sequence cycles continuously, the first display digit following on again as the fourth is switched off. To avoid flicker, the cycle of digits should repeat at least 100 times a second.

Since each display digit is switched on for only a quarter of the time, a four-digit multiplexed LED display has to be driven with roughly four times the normal current if it is to show up as brightly as a non-multiplexed display.

Some types of LED display digit have a common anode instead of a common cathode. The segments of this type of digit are turned on by making their voltage 'low' instead of 'high', the common anodes are connected to the positive of the supply instead of to ground, and the transistors in Figure 9.11 which make this connection must then be of *p-n-p* type instead of *n-p-n*.

The blocks shown in Figure 9.11 will not necessarily be found as separate components in a digital display instrument. Often they are incorporated within an integrated circuit which performs another function, such as a counter or an A/D converter, and which is therefore connected directly to the display.

## Sampling

An analogue signal is converted into a digital one by *sampling*, as shown in Figure 9.12. This is done by a *sample-and-hold amplifier*, an integrated circuit consisting of two op-amps connected by a solid-state switch as shown in Figure 9.13(a). The first op-amp is a wide-bandwidth amplifier which passes the analogue voltage to the second op-amp while the switch is closed (*sampling phase*). When the switch is opened, the second op-amp maintains the voltage it has at that instant, with as little droop as possible, for the circuits which follow (*holding phase*). Figure 9.13(b) shows the process as a voltage/time graph. With the switch closed, the output tracks the input. When the *hold* signal is applied to the switch control there is a slight delay, the *aperture time*, typically 50 ns, before the switch opens. This can be allowed for by advancing the *hold* signal by the same amount. After the switching transient has died out, the output stays constant except for the *droop*, typically 1 mV/ms. There is a similar delay

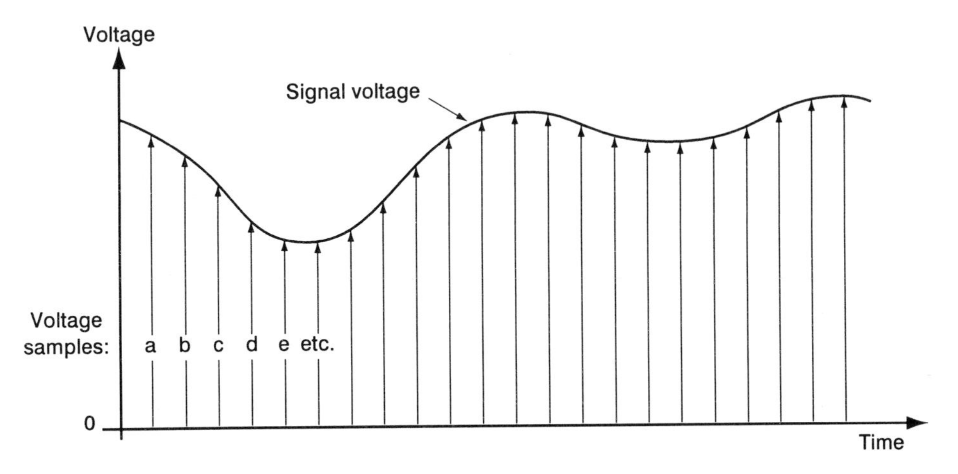


**Figure 9.12** Quantifying a signal by sampling

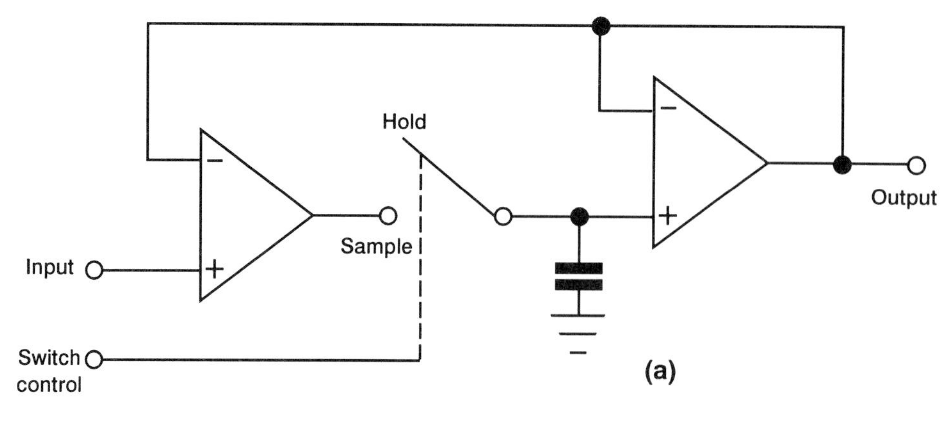


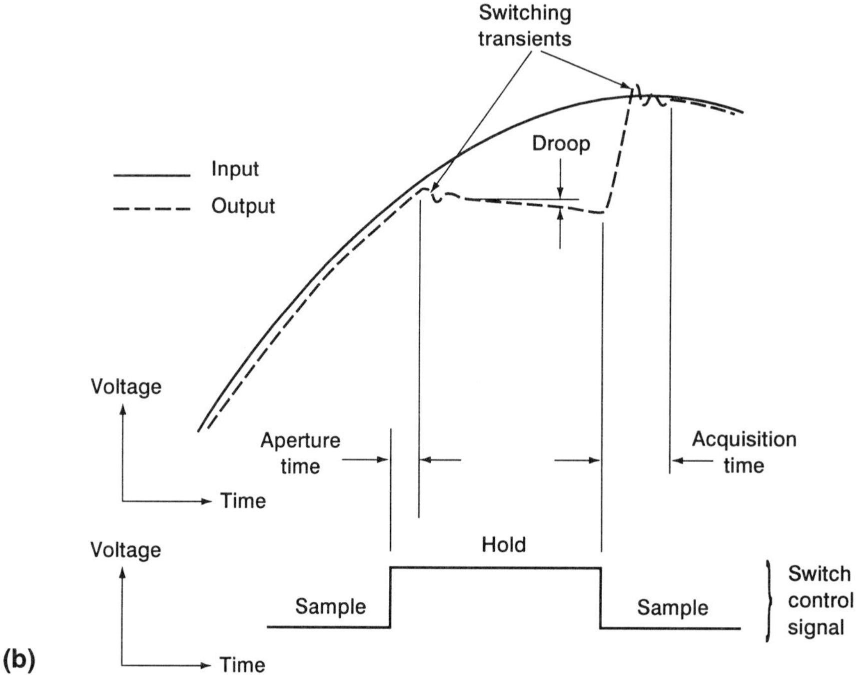


**Figure 9.13** Sample-and-hold amplifier: (a) circuit diagram, (b) voltage/time graphs

on switching from *hold* to *sample*, before the output can start tracking the input again. This delay, the *acquisition time*, is usually one or two microseconds from the instant of switching until the switching transient has died out.

## The sampling theorem

The sampling theorem states:

> *A continuous signal can be represented completely by, and reconstructed from, a set of instantaneous measurements or samples of its voltage which are made at equally spaced times. The interval between such samples must be less than one-half of the period of the highest-frequency component in the signal.*

In other words, we can convert an analogue signal into digital samples and convert them back into the original signal provided the number of samples per second is more than twice the highest frequency in the signal.

Suppose the signal contains a higher frequency than we expected, so that in fact the number of samples per second is less than twice that frequency. We then get what is known as *aliasing;* the samples convert back into a false lower frequency, as shown in Figure 9.14.

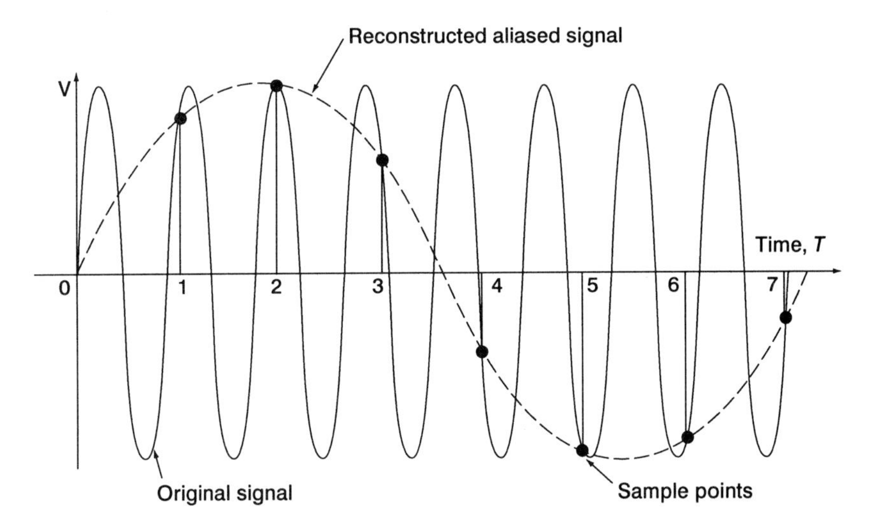


**Figure 9.14** The effect of sampling at less than twice the signal frequency

Of course an instrumentation signal is unlikely to be the nice tidy sine wave of Figure 9.14. It is more likely to be a varying voltage which is not a regularly repeating waveform. But by Fourier analysis, any such shape can be shown to be a summation of a constant term (in this case, a DC voltage) and sine waves of various amplitudes and frequencies.

An example of the application of sampling is the digital recording of music on compact disc (CD). The original design process was a compromise between disc size, playing time, maximum audio frequency which could be recorded, and the resolution of the analogue-to-digital conversion. For a given disc diameter (120 mm), maximum playing time (about 70 minutes – the duration of a symphony) and A/D resolution (14 bits), the bit capacity of the disc dictated the sampling frequency, which was set at 44.1 kHz (i.e. disc capacity = $70 \times 60 \times 44\,100 \times 14$ bits). To avoid aliasing, the highest allowable audio frequency had to be $<\frac{1}{2} \times 44.1$ kHz, so the input frequency was limited by a low-pass filter with 20 kHz cut-off.

## Analogue-to-digital conversion

The sample voltages from the sample-and-hold amplifier are converted into binary numbers by another integrated circuit, an analogue-to-digital converter (usually abbreviated to *A/D converter*). The main types of A/D converter, single slope, dual slope, successive approximation and flash, are described below.

### Single-slope or single-ramp A/D converter

This is shown diagrammatically in Figure 9.15. The 'ramp' is a steadily increasing voltage; its voltage/time graph would be an inclined straight line – hence the name. The ramp generator is

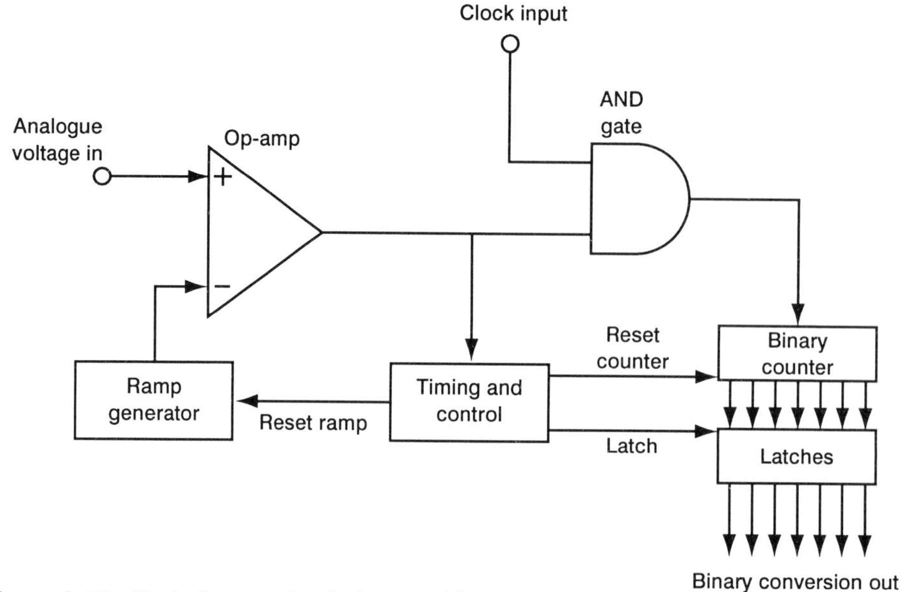


**Figure 9.15** Block diagram of a single-slope A/D converter

similar to the one shown in Figure 9.16(a): an op-amp with capacitive feed-back, which makes it integrate voltage with respect to time. At the start of the conversion operation the control circuit sets both the ramp voltage and the binary counter to zero. Any positive analogue voltage on the non-inverting input of the op-amp sends its output 'high'. This switches-on the ramp generator and at the same time enables the AND gate (see p.127) to pass cycles of square wave from the clock input to the binary counter. At the instant the ramp voltage exceeds the analogue input voltage, the op-amp output goes 'low'. This switches off the clock input to the binary counter and it *latches* (i.e. holds constant in a memory) the binary number it has counted to. The ramp generator and counter are then reset to zero, ready to convert the next sample.

This type of A/D converter is simple and cheap but comparatively slow, a 7-bit converter taking 1 ms or more to do a conversion. Another disadvantage is its inaccuracy if the clock frequency or the slope of the voltage/time ramp alter due to ageing or temperature-sensitivity of components.

## Dual-slope A/D converter

This cancels out any inaccuracy due to variations in clock frequency or ramp slope by using two ramps in a count-up, count-down process. The principle is shown in Figure 9.16 and explained in the following sequence:

1 The binary counter is set to zero and the analogue voltage is connected to the inverting input of the first op-amp, the integrator, which generates the ramps; its output is already at zero volts.
2 The positive analogue voltage applied to the inverting input of the first op-amp causes it to start generating a steadily increasing negative voltage at a rate proportional to the input voltage (i.e. a negative ramp with slope proportional to input).
3 The output of the second op-amp (the *comparator*) instantly goes 'high', enabling the AND gate to admit clock pulses to the binary counter.
4 Counting proceeds until the counter is full. The overflow causes the counter to be reset to zero and start counting again, and the inverting input of the first op-amp is switched to the negative reference voltage.

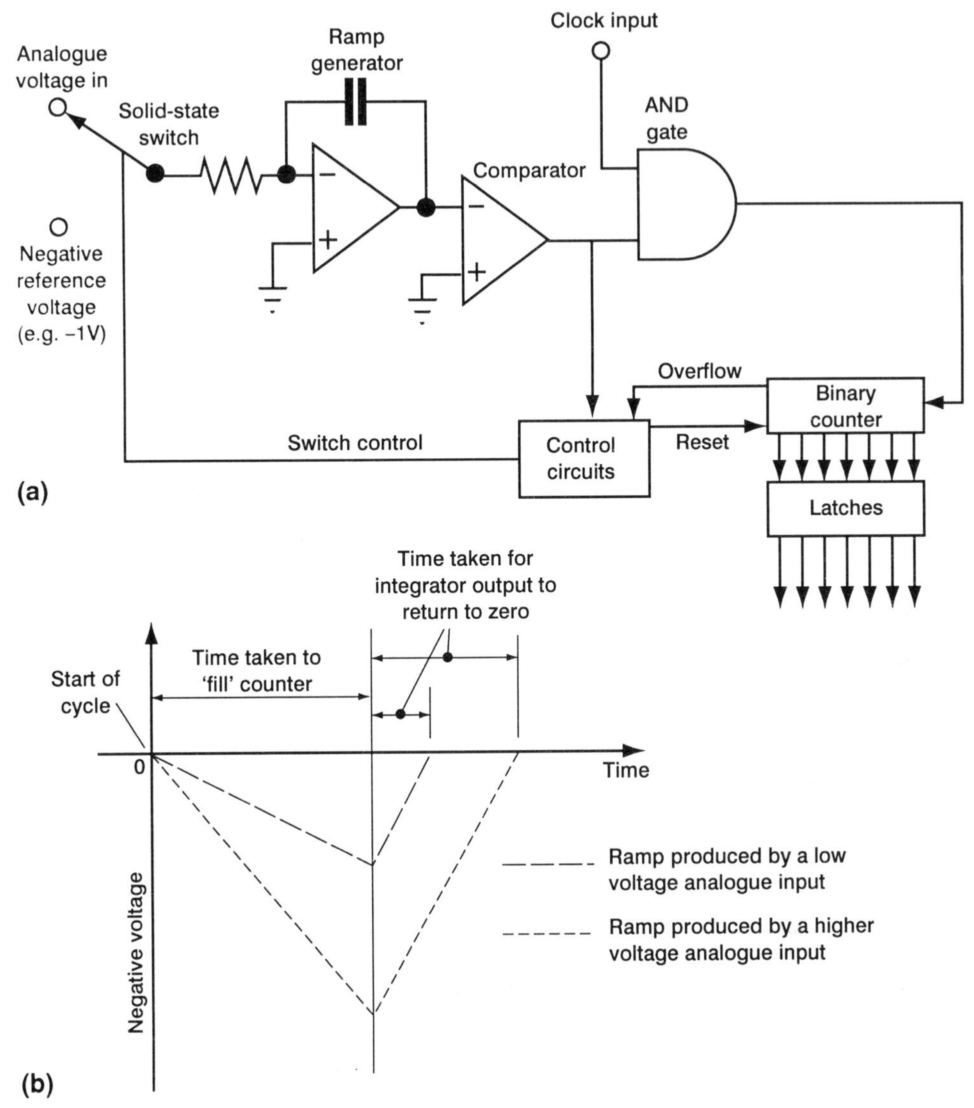


**Figure 9.16** Dual-slope A/D converter: (a) block diagram, (b) integrator output voltage/time graphs

5 This causes it to generate a positive ramp with slope proportional to the reference voltage. The negative output voltage of the first op-amp therefore starts to decrease; when it reaches zero the comparator output goes 'low', blocking the clock input, and the binary number in the counter is latched by the control circuit. This number is the digital equivalent of the analogue input voltage.

The cycle then recommences. The accuracy of a dual-slope A/D converter is unaffected by drift in the clock frequency or in the values of the resistance or capacitance in the integrator circuit, since the upward and downward ramps are affected equally. Also, high-frequency noise disappears in the integration, and changes in the analogue input signal during the integration period are averaged out. And if the duration of the integration is arranged to be a multiple of the period of a background noise such as 'mains hum', that noise is rejected.

The chief disadvantage of the dual slope converter is its slow speed. However it is used in many digital voltmeters, where its comparative slowness does not matter.

### Successive-approximation A/D converter

This type of A/D converter, shown diagrammatically in Figure 9.17, is very much faster than the ramp types. Its sequence of operations is as follows:

- Starting with the most significant bit (i.e. the left-hand binary digit) set to 1 and the remaining bits set to 0, it generates the equivalent analogue voltage to that number, by means of a built-in digital-to-analogue converter (see the next section in this chapter). That voltage is compared with the input voltage, in the comparator. If it is less than the input voltage the 1 is retained, otherwise it is altered to 0.
- The next most significant bit is then changed from 0 to 1, the equivalent voltage is again generated, and the comparison repeated. Again, this latest 1 is either retained or changed to 0, depending on whether the generated voltage is less than or greater than the input voltage.
- The process continues until, finally, the effect of a 1 as the least significant bit (LSB) has been tested. The digital conversion of the analogue input is then complete and the number is latched for output.
- The D/A converter output voltages are biased to increase them by the equivalent of $\frac{1}{2}$ LSB, so that the maximum error in conversion can only be $\frac{1}{2}$ LSB. Each comparison is operated by one clock pulse, and the complete conversion takes place in a few microseconds.

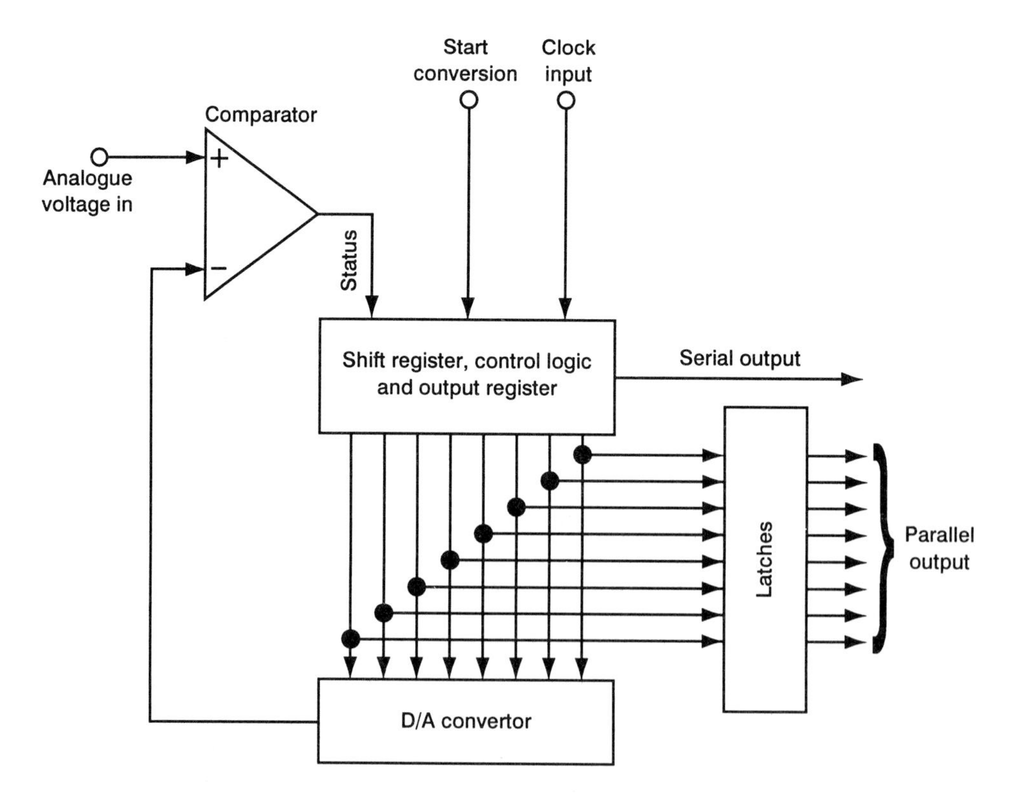


**Figure 9.17** Successive-approximation A/D converter

*Example 9.2*
Show the stages by which a 3-bit successive approximation A/D converter with an analogue span of 0 to 10 volts would convert 6.5 V to binary.

*Solution*
With reference to Figure 9.20, for a 3-bit D/A converter,

$$10 \text{ volts needs } \left(\frac{1}{2} \text{ step bias}\right) + (7 \text{ steps}) + \left(\frac{1}{2} \text{ step}\right) = 8 \text{ steps}$$

Therefore the stages of conversion are:

100 $\left(\frac{4.5}{8} \times 10 = 5.625 \text{ V} < 6.5, \text{therefore retain first 1}\right)$

110 $\left(\frac{6.5}{8} \times 10 = 8.125 \text{ V} > 6.5, \text{therefore replace second 1}\right)$

101 $\left(\frac{5.5}{8} \times 10 = 6.875 \text{ V} > 6.5, \text{therefore replace third 1}\right)$

$\therefore$ output is 100.

### Parallel, simultaneous or 'flash' A/D converter

This is the fastest converter of all. Figure 9.18 illustrates the principle, showing the circuit for a 2-bit 'flash' A/D converter with 4 volt power supply. By means of series resistors forming a potential divider across the power supply, a number of equal voltage steps is obtained. Each of these steps is applied to the inverting input of its own op-amp. The analogue input voltage is applied to the non-inverting inputs of all those op-amps. Where the analogue input voltage is higher than the voltage on the inverting input of an op-amp, the output voltage of the op-amp is 'high' (binary 1), otherwise it is 'low' (binary 0). These two strings of 0s and 1s are then turned into a corresponding binary number by means of encoding gates.

The 2-bit converter requires three op-amp comparators; an $n$-bit converter requires $(2^n - 1)$ comparators. Thus each additional bit virtually doubles the number of comparators required. All of the comparators and the encoding gates are included on the same integrated circuit chip. At present the practical limit is 11 bits; 12-bit resolution can be obtained by connecting two 11-bit chips in series. 8-bits is a more usual size; an 8-bit 'flash' A/D converter chip contains 255 comparators and is capable of making 20 million A/D conversions per second.

## Digital-to-analogue conversion

After transmission, processing etc., the digital signal may have to be converted back to analogue form. This is done by another integrated circuit: a *digital-to-analogue converter (D/A converter)*.

The D/A converter circuit, known as an *R-2R ladder*, is shown in Figure 9.19. It has as many inputs as there are digits in the binary numbers which represent the voltage values. Each input operates its own electronic switch (a logic gate) which connects that particular leg of the ladder to the reference voltage if the binary digit is a 1, or to earth if it is a 0. If all the legs but one are connected to earth, the one connected to the reference voltage produces a current which flows towards the inverting input of the op-amp and is halved by the resistance network at each junction through which it passes. Thus the current contribution of each leg is weighted to correspond to the position of that particular digit in the complete binary number; for example, in a 4-bit D/A converter the current produced by the most significant bit (MSB) will be eight

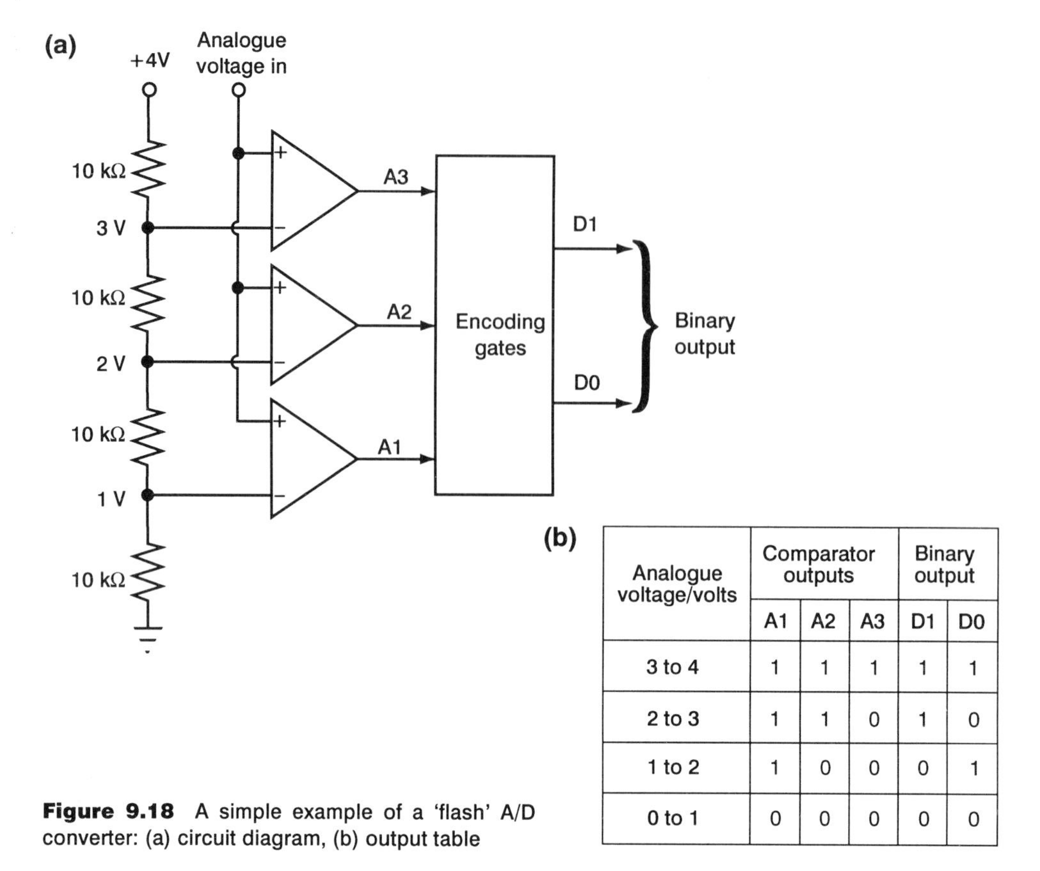


| Analogue voltage/volts | Comparator outputs | | | Binary output | |
|---|---|---|---|---|---|
| | A1 | A2 | A3 | D1 | D0 |
| 3 to 4 | 1 | 1 | 1 | 1 | 1 |
| 2 to 3 | 1 | 1 | 0 | 1 | 0 |
| 1 to 2 | 1 | 0 | 0 | 0 | 1 |
| 0 to 1 | 0 | 0 | 0 | 0 | 0 |

**Figure 9.18** A simple example of a 'flash' A/D converter: (a) circuit diagram, (b) output table

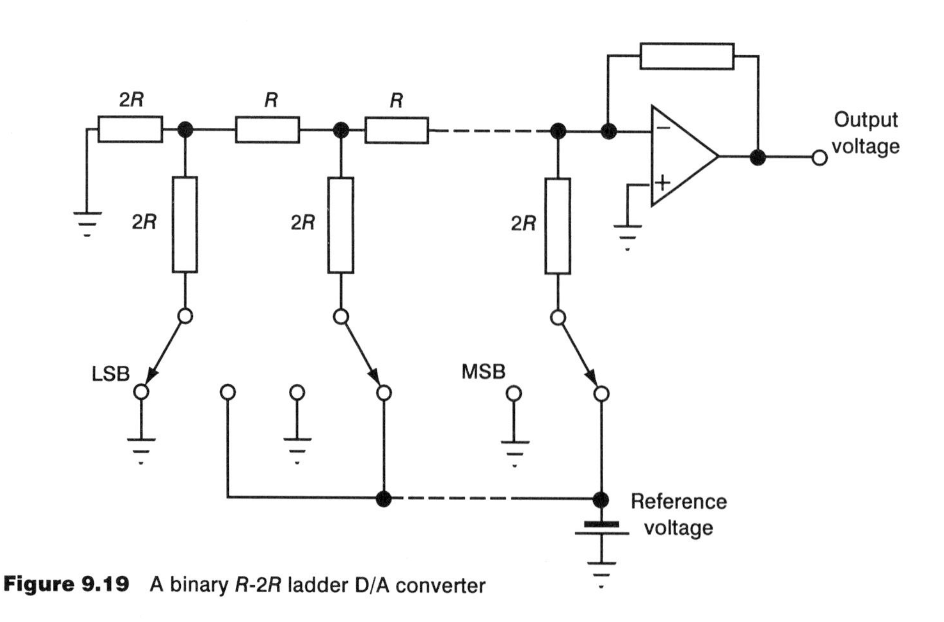


**Figure 9.19** A binary *R*-2*R* ladder D/A converter

times ($2^{(4-1)} = 8$) the current produced by the least significant bit (LSB). The op-amp produces an output voltage proportional to the sum of the currents.

There is a wide variety of D/A converter chips available with various output voltage limits and output current limits. There are also multiplying types, in which the output voltage depends on both the binary input and a varying reference voltage, and digitally buffered types for easy interfacing with a microprocessor, the 'buffer' being a memory which stores bits until the converter can deal with them.

In addition to deciding which of these types he needs, the system designer must also consider the speed at which he wants the chip to work, and the resolution he requires. The speed is specified as the *settling time*; the time it takes for a full scale input to be converted to within half of the least significant bit ($\frac{1}{2}$ LSB). Settling times range from a few microseconds to a few nanoseconds, depending on cost and the number of bits in the binary input. The number of bits also determines the resolution. Figure 9.20 shows the resolution ($\frac{1}{2}$ LSB) which would be obtainable from a 3-bit D/A converter. Each additional bit in the binary number doubles the number of steps in the graph, and so halves the (theoretical) maximum error in the output voltage. The actual maximum error depends also on the stability of the reference voltage applied to the chip and the stability of the resistors in the switching circuit.

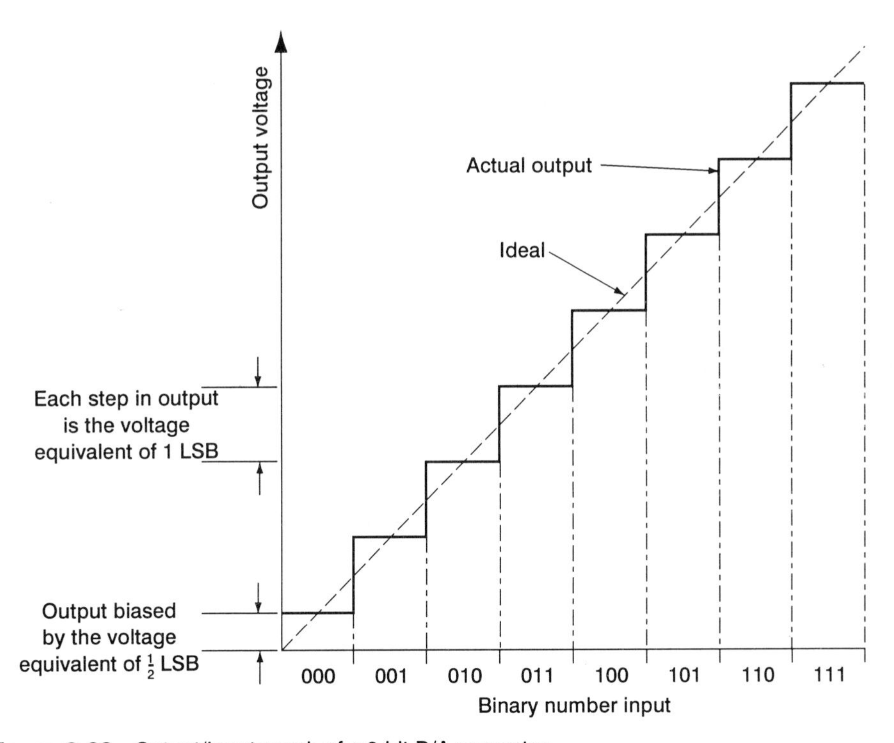


**Figure 9.20** Output/input graph of a 3-bit D/A converter

## A complete digital system

We have now considered the *essential* components of a digital instrumentation system: a sample-and-hold amplifier, an A/D converter and a D/A converter, such as might be found in a digital tape recorder for instance. And yet if the system consisted only of those components we should be very disappointed in the result, because the output of D/A converter fed with a sequence of digital voltage values will be a series of voltage level steps as shown in Figure 9.20. To smooth out the steps into a continuous curve, the high-frequency sinewave components which make up a step must be filtered out by a low-pass filter (see Filters in chapter 8).

Figure 9.21 shows another low-pass filter before the sample-and-hold amplifier, to filter out high-frequency noise in the signal and prevent aliasing, and a digital filter between the A/D and D/A converters. A digital filter is a microprocessor or logic circuit fulfilling the same function as an analogue filter but doing it much more precisely than the analogue version could, by performing addition, multiplication or delay operations on the stream of binary values. Even with all the filters in place, we shall never get absolutely perfect reproduction of the input signal at the output because the process of sampling changes a smooth analogue curve into an approximation composed of steps made up of horizontal and vertical lines. If we superimposed the approximation onto the original curve we should see that the divergence between them was a maximum at the sharp corners of the steps. It is as if by working from sample voltages of the signal we had introduced noise into the system. This noise is called *quantization noise*.

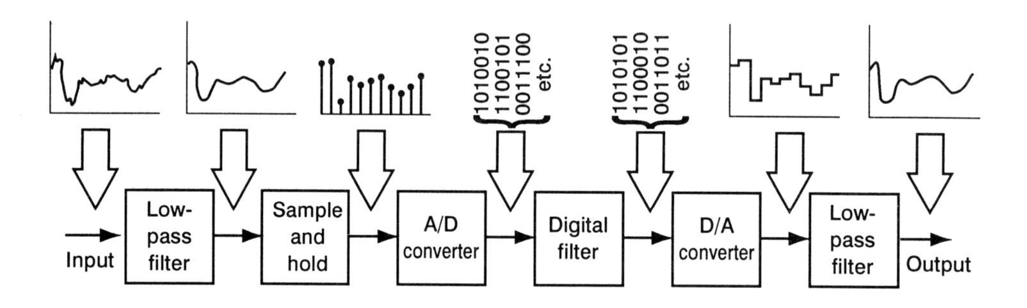


**Figure 9.21** Block diagram of a complete digital system, showing how the form of the signal is changed by each block

To reduce quantization noise we must reduce the height of the steps by increasing the number of voltage levels at which the LSB changes by 1, and this will require more digits in the digital conversion of the voltage samples.

*Example 9.3*

A digital system is to transmit an analogue voltage in the range 0 to 10 volts. Calculate the number of discrete voltage levels it can recognise and the minimum height of each step in the analogue output if it is (a) a 7-bit system, (b) an 8-bit system. (Neglect $\frac{1}{2}$ LSB biasing.)

*Solution*

a) 7 bits gives a maximum value of

$$2^6 + 2^5 + 2^4 + 2^3 + 2^2 + 2^1 + 2^0 = 64 + 32 + 16 + 8 + 4 + 2 + 1 = 127.$$

Adding in the value zero (binary 0000000), the system can recognise 128 levels of voltage. (i.e. $2^7$ levels). The minimum height of each step in the analogue output is:

$$\frac{10}{127} = 0.079 \text{ V}.$$

b) The corresponding values for an 8-bit system are 256 levels and 0.039 V.

## Polarity indication

Where an analogue signal voltage can have either a positive or a negative value (e.g. in a voltage range of ± 5 volts) an indication of polarity will be necessary after A/D conversion. This can take the form of an extra digit added to the binary 'word' in the MSB position, the extra digit being a 1 if the sample voltage is zero or positive or a 0 if it is negative. Thus, to take a very simple example, the quantization levels representing the range ± 5 V by 4-bit binary words (3 bits plus polarity indicator) would be:

| Volts | Binary number |
|---|---|
| 5.00 | 1111 |
| 4.29 | 1110 |
| 3.57 | 1101 |
| 2.85 | 1100 |
| 2.14 | 1011 |
| 1.43 | 1010 |
| 0.71 | 1001 |
| 0 | 1000 |
| –0.71 | 0001 |
| –1.43 | 0010 |
| –2.14 | 0011 |
| –2.85 | 0100 |
| –3.57 | 0101 |
| –4.29 | 0110 |
| –5.00 | 0111 |

## Companding

Whether the quantization noise of a given system will seriously interfere with a signal depends on the amplitude of the signal. For example, in the 8-bit system of the previous worked example, the ratio:

$$\frac{\text{signal amplitude}}{\text{quantization step}} = \frac{10}{0.039} = 256$$

if the signal amplitude is 10 V but only 25.6 if the amplitude is 1 V. To improve the signal-to-quantization noise ratio of low level signals the technique known as *companding* may be used. (The word is a combination of *compressing* and *expanding*.) Before going to the sample-and-hold amplifier the signal is passed through a *compressor amplifier*. As Figure 9.22 shows, this has a gain which varies as the input voltage alters. At low input voltages the gain is high, so that the quantization steps are closer together. As the input voltage increases the gain decreases, to space the steps further apart. To cancel out the resulting distortion of the signal, the output of the complete digital system is finally passed through an *expander amplifier*, which has the opposite gain characteristic to that of the compressor, as shown in Figure 9.23. One example of the use of companding is in digital telephone transmission systems, where loudness of speech may vary considerably between different phone users.

As an alternative to compressor and expander amplifiers, non-linear A/D and D/A converters may be used to obtain the same effect.

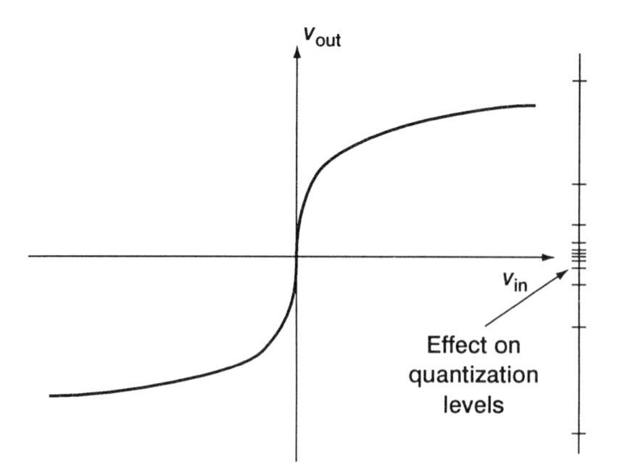


**Figure 9.22** Input/output relationship of a compressor amplifier

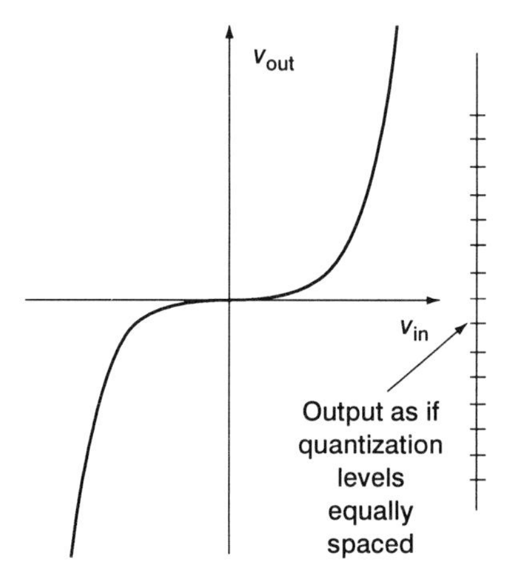


**Figure 9.23** Input/output relationship of an expander amplifier

## Pulse code modulation

This is the modulation of a digital signal on to a carrier wave, magnetic tape or disc. The digits 1 and 0 are represented by opposite voltage levels, or by opposite magnetization in the case of magnetic recording. Figure 9.24 shows two alternative formats; RZ (*return to zero level*) and NRZ (*non-return to zero level*). NRZ is the usual format; RZ enables individual bits to be distinguished but needs twice as much bandwidth because each transition between 1 and 0 entails two changes of level instead of one.

For NRZ, the upper limit of the frequency range of the digital signal is determined by assuming the worst case: 0s and 1s alternating continuously at the bit-stream frequency. As Figure 9.25 shows, a frequency of half the bit rate, sampled at minimum sampling frequency would (just) give the necessary digital information. With NRZ the theoretical lower limit of the frequency range is 0 Hz – that is DC, due to long streams of 1s or of 0s. DC is difficult to record properly, so special NRZ codes may be used; for example in a stream of 8-bit data words, bits 2, 3, 6 and 7 may be inverted to ensure that transitions between the two levels occur with reasonable frequency.

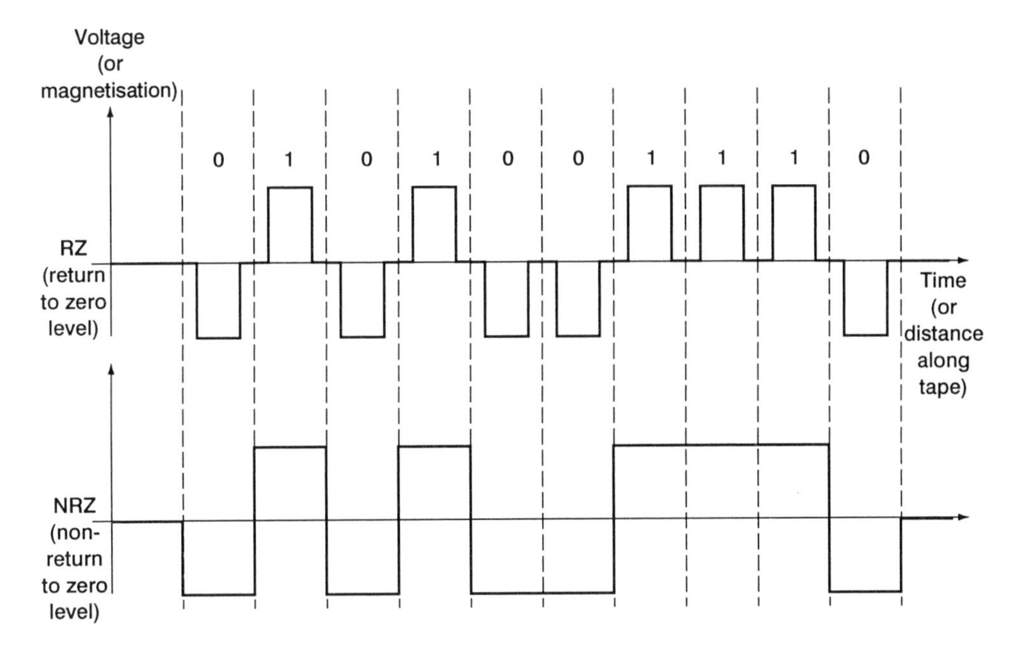


**Figure 9.24** Pulse code modulation formats

Digital data can be recorded in serial, parallel or serial-parallel form. This corresponds to an 8-track head, recording 8-bit words in parallel on eight separate tracks, or a 2-track head recording alternate bits on alternate tracks. The unit of digital data transfer rate is the *baud* (abbreviation: Bd). 1 Bd = 1 bit per second.

Pulse code modulation is much less likely to be corrupted by noise than an analogue signal is, since only two levels have to be recognised. To guard against the occasional mis-reading of a bit in the data stream, a parity check digit or *parity bit* may be included in each word. For example, the first 7 bits of an 8-bit word may be data, with the eighth bit the parity bit. This is either a 0 or a 1 to make the total of number of 1s in the word an odd number if the system uses *odd parity*, or an even number for *even parity*.

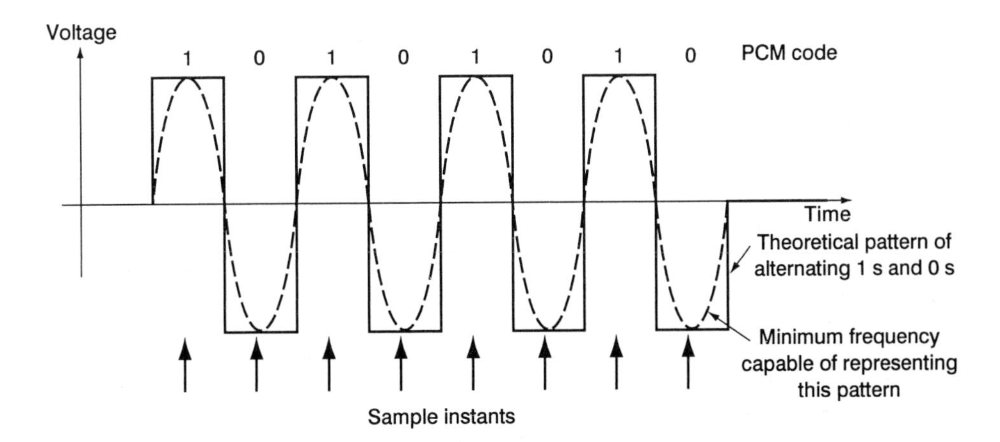


**Figure 9.25** Minimum upper frequency limit of PCM bandwidth

*Example 9.4*
Make the following into 8-bit words in which the eighth bit is the parity bit, using odd parity.

0011111 0100010 1111101 1100010

*Solution*

00111110 01000101 11111011 11000100

A parity bit shows up a single error in a binary word, but if there are two errors in the same word, parity will still be satisfied and the errors will go undetected. Error-detecting and error-correcting codes have been devised which can deal with more than one error in a word, but at the cost of greater complication in coding and decoding.

## Exercises on chapter 9

1 Compare the TTL and CMOS versions of a given type of integrated circuit by stating which of the two versions is the better in respect of each of the following considerations:
  a) high input impedance
  b) low power consumption
  c) ease of handling during circuit construction.
2 A logic circuit consisting of an inverter, a two-input NAND gate and a two-input OR gate receives three digital signals, A, B and C. Signal A passes through the inverter to one of the inputs of the NAND gate; signal B is connected directly to the other input. The output of the NAND gate is connected to one of the inputs of the OR gate; signal C is connected directly to the other input.
  a) Draw a diagram of the logic circuit, showing the inputs A, B, and C.
  b) Allot any other identifying letters necessary, and draw up a truth table showing the output of the OR gate for any combination of A, B and C.
3 Which of the following statements about serial data transfer is/are true?
  a) 'It is faster than parallel data transfer.'
  b) 'It needs only one wire.'
  c) 'It is unsuitable for BCD.'
  Explain the error in any untrue statements.
4 The logic gate shown in Figure 9.26 on the next page has input waveforms A, B and C. Which of the waveforms $X_1$, $X_2$ or $X_3$ represents its output? Explain why the waveform you have chosen is the correct one.
5 Which of the following arrangements will enable a three-input NAND gate to be function as an inverter?
  a) All inputs connected together to form a single input.
  b) A permanent 'high' on the two unused inputs.
  c) A permanent 'low' on the two unused inputs.
6 A particular Schmitt trigger circuit gives an output voltage corresponding to binary 1 when the input voltage reaches +1.6 V while rising, and an output voltage corresponding to binary 0 when the input voltage reaches +0.8 V while falling. If the input is a sinusoidal voltage with amplitude ±7.5 V, for how many electrical degrees of this input will the output be 1 and for how many degrees will it be 0?
7 A binary counter's output pins are denoted $Q_1$ to $Q_{10}$.
  a) Write down in binary digits the logic levels at the pins when 189 pulses have reached the clock input since it was reset, taking the pins in order from $Q_1$ (LSB) on the right to $Q_{10}$ (MSB) on the left.
  b) Write down the binary digits of the NBCD code for this number.
  c) Sketch a seven segment digit, labelling the segments a to g in the standard notation.

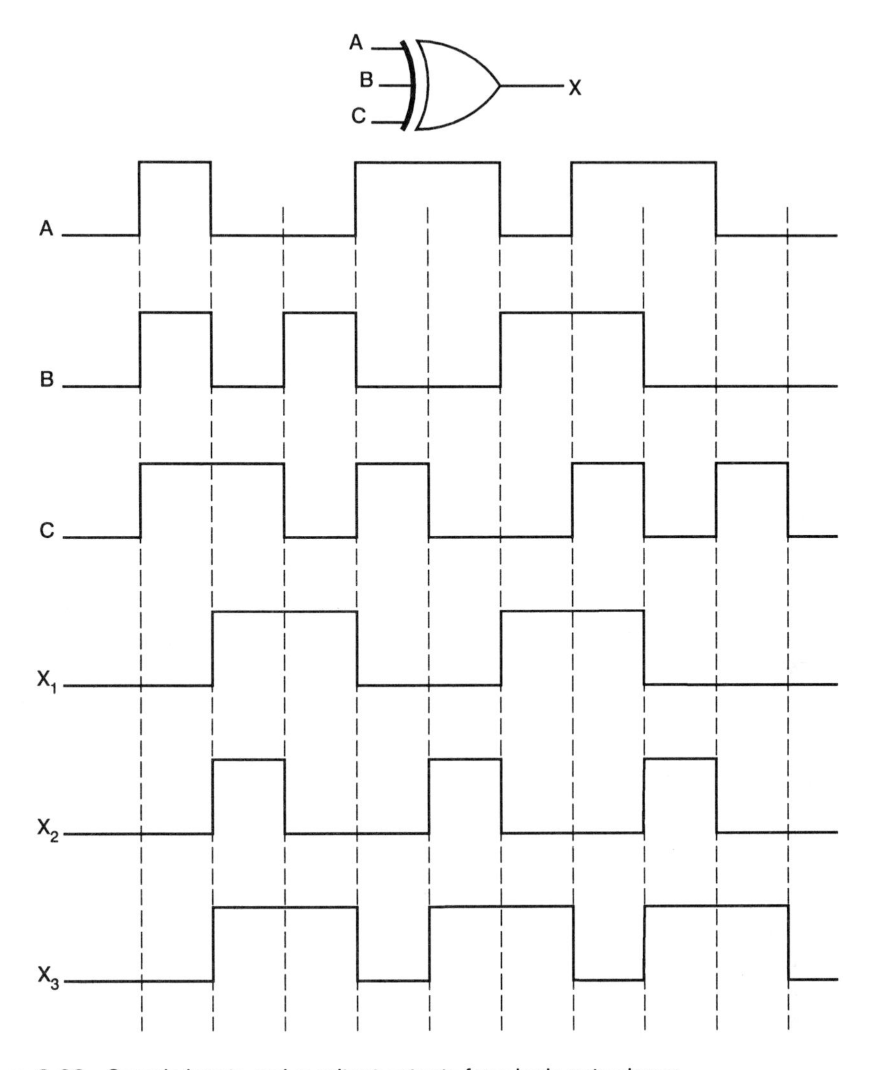


**Figure 9.26** Sample inputs and resultant outputs from logic gate shown

d) For each of the digits 1, 8 and 9, show as a 1 or a 0 the logic level required on each of the segments a to g.

8 a) Explain how a counter-timer measures
  i) the frequency
  ii) the period of a waveform.
b) List the possible sources of error in a reading obtained using a counter-timer.

9 a) The timebase of a counter-timer has a 10 MHz oscillator. The instrument, which has an 8-digit display, is to be used to measure a frequency of 5 kHz
  i) in the frequency mode with a gate time of 1 s
  ii) in the period mode. What will the display show in each case?
b) Considering only the normal display error of ± 1, what is the corresponding percentage error in the displayed value, in each case?
c) What is the 'break-even' frequency at which the two modes would give equal percentage errors?

10 a) Explain, with the aid of a simple circuit diagram, the operating principle of a sample-and-hold amplifier.
b) Sketch a graph with time axis horizontal, showing a changing input voltage and the corresponding output voltage when the commands: SAMPLE, HOLD, SAMPLE are applied to this device. Indicate *aperture time* and *acquisition time* on your sketch.

11 a) According to the sampling theorem, what is the minimum number of samples per second at which a signal should be sampled?
b) Name and explain the effect of sampling at a rate less than that minimum frequency.
c) Why is it desirable to sample at a rate much higher than the minimum rate specified by the sampling theorem?

12 a) Draw a simple circuit diagram to illustrate the principle of the $R$-$2R$ ladder in a D/A converter, and briefly explain how it gives an output voltage proportional to the digital input.
b) Sketch a graph showing, for a 3-bit D/A converter, the relationship between the binary input and the output voltage. Indicate on the graph:
i) the biasing voltage necessary to minimise the divergence from the ideal output
ii) the value of that minimum divergence in terms of the least significant bit.
c) What other factors may cause error in the output voltage?

13 a) Explain how a dual-slope A/D converter converts an input voltage into its binary equivalent. Illustrate your answer with a sketch of voltage-time graphs of the process.
b) What is its main disadvantage compared with other types of A/D converter?
c) In what application is this type of converter most likely to be found?

14 a) Explain briefly the process of conversion in a successive approximation A/D converter.
b) Show the stages by which a 4-bit successive approximation A/D converter with an analogue span of 12 V would convert a voltage of 5.2 V to binary.

15 Explain briefly, with the aid of a diagram of a 3-bit version, the principle of operation of a 'flash' type of A/D converter. What is its chief advantage over other types of A/D converter?

16 Draw the block diagram of a system in which an analogue input voltage is converted into digital form for transmission and processing, then reconverted into an analogue output voltage. Indicate, by simple diagrams, the form of the signal after each stage in the process.

17 a) What is:
i) quantization noise?
ii) companding?
b) Sketch graphs showing the output/input voltage relationship for:
i) a compressor amplifier
ii) an expander amplifier.
c) By means of a simple block diagram show where each amplifier should be placed in relation to an existing digital system.

18 a) The maximum component frequency in a signal is 3 kHz. What is the minimum frequency at which the signal should be sampled if it is to be converted from analogue to digital form?
b) The voltage range of 0 to 2.5 V is to be expressed in terms of 6-bit binary quantities. What resolution will this give in terms of volts?
c) If the signal is actually sampled and converted into a 6-bit quantity 20 000 times per second, specify a minimum value for the upper limit of the transmission system's bandwidth.

19 a) What is the purpose of a parity bit and what limits its usefulness for this purpose?
b) Taking the binary number 0011000 as an example, add a parity bit, giving the two alternative versions of the resulting 8-bit word, and stating which type of parity each version is using.

# 10
# Display and recording of output

## The moving coil meter

Figure 10.1 shows the usual arrangement of a moving coil meter. It consists of a coil of very fine wire, the 'moving coil', wound on a rectangular former, suspended in the field of a permanent magnet by two conical pivots. A cylindrical soft iron core is fixed in the space at the centre of the coil to concentrate the magnetic field and shape it so that it is radial in the gap between core and magnet, as shown in Figure 10.2. By Fleming's left-hand rule, the current flowing through the coil causes a proportional force, $F$, to act on each side of the coil. This rotates it to a position where this magnetic torque is balanced by the restoring torque from a torsion spring. A pointer carried by the moving coil thus indicates on an analogue scale the current through the coil.

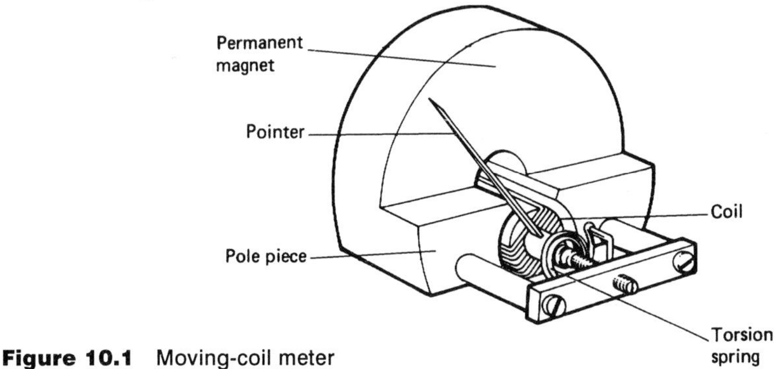


**Figure 10.1** Moving-coil meter

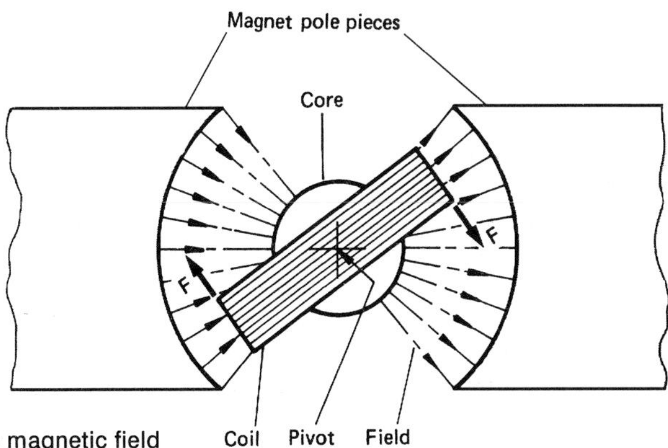


**Figure 10.2** The magnetic field

On the casing of the meter, a small plastic button with a screwdriver slot, close to the pivot, can be rotated to adjust the position of the fixed end of the torsion spring. This is used to set the pointer to zero before measurements are taken.

A typical high-sensitivity meter would have a coil resistance of 2000–3000 Ω, and would give a maximum reading of 50 μA (specified as '50 μA FSD', *FSD* being the abbreviation for *full scale deflection*). A more rugged meter might have a coil resistance of 100 Ω and an FSD reading of 1 mA. Currents greater than the FSD current are measured by connecting a resistance in parallel with the moving coil as shown in Figure 10.3, so that the coil has only to carry a small proportion of the total current. Such a parallel resistance is called a *shunt*.

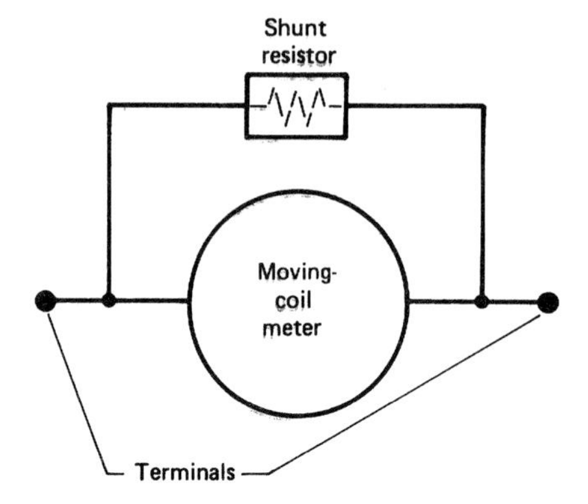


**Figure 10.3** Construction of an ammeter

## Current meters

Current meters (*ammeters*) are connected in series with the rest of the circuit, so that the whole current flows *through* them. Their resistance is usually such a small proportion of the total resistance around the circuit that it is often treated as negligible, although, of course, there is always some voltage drop across the terminals of the meter.

## Voltmeters

If we know the resistance of a current meter we can use it as a voltmeter by applying Ohm's law: $V = I \times R$, where $I$ is the current shown by the meter and $R$ is its resistance. Thus a moving coil meter could have two scales on its dial: a scale of current and a scale of voltage. To use a meter for voltages greater than its FSD voltage it is connected in *series* with a resistor as shown in Figure 10.4, the greater total resistance serving to reduce the current through the meter. Thus voltmeters are high-resistance instruments. They are applied in *parallel* with a component to

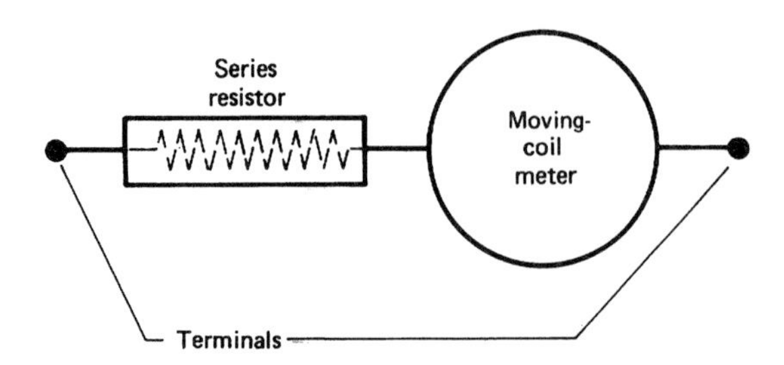


**Figure 10.4** Construction of a voltmeter

measure the voltage drop across it, or they are applied across the output terminals of an instrumentation system to show the signal voltage. When they are applied to ordinary electrical circuits the current they take is negligible, but when they are applied to electronic circuits the current is comparable with the currents in the circuit, so a moving coil voltmeter can sometimes virtually short-circuit the two points to which it is applied. To measure voltages in semiconductor circuits accurately requires an instrument with a much higher input resistance: a digital voltmeter or a cathode ray oscilloscope.

## Moving-coil multimeters

A moving-coil multimeter uses a single moving coil meter to measure several ranges of current, voltage and resistance. It does this by switching appropriate shunt or series resistances into circuit by means of a rotary switch.

While most multimeters have some form of overload protection to prevent the moving coil from being burnt out, it is inadvisable to rely on it, so before connecting the meter to measure an unknown voltage or current, set it to the *highest* (least sensitive) range, then switch down through the ranges, step by step, to the one which gives a reading in the middle third of the scale.

**To measure resistance**, the multimeter must contain a battery to drive current through the resistance being measured. The current, or the voltage drop across an internal resistor, is then a measure of that resistance. The simplest arrangement is shown in Figure 10.5; a 1.5 V cell

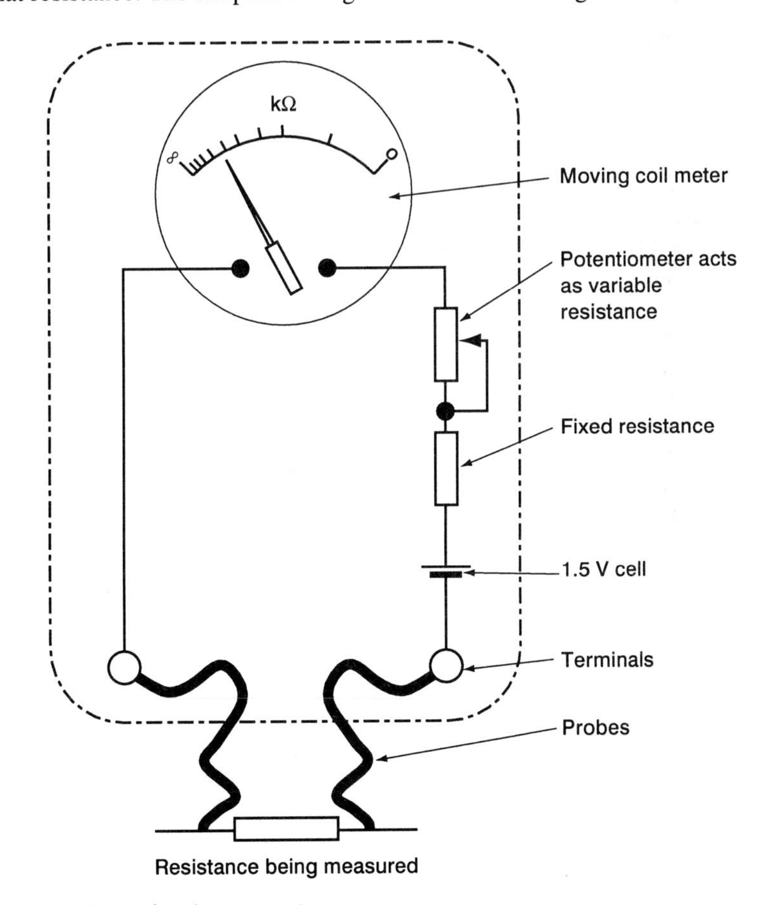


**Figure 10.5** A simple form of resistance meter

(battery) and a variable resistor in series with the meter. The variable resistor is used to adjust the current through the meter to the FSD current when the resistance being measured is zero (i.e. when the probes are touching). The diagram also shows a fixed resistor in series with the variable resistor. This is necessary to keep a certain minimum resistance in the circuit so that the meter cannot be damaged by too much current if the variable resistance is carelessly adjusted. The scale of resistance on a simple moving coil multimeter has '0 ohms' on the right-hand side and 'infinity' (i.e. a gap in the external circuit between the probes) on the left, and it is non-linear, with the higher-resistance graduations crowded together on the left.

Most multimeters offer more than one range of resistance, e.g. 'Ω', 'Ω × 10', 'Ω × 100', and this necessitates something more than the simple circuit described above. In such cases the meter, while still indicating the external resistance in ohms, is used as a voltmeter, responding to the voltage drop across resistors which are switched in series with the resistance to be measured.

Using a multimeter to measure resistance: always touch the probes together as a first operation, to check that the meter shows '0 ohms' correctly, adjusting the scale reading, if necessary, by rotating the appropriate zeroing button. If more than one resistance range is available, the one to use for a particular measurement is the one which gives a reading on the middle third of the scale.

Note that to measure a resistor which is part of an existing circuit it is essential to disconnect one end of the resistor – otherwise what we will get is the resistance equivalent to the resistor with the rest of the circuit in parallel.

Note also that when a multimeter is switched to the resistance scales, the polarity of the probes may be opposite to the polarity indication on the terminals; it depends on the internal circuitry of the multimeter. If necessary, it can be checked by touching the probes on to another voltmeter.

Probably the most common use of the resistance scale is simply to check whether the resistance between the probes is 'zero', 'infinity' or 'something in between'. We should have 'zero' if we are checking a soldered joint in a circuit, 'infinity' between two adjacent conductors checks that we have not accidentally joined them by careless soldering, while 'something in between' verifies that there is not a break in a suspect component:

- **Diodes** can be checked for damage by applying the probes to pass current through the diode first in one direction, then in the other. If the diode is satisfactory, its resistance should be 'something in between' in one of the two directions and 'infinity' in the other.
- **Transistors** can be similarly checked by treating them as two diodes connected anode-to-anode (n-p-n transistor) or cathode-to-cathode (p-n-p transistor), the base being the point where they are connected together, the emitter and collector their opposite ends. Each of the two 'diodes' can be checked for damage, as above; also the resistance between emitter and collector should be 'infinity' in both the positive-probe-to-emitter and negative-probe-to-emitter configurations. In order not to damage semiconductors by excessive currents during these tests, only the Ω × 1 or, preferably, Ω × 100 ranges of the multimeter should be used.

*Example 10.1*
A moving coil meter has a resistance of 2000 Ω and an FSD reading of 50 μA. Calculate:

a) its FSD voltage reading if without any extra resistance it is used as a voltmeter
b) the shunt resistance which would give it an FSD reading of 2 A
c) the series resistance which would give it an FSD reading of 10 V

*Solution*
a) $V = I \times R = 0.00005 \times 2000 = 0.1$ V

b) Referring to Figure 10.3, and using the answer to part a): at FSD, the voltage across both meter and shunt is 0.1 V. Also, the current carried by the shunt is 2 – 0.00005 = 1.99995 A.

$$\therefore \text{shunt resistor} = \frac{V}{I} = \frac{0.1}{1.99995} = 0.0500125\ \Omega$$

i.e. 0.05 Ω for practical purposes.

c) Referring to Figure 10.4, again the FSD current is 0.00005 A and the voltage drop across the meter 0.1 V.

$$\therefore \text{series resistor} = \frac{V}{I} = \frac{10 - 0.1}{0.00005} = 198\ 000\ \Omega$$

## Digital display multimeters

Analogue and digital multimeters are two completely different types of meter. The analogue meter is a current-operated device, based on the d'Arsonval moving coil. The digital meter is a voltage-operated device, and as such it has a very high input impedance, usually 10 MΩ, so, as a voltmeter, it usually does not overload the circuits to which it is applied. An analogue meter gives a continuous reading which can follow variations of the input quantity at frequencies up to about 1 Hz; a digital meter displays an instantaneous sample of the input at half-second intervals so if the input is fluctuating the result is a confusing sequence of random numbers. Each of the two types of meter has its particular advantages; a comparison is given below.

| Analogue | Digital |
|---|---|
| Simple | More accurate |
| Inexpensive | Better resolution |
| No power supply needed | Very high input impedance |
| Shows trends better | Impossible to misread |
| Filters noise in signal | Readable at long distance |

A digital display multimeter is based on a digital display voltmeter, a block diagram of which is shown in Figure 10.6. The sample-and-hold amplifier, the A/D converter (usually

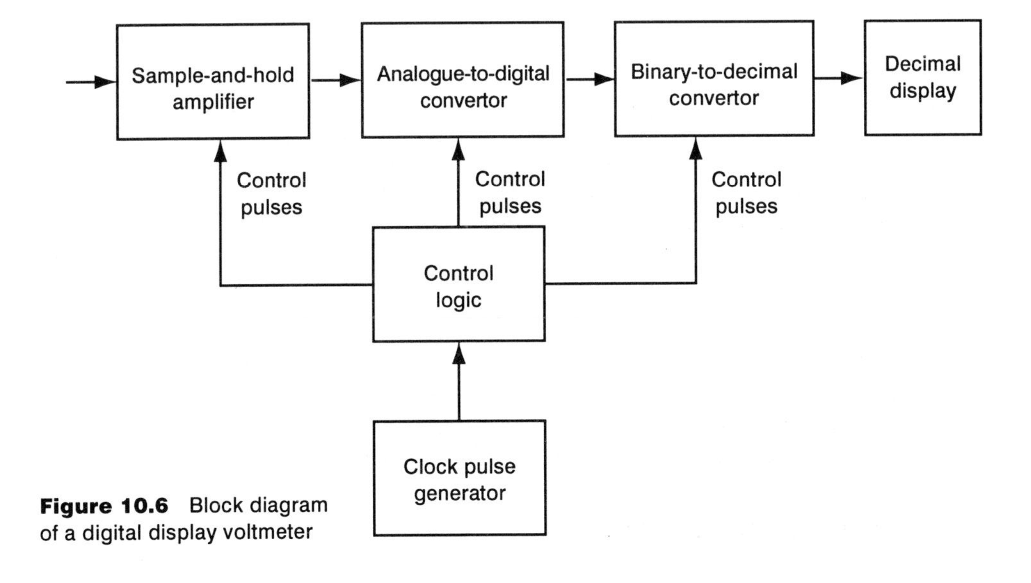


**Figure 10.6** Block diagram of a digital display voltmeter

the dual-slope type is used) and the principles of binary-to-decimal conversion and multiplexing to a row of 7-segment display digits have already been explained in Chapter 9.

A digital multimeter is usually specified as having a $3\frac{1}{2}$ or $4\frac{1}{2}$ digit display. The '$\frac{1}{2}$' indicates that the most significant digit can only be a 0 or a 1, so the reading on a $3\frac{1}{2}$ digit display can only range between +1999 and −1999 (a decimal point can be inserted after any of the digits).

The segments of a 7-segment digit can be formed by *light-emitting diodes* (*LED*), or by a *liquid crystal display* (*LCD*).

- A **light-emitting diode** gives an area of coloured light which can be produced in any required shape. It has the advantage of being visible in the dark, but it has a low intensity of illumination which can make it difficult to see in bright sunlight. Also it takes a few milliamps of current, which is a serious disadvantage for battery-powered equipment so any battery-powered instrument usually has a liquid crystal display.
- A **liquid crystal display** uses about one-thousandth of the current of a light-emitting-diode display. A diagram of a single-digit display is shown in Figure 10.7. It consists of two glass plates with a very thin layer of liquid crystal material sandwiched between them. Over its working temperature range, this material is a transparent liquid in the form of molecular crystals. (At lower temperatures it is solid; at higher temperatures it is non-crystalline.) Electrodes of tin oxide are deposited by evaporation on to the inside faces of the glass plates, the coating being so thin as to be effectively transparent. The coating on the upper plate is in the shape of the segments or other characters to be displayed; that on the lower plate is a continuous electrode, the *backplane*, common to all the characters. When a voltage is applied to the electrodes, the molecular crystals lose their random orientation and rotate so that their electrical axis is aligned with the electric field between the two plates. This causes the liquid to become opaque, causing the area covered by the electrode to appear black. The liquid becomes transparent again when the voltage is no longer applied. For viewing in normal lighting, the display is backed by a white reflecting coating; for viewing in poor lighting it can be backlit by a lamp.

The electrodes and the liquid crystal material form an electrolytic cell, and a DC voltage across them would eventually erode one of the electrodes. For this reason, the applied voltage has to be AC. The source of the AC is an oscillator generating a square-wave alternating between 0 and (e.g.) +15 V, at a frequency in the range 30 to 400 Hz. Each of the

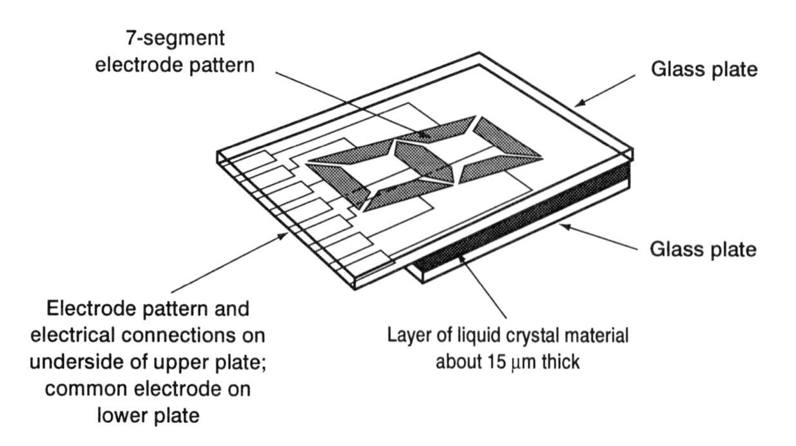


**Figure 10.7** Single-digit LCD assembly

seven electrodes of a 7-segment LCD digit is connected to the output of its own 2-input EXCLUSIVE-OR gate, and the oscillator output is connected both to the backplane and to one of the EXCLUSIVE-OR gate's inputs. The other input receives a logic 0 or logic 1 voltage from the 7-segment driver circuit's output to that segment. The truth table of a 2-input EXCLUSIVE-OR gate is:

| **input** | | **output** |
|---|---|---|
| *x* | *y* | |
| 0 | 0 | 0 |
| 0 | 1 | 1 |
| 1 | 0 | 1 |
| 1 | 1 | 0 |

This shows that when the control voltage, on input *x*, is logic 0, the output voltage is in phase with the oscillator on input *y*; thus there is no voltage difference between electrode and backplane and the segment remains transparent. But when the driver applies logic 1 to input *x*, the output is antiphase to that of the oscillator. Thus there is then a square-wave voltage between the segment electrode and the backplane, alternating between +15 V and −15 V, which rotates the crystals between them, causing the segment to appear black.

The change in optical density from transparent to black and back again to transparent takes place relatively slowly; Figure 10.8 shows how it relates to the energising signal. Typical rise and fall times are 20 ms and 100 ms, respectively.

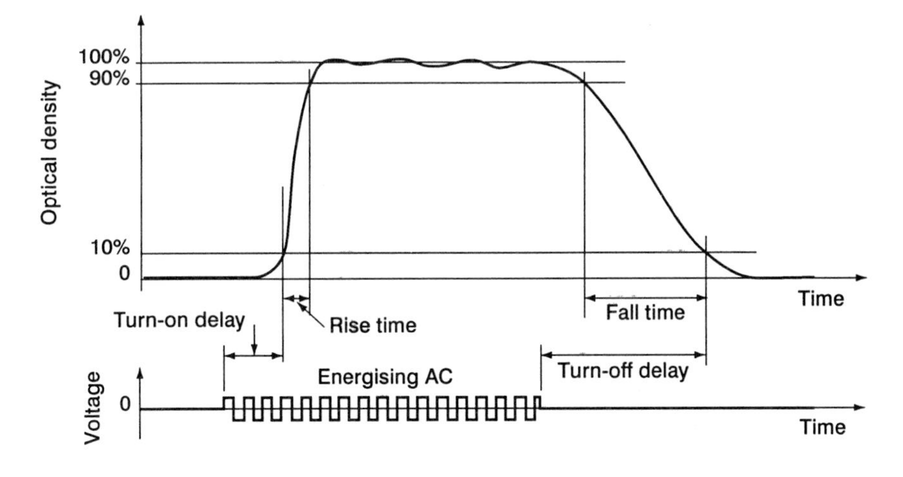


**Figure 10.8** Relationship between energising signal and liquid crystal display

The essential components of a digital voltmeter have now been described, though it is unlikely that they will be found as separate items in the actual instrument; almost certainly they will be incorporated in a single large integrated circuit. This will be combined with one or more rotary switches and some resistors and capacitors to turn it into a multi-range, multi-function meter. The following is a brief description of the main features of these extra circuits. They cannot be the same as the circuits used in a moving coil multimeter because the digital voltmeter is a voltage-operated device, whereas the moving coil meter is current-operated.

**Voltage** To extend the range of voltage measurements, the potential divider circuit of Figure 10.9 is used. As drawn, the circuit gives four ranges of voltage. The lowest range is

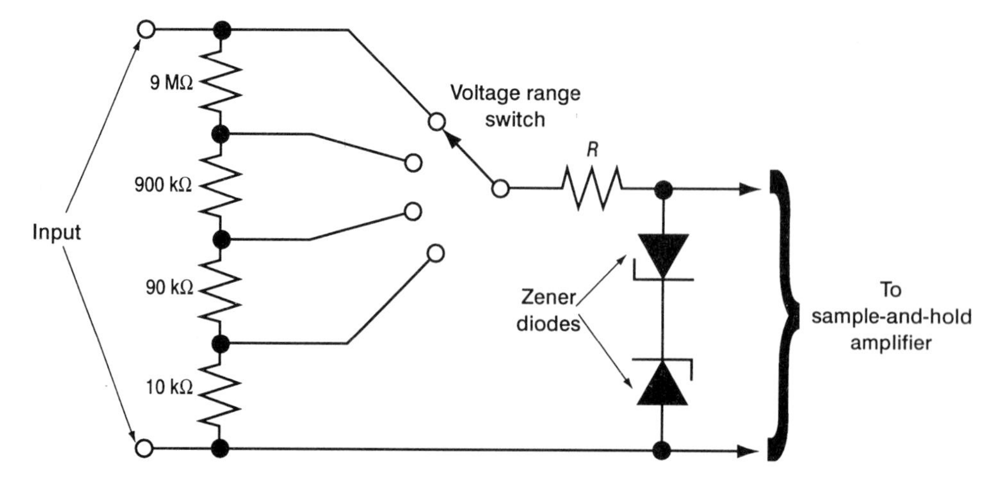


**Figure 10.9** Potential divider circuit to extend the voltage range of the digital voltmeter

obtained with the range switch contact in the position shown. There is then a total resistance of:

$$9\ \mathrm{M\Omega} + 900\ \mathrm{k\Omega} + 90\ \mathrm{k\Omega} + 10\ \mathrm{k\Omega} = 10\ \mathrm{M\Omega}$$

across the input terminals; this is the input resistance of the voltmeter. (The input resistance of the sample-and-hold amplifier, connected in parallel, is greater than 1000 MΩ, so it makes negligible difference to the multimeter's input resistance.) Each step of the range switch attenuates the voltage applied to the sample-and-hold amplifier by a factor of 10. For example, at the step below the top, the fraction of the input voltage which is applied to the sample-and-hold circuit is

$$\frac{900\ \mathrm{k\Omega} + 90\ \mathrm{k\Omega} + 10\ \mathrm{k\Omega}}{10\ \mathrm{M\Omega}} = \frac{1\,000\,000}{10\,000\,000} = \frac{1}{10}$$

Thus if the basic range of the voltmeter was 0 to 2 V, the step below the top would give a range of 0 to 20 V. High-precision metal-film resistors are used throughout. The back-to-back zener diodes conduct to protect the sample-and-hold amplifier if an excessive input voltage of either polarity is applied to the meter. The current they have to conduct is limited to a safe value by the resistance $R$ in series with the switch.

- **Current** Current must be converted into a voltage drop by passing it through the current-measuring resistor $R_c$, as shown in Figure 10.10. This voltage is then amplified by the

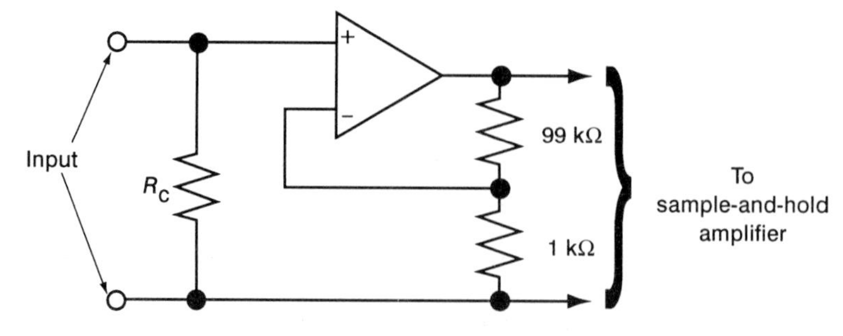


**Figure 10.10** Current-to-voltage converter

op-amp before it is passed to the sample-and-hold circuit. The resistance values shown give a voltage gain of

$$\frac{99\ \text{k}\Omega + 1\ \text{k}\Omega}{1\ \text{k}\Omega} = 100$$

Thus $R_C$ need only be 1/100 of what it would have to be if the op-amp were omitted. (An ideal current meter would have zero resistance.) Other current ranges are obtained by switching alternative values of $R_C$ into circuit.

- **Resistance** Figure 10.11 shows one way of measuring resistance. The resistance to be measured is connected so that it is the negative feedback resistance of the op-amp. An op-amp with field-effect-transistor input (*FET input op-amp*) is used because having a very high input resistance, it draws negligible current from its inputs. The voltage passed to the digital voltmeter is then:

$$\frac{\text{(Unknown resistance)}}{R_1} \times \text{(Reference voltage)}$$

Switching-in alternative values of $R_1$ and $R_2$ gives alternative ranges of resistance.

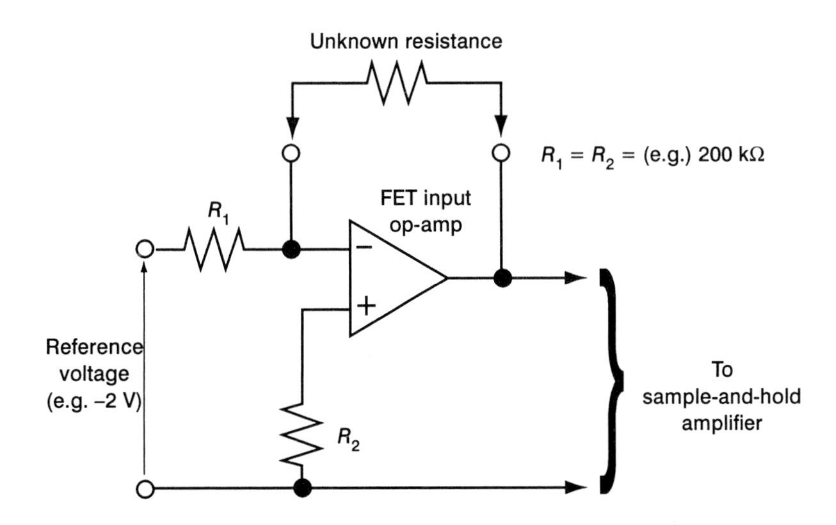


**Figure 10.11** Resistance-to-voltage-converter

- **AC Ranges** The AC must be converted into a DC voltage before it can be measured. For an approximate value it would be sufficient to measure the DC voltage of a capacitor charged from the AC through a diode, but the usual 0.6 V voltage drop across the diode would make this method unsuitable for precision measurements.

  An *accurate* AC to DC voltage conversion can be obtained in various ways. In one circuit, the input AC is applied to an op-amp circuit which converts it to half-wave rectified DC without any voltage drop. This is input to another op-amp circuit which smooths it and amplifies the resulting pure DC to the exact RMS (root mean square) voltage of the original AC sinewave.

## Means of recording data

### The ultraviolet recorder (UV recorder)

This applies the principle of the moving coil meter (the *d'Arsonval movement*) to the simultaneous recording of a large number of analogue signals (current or voltage) by projecting rays of ultraviolet light on to a continuously unrolling sheet of ultraviolet-sensitive paper. Each signal is applied to the moving coil of a galvanometer. Figure 10.12 shows an external view of a typical galvanometer and Figure 10.13 its internal construction. The moving coil is suspended between bobbins and pulled taut by the upper and lower suspension bands – narrow strips of metal foil which also act as the electrical connections to the coil and as the torsion spring against which the magnetic torque acts. Instead of a pointer, the upper suspension carries a tiny mirror to reflect a ray of ultraviolet light on to the paper.

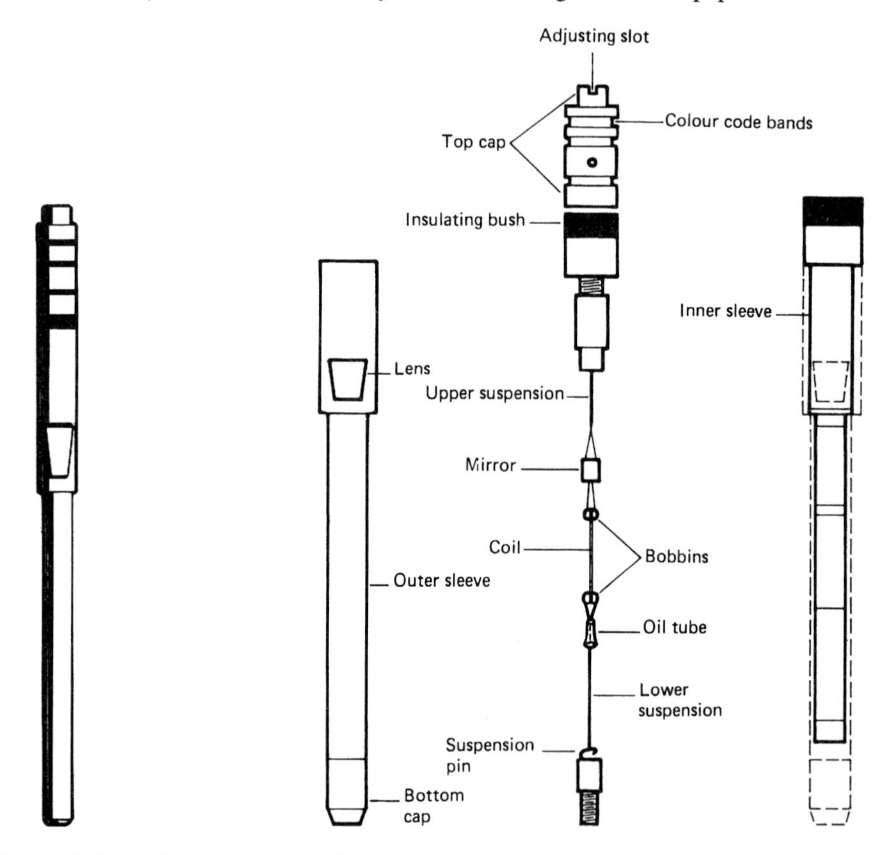


**Figure 10.12** Typical galvanometer

**Figure 10.13** Diagram of the construction of a typical galvanometer

The galvanometers are inserted into a block magnet which provides the magnetic field in which the galvanometer coils operate. An optical system projects a 'wedge' of ultraviolet light on to all the galvanometer mirrors, each mirror reflecting a portion of this light so that it converges to a spot on the paper. The essential features of an ultraviolet recorder are as shown in Figure 10.14. Further systems of lenses and mirrors project ultraviolet light to draw grid lines along the paper and timing lines across it, and each trace in turn is interrupted

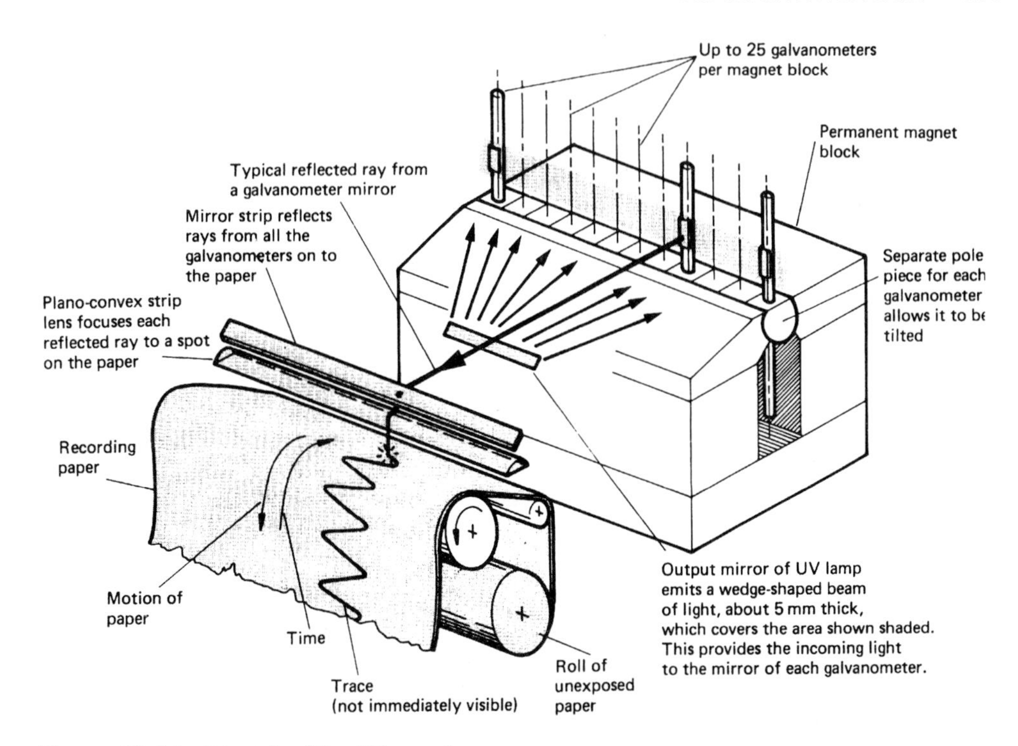


**Figure 10.14** Principle of the UV recorder

automatically so that the traces can be identified one from another. This identification is essential because the individual rays of light from the galvanometer mirrors can pass through each other unhindered, so each trace can traverse the full width of the paper if required. Because the 'pointer' of each galvanometer is a ray of light, which has no mass and therefore no moment of inertia, very high frequencies can be recorded – an upper frequency limit of 13 kHz is typical. Manufacturers of ultraviolet recorders can usually supply a 'family' of interchangeable galvanometers of the type shown in Figure 10.12, ranging from high frequency types, which unfortunately have low sensitivity, to high sensitivity types, which have a low frequency-limit. Their movements are damped to a damping ratio of about 0.65, which gives the fastest possible response with minimum overshoot, electromagnetic damping being used in the high-sensitivity types, fluid damping in the high-frequency types. The ultraviolet light-sensitive paper can be driven at various speeds, the highest setting being about 4 m/s – a speed which can submerge one in paper in a few seconds! The traces drawn by the spots are not immediately visible but develop in a few minutes under ordinary fluorescent room lighting. They remain readable for years if kept away from sunlight, and, if necessary, can be made permanent by a simple liquid fixing process.

## Servo recorders

In this type of recorder a pen, driven by some form of electrical drive, draws a trace on paper unrolling at constant speed, the displacement of the pen across the paper being proportional to the input signal voltage. In *open-loop servo recorders* the pen position is not measured; the incoming signal is amplified and applied directly to the pen positioning device. In *closed-loop servo recorders* the pen position is fed back from a displacement transducer and corrected by comparison with the input signal, in a closed-loop control system.

The traces of up to six input signals may be drawn simultaneously if the displacement of the pens is limited to enable them to work side by side, or if some other arrangement is made to avoid mechanical interference between them. The usual method of marking the trace on the paper is by lightweight fibre-tip pens using coloured inks, but some recorders use a heated stylus and heat-sensitive paper. Pressure-sensitive, spark-sensitive, and electrolytic papers have also been used. Whatever the method of trace-marking adopted, the paper must be dimensionally stable, so that traces remain undistorted.

If the pen or stylus is carried by a radial arm, it traverses a curved path; this can slightly distort the shape of a waveform, so the recording paper should be printed with a correspondingly curved grid of lines for accurate measurements.

Most closed-loop servo recorders use a potentiometer as the feedback transducer; these instruments are therefore sometimes called *potentiometric recorders*. Figure 10.15 shows the arrangement of pen carriage and feedback transducer in one type of potentiometric recorder. The pen 'motor' consists of a coil, free to slide along a rod. The rod is part of a system of magnets, arranged to provide a permanent magnetic field at right angles to the rod throughout its length. Current through the coil causes a magnetic force to act on it in the direction given by Fleming's left-hand rule for motors. This propels it along the rod to a position where the error voltage in the closed-loop system is zero.

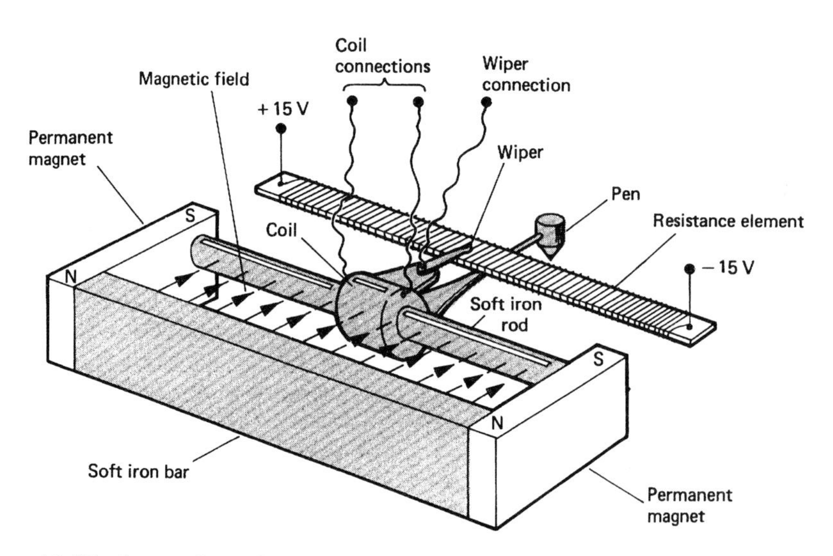


**Figure 10.15** A potentiometric recorder

## The XY plotter

This is an instrument which draws a graph of two inputs ('X' and 'Y') on a fixed sheet of paper which is usually held down to the bed of the machine by vacuum. Figure 10.16 shows the layout of the instrument.

The 'X' input is applied to a closed-loop system which positions a gantry along which the pen carriage moves. The 'Y' input, by means of another closed-loop system, positions the pen carriage on the gantry. The pen carriage and gantry may be driven by pulley and cable systems or by small electric motors carried on the gantry. Most XY plotters can also be used as Y-t plotters to show the variation of an input signal with time, the gantry being driven at various set speeds from left to right across the paper by an internal timebase.

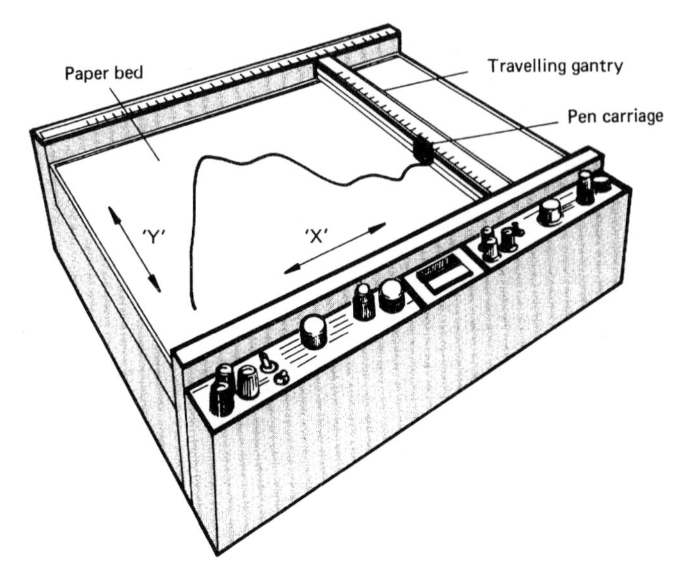


**Figure 10.16** XY plotter

## Chart recorders

Figure 10.17 shows a simplified diagram of a chart recorder. The 'chart' is a disc of paper carried on a metal disc which is rotated very slowly about a horizontal axis. The disc is driven by an electric or clockwork motor through reduction gearing so that it makes one revolution per day or one revolution per week. The recording pen is carried by a pointer which hangs down across the face of the disc. A transducer rotates the pointer so that an increase or decrease in the input to the transducer causes a corresponding change in the angular position of the pointer.

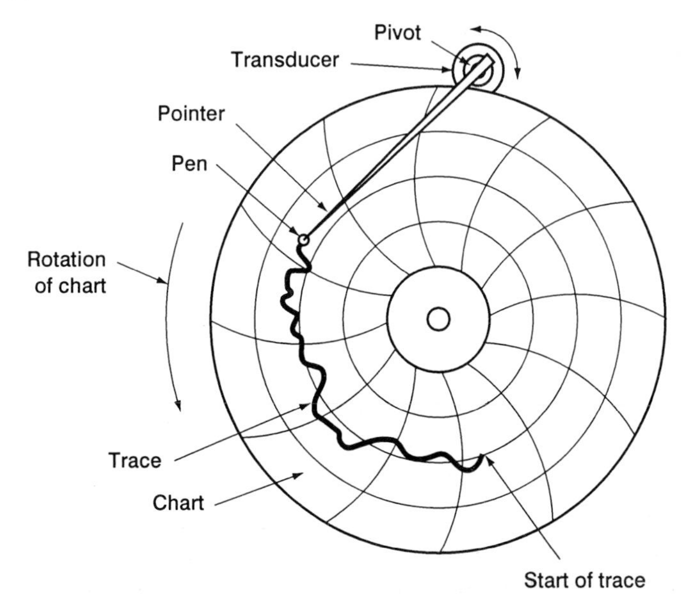


**Figure 10.17** The essential features of a chart recorder

The transducer used is usually Bourdon tube, a bimetallic strip or the movement of a moving coil meter, depending on whether the signal to be recorded is a pressure, a temperature or an electrical signal. The chart recorder draws a circular graph of the long-term variation in a measured quantity. Concentric circles on the paper represent graduations in the measured quantity, curved lines 'radiating' from the centre represent time divisions, the curvature of these lines corresponding to the curvature of the arc traced out by the pen. Up to four transducers, arms and pens can be accommodated in the same recorder, recording four different signals simultaneously in different coloured inks.

## The cathode-ray oscilloscope

This is based on the cathode-ray tube, a special kind of thermionic valve in which electrons thrown off by the heated cathode are accelerated and focused by anodes into a beam or 'ray' which strikes a flat face at the end of the tube. This produces a glowing spot on a phosphor coating on the inside of the tube. A diagram of the cathode-ray tube electrodes is shown in Figure 10.18.

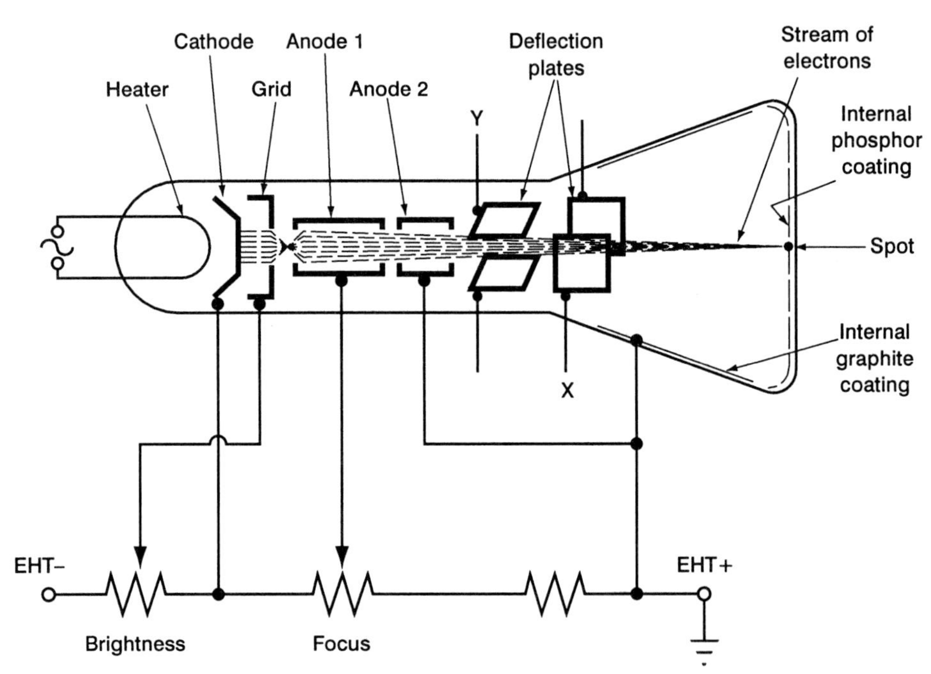


**Figure 10.18** The cathode-ray tube

The brightness of the spot is controlled by increasing or decreasing the negative voltage of the grid, relative to the cathode, by means of the *brightness potentiometer*. This decreases or increases the flow of electrons along the tube. The *focus potentiometer* adjusts the positive voltage on anode 1 to make the beam of electrons converge to a point at the phosphor coating. Each of these controls affects the other: an alteration to the brightness control causes the beam of electrons to converge on a point in front of or behind the screen, so it then usually has to be re-focused to give as sharp an image as possible. *A spot which is too bright cannot be focused sharply and may burn away the phosphor coating of the screen*, especially if it is allowed to dwell on one point for a long time.

A high DC voltage is needed to accelerate the electrons along the tube; this is denoted *EHT* (extra high tension). Typical voltages within the tube are:

| Electrode | Voltage |
|---|---|
| Cathode | –1900 V |
| Grid | –1900 V to –2000 V |
| Anode 1 (the focusing anode) | –1500 V to –1300 V |
| Anode 2 (the accelerating anode) | 0 V (i.e. earthed) |

The electron beam (and thus the spot) is deflected horizontally by voltages applied to the X-plates and vertically by voltages applied to the Y-plates. The X-plates are connected to the *time base*, which generates a steadily increasing voltage to drive the spot at constant speed from left to right across the screen. The voltage applied to the Y-plates is the input voltage, attenuated or amplified in accordance with the Y-sensitivity setting. The spot therefore draws a 'graph' of voltage (vertically) against time (horizontally) and the cathode-ray oscilloscope is in this respect basically an analogue voltmeter. However, it has the advantage that the beam of electrons can be deflected very rapidly in any direction, so it can display very high frequency waveforms – most oscilloscopes can display frequencies up to 20 MHz. Also it is a voltmeter with a very high input impedance (typically 1 MΩ with 40 pF in parallel), so it does not appreciably load the circuit it is applied to.

For capturing the waveform of an isolated event, a *digital storage oscilloscope* may be used. This stores voltage/time co-ordinates of the waveform in binary digit form so that it can be displayed continuously on the oscilloscope screen. A storage 'scope is also useful for displaying an event in which the spot is moving too slowly to produce the usual trace we normally see on the screen.

Most oscilloscopes can display two independent voltage signals simultaneously. This used to be done by splitting the electron beam into two, so that the two spots were in vertical alignment as they moved across the screen (essential because horizontal distances represent elapsed time from the start of the trace). Beam splitting is now superseded by the time-sharing of the single beam, by chopping or by alternating its use. In the **'chop' mode**, each signal in turn is connected to the Y-plates, at a frequency of about 100 kHz. Thus the two traces are drawn simultaneously across the screen, the spot being chopped from one trace to the other, as shown in Figure 10.19. The vertical movement of the spot between the two traces is so rapid that the vertical lines are invisible, and the traces appear continuous and independent.

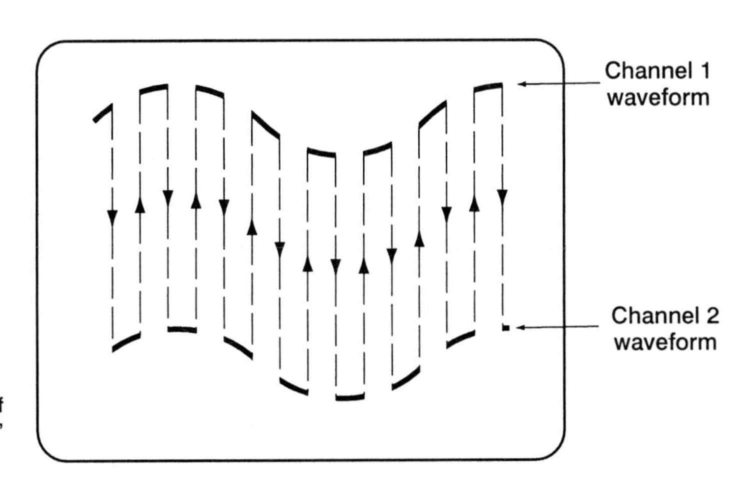


**Figure 10.19** Motion of the spot in the 'chop' mode

In the **'alternate' mode**, the two traces are drawn alternately, each signal in turn being connected to the Y-plates for one complete traverse of the spot across the screen.

Chop mode is better for low timebase speeds; its use is limited to signal frequencies well below the chopping frequency. 'Alternate' is better for high timebase speeds, where the persistence of glow from the phosphor and the persistence of one's vision make it appear that the two traces are being drawn simultaneously.

### Triggering

Having drawn a voltage-time graph of the input signal, the spot disappears into the right-hand edge of the screen. If the timebase is free-running, it then returns very quickly (*flyback*) to the left-hand side and repeats its steady traverse from left to right, this cycle continuing indefinitely. The spot is not seen during flyback, because the electron flow in the tube is cut off by a blanking pulse at that time.

A free-running timebase will draw voltage-time 'graphs' starting, each time, from a random point in the waveform, and the result will be a jumble of traces. For a single 'graph' of the waveform, the timebase must start its traverse from the same point in the waveform each time, so that successive traces lie exactly on top of one another. Thus, after flyback is completed, the timebase must wait until it is *triggered* by some event.

The timebase may be triggered by some feature of the signal itself (*internal triggering*) or by a time-related signal applied to a triggering input (*external triggering*). With internal triggering, the timebase is triggered by a particular voltage level in the waveform. This is set by the *trig-level* control knob. There is also a '+/-' push button switch which determines whether the triggering at this voltage level occurs when the waveform has a positive or a negative slope.

On many oscilloscopes, there is an 'auto/normal' switch. When 'auto' is selected, the trig-level control is inoperative, the triggering being set automatically by the oscilloscope to give a steady display of the waveform. This simplifies the use of the oscilloscope, and it can be used in this mode for most routine measurements. Another mode of triggering is *single shot*, which is used for capturing one-off events. In this mode the timebase, instead of running continuously, gives one traverse of the screen and a flyback. Then it waits to be triggered again for the next single shot. Yet another mode of triggering is *line triggering*, in which the timebase is triggered at power supply frequency (50 Hz). This is used to examine ripple in DC power supplies.

### Input coupling

Most oscilloscopes have a Y1 input and a Y2 input. Each of these is connected to its Y-amplifier via a 3-position switch marked 'DC-GND-AC'. In the 'DC' position the signal is connected directly to the Y-amplifier. In the 'GND' position the Y-amplifier input is grounded (i.e. earthed) so that the trace can be shifted to a level on the screen which will indicate zero volts for DC voltage measurements. In the 'AC' position, the signal is connected to the Y-amplifier through a large capacitance, which blocks any DC component in the signal but lets the AC component through.

### Probes

A probe is used to pick off a voltage from a circuit and convey it to the oscilloscope, so the probe is combined with a length of co-axial cable with a plug on the other end which plugs into the Y-input socket of the oscilloscope. The probe usually gives a 10 to 1 attenuation of the signal voltage, so the value shown on the 'volts/division' setting of that channel needs to be multiplied by ten to give the true voltage applied to the probe. The reason for using a probe of this kind is to eliminate the integrating effect caused by the capacitance of the oscilloscope's Y-input and co-axial cable, which are in parallel with the oscilloscope's 1 MΩ input resistance. Figure 10.20(b) shows the distortion caused to a square-wave input by such

**Figure 10.20** Oscilloscope displays of a square-wave: (a) undistorted, (b) distorted by integration, (c) distorted by differentiation

integration. High-frequency sinewaves are similarly affected, being reduced in amplitude by integration.

For accurate measurements, the integrating effect must be cancelled out by a differentiating effect, shown in Figure 10.20(c). This is provided by the probe's 9 MΩ resistor and trimming capacitor in parallel (see Figure 10.21). To set the trimmer, a square-wave voltage is applied to the probe at a frequency which gives one of the types of distortion shown in Figure 10.20 and the trimmer is adjusted until the correct waveform of Figure 10.20(a) is obtained.

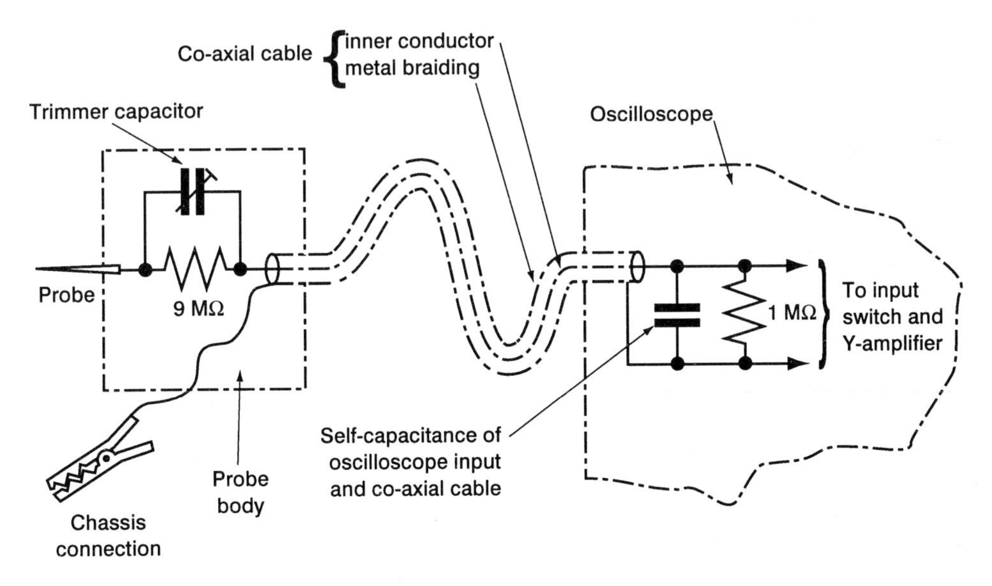


**Figure 10.21** Oscilloscope input with 10-to-1 attenuation probe

**The X-Y mode**

This enables the oscilloscope to display a graph of one voltage vertically, against the other voltage horizontally, instead of the usual two separate voltage/time 'graphs'. In this mode, the X-plates are disconnected from the timebase and connected to the output of one of the Y-amplifiers. If the two voltages are sinewaves with the same frequency but with a phase difference between them, the display is a *Lissajous figure* in the form of an ellipse. The phase difference can be calculated from the proportions of the ellipse, though it is more usual to calculate it from the normal voltage/time display, as in the worked example which follows.

**The measurement of current**

Like any other voltage-measuring instrument, an oscilloscope can be used to display waveforms of current, by passing the current through a resistor and using the oscilloscope to show the corresponding voltage drop across the resistor.

*Using an oscilloscope – a worked example*

An AC voltage at a frequency of about 700 Hz is applied to a 1 kΩ resistor and 0.1 μF capacitor in series. The following measurements are to be taken by means of an oscilloscope:

a) the RMS value of the applied voltage
b) the amplitude of the resulting current
c) the exact frequency of the applied voltage
d) the phase difference between the applied voltage and the current.

Figure 10.22 shows the resistor and capacitor, with the connections which have to be made to the oscilloscope.

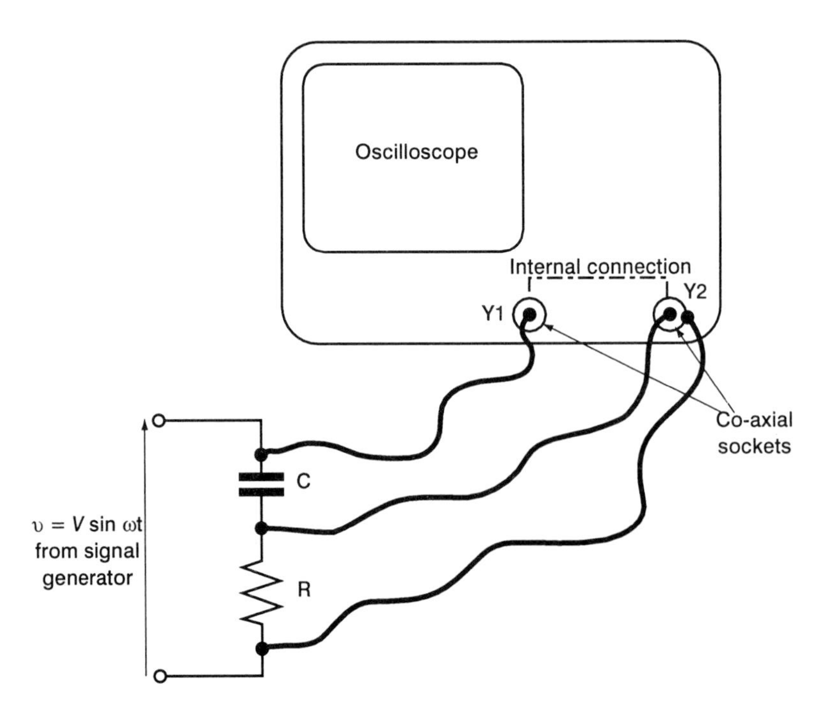


**Figure 10.22** Worked example on use of the oscilloscope

*Procedure and solution*

If we are to take measurements from an oscilloscope screen, we must first make sure that the variable controls are all in the 'calibration' position. There are three variable controls: one to vary the set value of the 'time per division' switch, the other two to vary the set values of their respective 'volts per division' switches. The control knobs are usually concentric with their switches; before any readings are taken, their arrows must be aligned with arrows marked 'CAL' adjacent to the knobs, and they must stay in those positions throughout the measuring process. They usually 'click' into the CAL position.

When we have obtained steady traces on the screen, the display will probably be something like Figure 10.23. For maximum accuracy, the trace we are to measure should be as large as possible, so we enlarge it, like this:

1 use the 'time per division' switch to obtain the smallest number of *complete* cycles we can get on the screen
2 move the Y2 trace off-screen (up or down) by means of the Y2 position control
3 use the Y1 'volts per division' switch and the Y1 position control to obtain the tallest Y1 sinewave we can get on the screen
4 use the Y1 position control, the X position control and (possibly) the trig level control and +/- switch to obtain something like the display shown in Figure 10.24, with two minima tangential to one of the horizontal lines of the graticule (at points A and B) and the maximum between them on the graduated vertical centre line of the screen (point C).

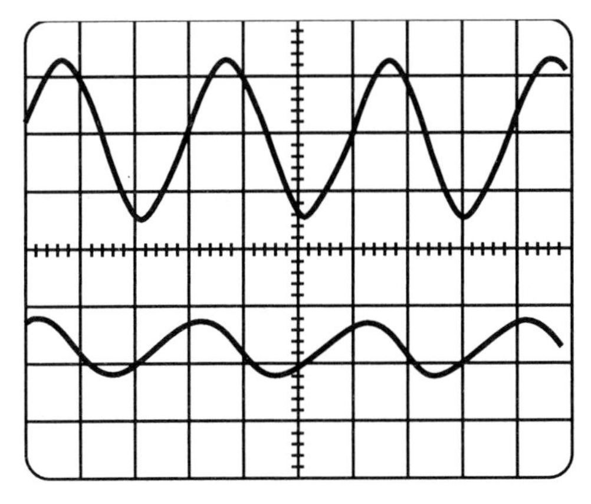

**Figure 10.23** Oscilloscope display of two AC voltages with the same frequency but with a phase difference between them

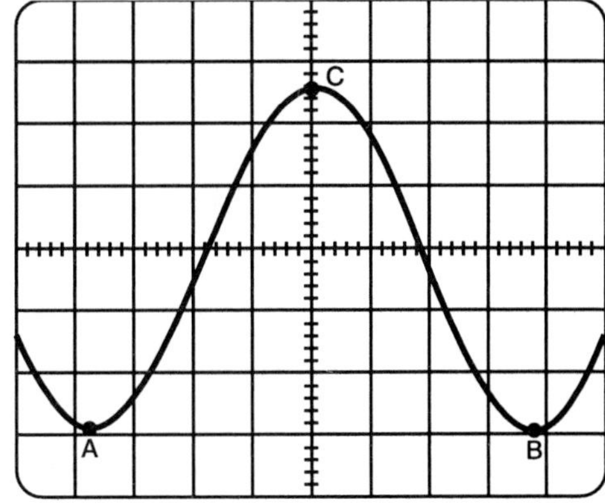


**Figure 10.24** Measuring the peak-to-peak voltage of one of the traces

The switch settings are now found to be:

- Y1: 0.5 volts per division
- time per division: 0.2 ms.

Reading from the screen, point C is 5.6 divisions vertically above AB. Therefore:

- peak-to-peak voltage is $0.5 \times 5.6 = 2.8$ V
- amplitude is $\frac{2.8}{2} = 1.4$ V;
- **root mean square voltage is** $\frac{1.4}{\sqrt{2}} = 0.99$ V .

We now move the Y1 trace off the screen and bring the Y2 trace back on, by means of the Y position controls. We then repeat steps (3) and (4) with the Y2 controls, which gives us a peak-to-peak measurement of 5.4 divisions at a Y2 switch setting of 0.2 volts per division. Therefore:

- peak-to-peak voltage is $0.2 \times 5.4 = 1.08$ V
- amplitude is $\frac{1.08}{2} = 0.54$ V

$\therefore$ from Ohm's Law $\left(I = \frac{V}{R}\right)$

- amplitude of current is $\frac{0.54}{1000} = 0.54$ mA.

Now using the Y position controls we bring the Y1 trace back and arrange the two traces so that they are exactly symmetrical about the horizontal centre line of the screen, as in Figure 10.25. This symmetry can be checked by using the X-position control to move the whole display sideways to see that the maxima and minima of each trace are an equal number of graduations from the midpoint of the screen, on the graduated vertical centre-line.

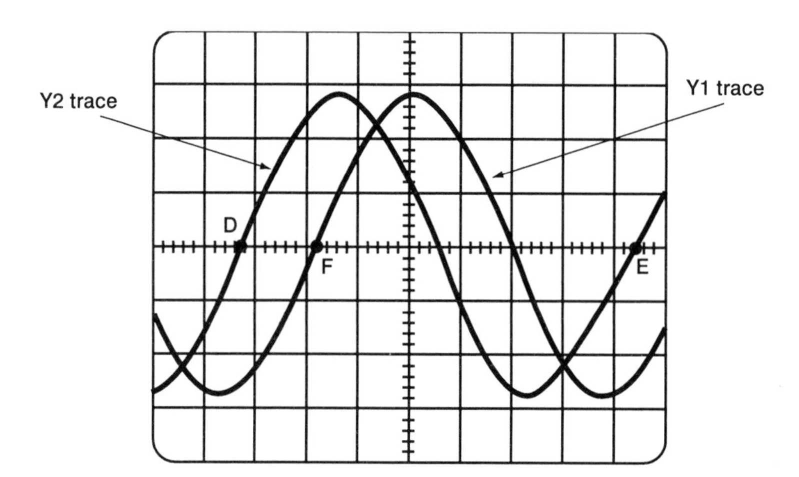


**Figure 10.25** Measuring the periodic time of the AC and the time difference between the two voltages

The periodic time is given by the distance DE between one point where a trace crosses the line and the next point where the same trace crosses the line in the same direction.

From Figure 10.25:

- DE = 7.5 divisions
- $\therefore$ periodic time = 7.5 divisions $\times$ 0.2 ms/division = 1.5 ms
- Frequency $= \frac{1}{\text{periodic time in seconds}} = \frac{1}{0.0015} = 667$ Hz

The phase difference between the two traces is the ratio

$\frac{DF}{DE}$ of 360°.

From Figure 10.25:

- DF = 1.4 divisions

  ∴ Y2 (current) leads Y1 (applied voltage) by $\frac{1.4}{7.5} \times 360° = 67°$

Readers with access to a signal generator and an oscilloscope could set up the circuit of Figure 10.22 using $R = 1000\ \Omega$ and $C = 0.1\ \mu F$, obtain displays similar to Figures 10.23–10.25 and work through their own versions of the above example as an exercise. You may, however, find considerable differences from the values obtained in the example, because:

a) some types of capacitor have very large tolerances (e.g. +50%, –25%)
b) the oscilloscope calibrations may have gradually changed (*secular change*).

Some oscilloscopes provide an output in the form of a mains-derived 50 Hz square-wave, of fixed amplitude, which can be connected to the Y-inputs for the purpose of checking and adjusting calibrations.

## Magnetic tape recording

This stores a time-varying electrical signal by recording it magnetically on to suitable tape. The magnetized tape can then be played back to re-convert the magnetic recording into a reproduction of the original electrical signal. Playback does not erase the recording, so it can be played back as often as required. Most readers will be familiar with this process in cassette players and video recorders.

The tape is a polyester ribbon coated on one face with the magnetic material and on the other face with a thin layer of backing material to improve driving friction and reduce static electricity effects. The magnetic material consists of a 10 μm layer of ferromagnetic or chromium dioxide powder bonded to the tape, the grains of powder acting as microscopic bar magnets with their axes in random directions. When the tape is magnetized during a recording, waves of magnetism are set up along the tape. Within each half-wave of magnetism, the microscopic magnets have been wrenched round by an applied magnetic field so that their north poles lie more or less in the same direction along the tape. The strength of the magnetism (*the remanent magnetic flux density*), and hence the strength of the signal, depends on how closely they are aligned with the direction of travel of the tape. If the magnetizing field was so strong that all the microscopic magnets are completely aligned, the tape is *magnetically saturated*, and no further increase in magnetic signal strength can be obtained.

### The recording head

Figure 10.26 shows a tape recording head. The electrical signal, after suitable processing, is passed through the head windings, magnetizing the pole pieces. These are separated by a very thin wedge of non-magnetic material, creating a magnetic field which bulges out from the gap, penetrating the magnetic layer on the tape and magnetizing it.

An instrumentation tape recorder may record as many as 14 tracks simultaneously on the same tape; each track requires its own recording head, the heads being stacked on top of one another with magnetic shields between them as shown in Figure 10.27. The head gaps must be in exact alignment vertically, otherwise simultaneous events would appear to have a time difference on the tape (*static skew*).

### The playback head

A tape recorder has two such head stacks, as shown in Figure 10.28. The tape first passes over the *record head*, then over the *playback head*, which has the same construction as the record head but smaller head gaps. The heads are manufactured to very close tolerances from

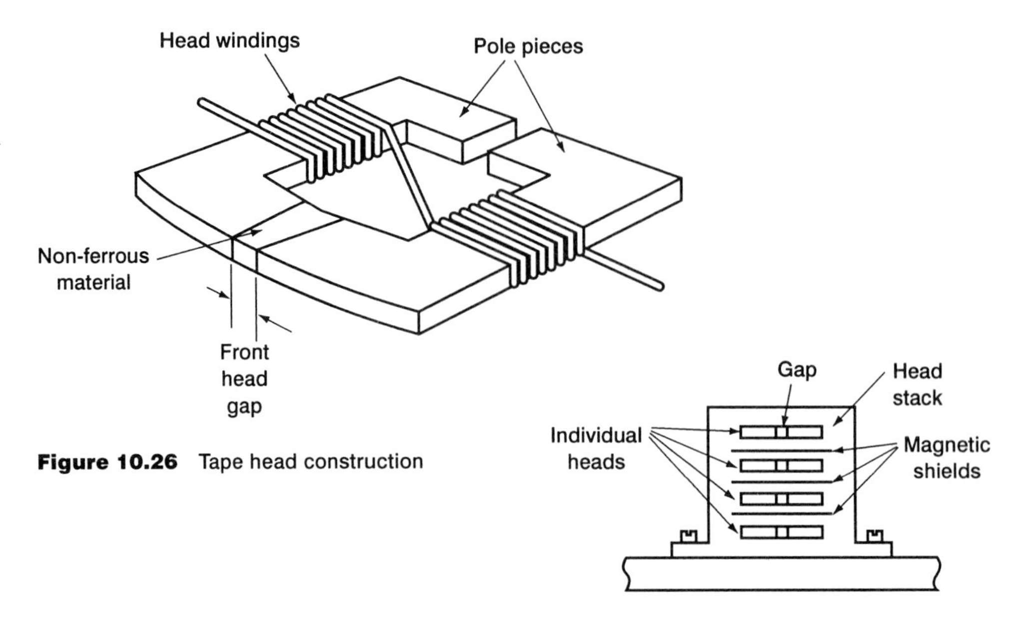


**Figure 10.26** Tape head construction

**Figure 10.27** A multi-track head

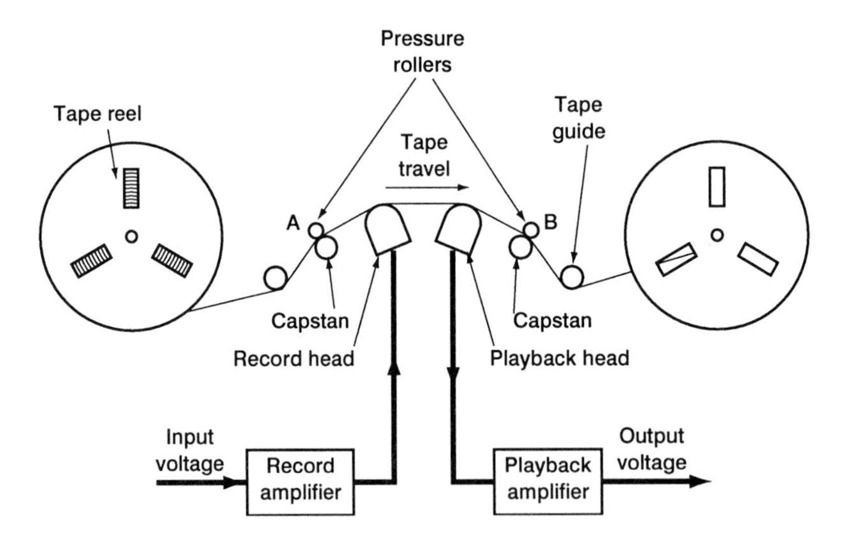


**Figure 10.28** Instrumentation tape recorder

hard-wearing material, and are therefore quite expensive. The record head will have a gap of a few micrometres between the pole pieces, to give a high flux density for magnetising the tape. The playback head has about half as much gap, to give maximum frequency response.

In the playback head, the waves of magnetism in the tape, passing the gap, create a changing magnetic field in the pole pieces, which creates a changing voltage in the head windings. This output voltage is proportional to the rate of change of the magnetic field, so zero frequency (DC) gives no output, and the *extinction frequency*, at which the magnetic wavelength on the tape equals the head gap, also gives no output. In between these two extremes, the relationship between output voltage and frequency is as shown in Figure 10.29. By amplifying the linear part of the curve with an amplifier which has the opposite gain characteristic (its gain decreasing as the frequency increases) the output can be maintained constant over the usable frequency range.

$$\text{extinction frequency} = \frac{\text{tape speed}}{\text{head gap}}$$

Thus a tape speed of 1524 mm/s (= 60 inches per second) with a head gap of 2 μm gives an extinction frequency of

$$\frac{1.524}{0.000002} = 762\ 000 \text{ Hz or } 762 \text{ kHz}$$

Actually, as Figure 10.29 shows, the practical frequency limit will be about half this, say 400 kHz.

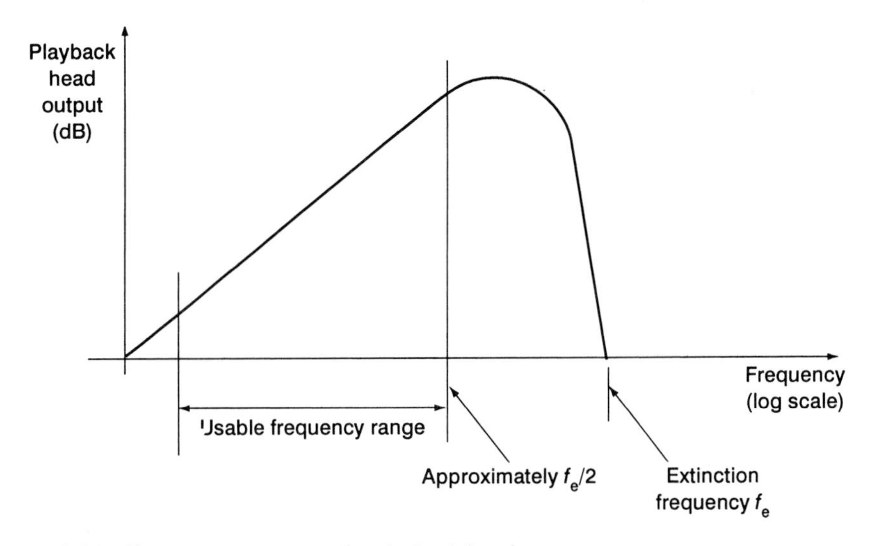


**Figure 10.29** Frequency response of a playback head

## The erase head

In a domestic tape recorder the tape passes over an erase head before it reaches the record head. The erase head has the same construction as the other two heads but larger head gaps. It is fed with high-frequency AC so that as the tape passes it, it applies an alternating magnetic field, the effect on the tape diminishing to zero as the tape moves away from it. This demagnetizes the tape, getting rid of any previous recording before the recording head remagnetizes the tape with a new recording. Instrumentation tapes are demagnetized in a similar way before they are used, by enclosing a whole reel in a *bulk eraser*, which subjects it to an alternating magnetic field which is gradually decreased to zero.

## Making a recording

The recording technique which has been described above is known as *direct recording*. Its advantage is the fact that the upper limit of its frequency range is very high (up to 1.5 MHz at a tape speed of 240 inches per second). Its disadvantages are (1) the lower limit of its frequency range is about 20 Hz, and (2) its accuracy is limited by electrical 'noise' generated by microscopic defects in the magnetic coating of the tape. When greater accuracy or the ability to record DC voltages is required, FM recording is employed (see chapter 8). With frequency modulation, the signal is immune to tape-generated noise but there is a penalty; the carrier frequency is limited by the upper frequency limit for recording, and with FM the maximum signal frequency can only be about one-fifth of the carrier frequency, so we lose high frequency response compared with direct recording. When the maximum frequency response is not needed, tape is conserved by reducing both tape speed and carrier frequency so that they are always in the same ratio. FM recorders offer a range of tape speed settings in which this is done automatically.

### Errors

Errors may be introduced into a recording by variations in the driving motor speed, and by uneven stretching of the tape as it passes the heads (*dynamic skew*). Cyclic variations in tape speed are eliminated as far as possible by precision manufacture of the tape drive. Variations in the driving (*capstan*) motor speed may be reduced by an automatic speed control circuit in which pulses from a crystal-controlled oscillator are compared with pulses derived from a phototransistor illuminated through holes in a disc carried by the driving motor spindle. To ensure that a tape is played back at exactly the speed at which it was recorded, a constant-frequency sinewave from a crystal-controlled oscillator may be recorded on one of the tracks, and its playback frequency compared with that of the oscillator.

Tape stretch may be controlled by virtually 'clamping' the tape to drive-motor capstans at points just before the record head and just after the playback head – points A and B in Figure 10.28. The drive motor at A is speed-controlled; the drive at B is given a slightly higher speed and allowed to slip, to keep the tape under constant tension.

### Digital recording

This is the sampling of a signal at regular intervals, as described in chapter 9, converting the sample value into a binary number, and recording the 0s and 1s on the tape as pulses, by direct recording. The binary digits may be recorded on a single track (*serial recording*), or an 8-digit number may be recorded simultaneously on eight tracks (*parallel recording*). Digital recording gives complete immunity to noise and very much greater precision than analogue recording, but it needs a great deal more tape – perhaps 50 times more to record a sinusoidal type of signal.

## Data acquisition systems

Throughout this book we have been looking at ways of measuring a single property. Whether it be a force, a temperature, an acceleration, a displacement or whatever, we have considered it in isolation. But in, for example, the acceptance tests of a complex system such as an aircraft or an oil refinery, it may be necessary to continuously record the variations in output of many different transducers simultaneously. Hence the need for data acquisition systems.

A data acquisition system, which may deal with hundreds of simultaneous transducer signals, is put together from *signal conditioning modules*, *signal processing modules* and *readout modules*, all controlled by the system's own computer.

**Signal conditioning modules** provide, for the transducers, whatever power supplies, reference voltages, amplification, and linearization circuits they need so that each transducer produces a standardised output signal with full-scale range of for example ±5 V DC. Since the transducers may include strain gauge bridges, thermocouples, LVDTs, pulse generators, etc., each different type of transducer will require a different form of signal conditioning. Where amplification is required, each channel has its own dedicated amplifier.

A signal conditioning channel may also provide its own push-button calibration check. This is not to check the calibration of the transducer, which would be difficult or impossible once it has been installed, but to check that the gain of amplifiers and other circuits between input and output of the signal conditioning unit have not changed due to change of ambient conditions. The push-button applies a simulated transducer output, and the corresponding output of the signal conditioning channel should be within some allowable tolerance. The transducer itself is calibrated by the manufacturer before installation.

A **signal processing module** receives the outputs from two transducers and performs analogue operations to obtain either their product, ratio, sum, difference or to make logical decisions. For example, an engine test facility may have a signal processor which receives inputs from a load cell and from a toothed-wheel magnetic transducer, and gives a continuous output signal of mechanical power in kilowatts.

A **readout module** takes the output from a signal conditioning or signal processing channel and makes it suitable for the display or recording device which is to receive it. Display devices range from simple three-light systems indicating 'low' – 'OK' – 'high', to sophisticated devices such as cathode-ray tube monitors. Recording devices may include UV recorders, servo recorders, magnetic tape recorders or computers. For recording by computer, channels may be scanned and their data digitally recorded at rates of the order of 1500 channels per second.

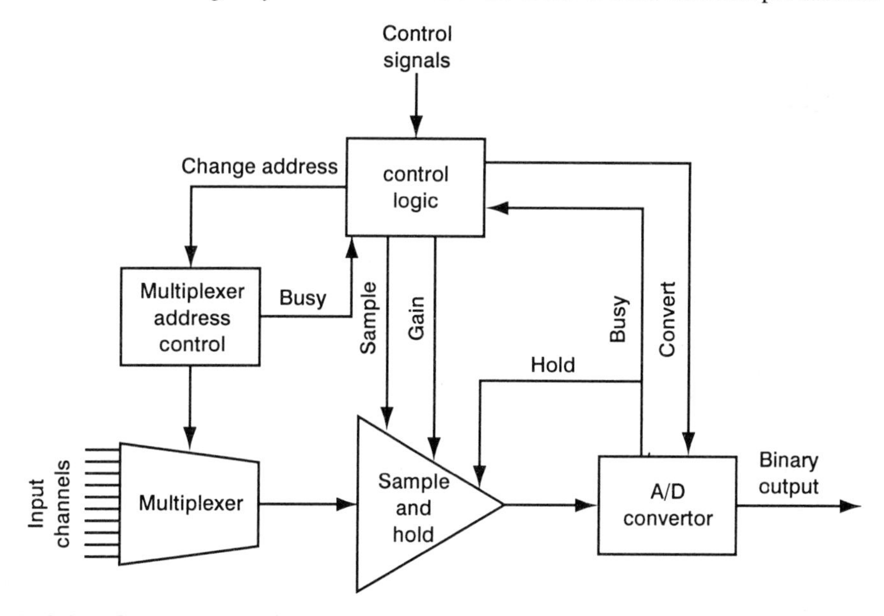


**Figure 10.30** Block diagram of a data logger

**Data loggers** are smaller standardised data acquisition units, usually dealing with up to 64 channels of data, with a built-in microprocessor to control internal operations. After signal conditioning, the input channels are scanned in sequence and multiplexed to a single sample-and-hold amplifier and an analogue-to-digital converter. The A/D converter gives a digital output suitable for display by digital indicator, or recording by magnetic tape, punched tape, teletype or line printer, or for input to a computer. Figure 10.30 shows a block diagram of a data logger. To read the signal on one input the following sequence of operations takes place:

1 The control logic sends a signal to the multiplexer address controller to select the address (a 6-bit binary code) of the next input channel to be read. As this can take a few microseconds, the address controller sends a 'busy' signal to control logic until the operation is completed.

2 The control logic then sends a 'sample' signal to the sample-and-hold amplifier, and by means of a 2-bit binary signal sets the amplifier gain which will be required for this particular input.
3 The control logic sends a 'convert' signal to the analogue-to-digital converter. As A/D conversion takes time, the converter sends a 'converter busy' signal back to control logic, and also to the sample-and-hold amplifier, which switches it to the 'hold' mode.
4 At the end of the conversion, the A/D converter sends a 'conversion complete' signal to the control logic, which then synchronises the transfer of the binary output to the recording device.

Readout by digital indicator or printer requires a low scanning rate but this may give erroneous results due to the time difference between the recording of signal inputs at the beginning and end of the scanning sequence. Thus, although a data logger is much cheaper than a 'made-to-measure' data acquisition system, the time-sharing of channels of the data logger may make it unsuitable for fast-changing signals. A data acquisition system in which all channels are continuously monitored does not suffer from this disadvantage.

## Exercises on chapter 10

1 A moving coil meter has a 100 μA movement with a coil resistance of 1300 Ω.
   a) Calculate the resistance required, and show it in relation to the meter in a simple circuit diagram for:
      i) an ammeter to measure 0 to 1 A
      ii) a voltmeter to measure 0 to 100 V.
2 Draw a block diagram of a digital voltmeter.
3 As the means of displaying numerical values, discuss the relative advantages and disadvantages of:
   a) light-emitting diodes
   b) liquid crystal.
4 a) Sketch the construction of an LCD 7-segment digit, showing the electrical connections to the segments.
   b) Explain how the LCD segments become visible.
   c) Why is DC unsuitable for driving LCD segments?
   d) Show, on a horizontal axis representing time, the relationship between the start and finish of the controlling waveform and the appearance and disappearance of an LCD segment. Insert typical values of the rise and fall times of optical density.
5 a) A digital voltmeter has a basic range of 0 to 1.999 V. Draw the circuit diagram of a resistance chain and rotary switch arrangement to give it an input resistance of 10 MΩ and additional ranges of 0 to 19.99 V; 0 to 199.9 V; 0 to 1999 V.
   b) Explain how the instrument may be protected from an over-range input voltage of either polarity.
6 Explain the essential features of an ultraviolet recorder. How does it produce a record of the variations in the input quantity?
7 Without specifying the means by which the pen carriage is moved, show, by means of a simple labelled sketch, the arrangement of pen carriage, chart roll and feedback transducer in a potentiometric recorder.
8 a) Make a sketch of an XY plotter and indicate on it the gantry and the pen carriage. Show, also, the respective directions in which the X and Y signals are measured.
   b) What additional feature is needed to convert an XY plotter into a Y-t plotter, and what is its purpose?

9 Show by a simple sketch the pen and chart arrangement of a chart recorder. Why does the grid of the chart consist entirely of curved lines?

10 Draw a diagram of the arrangement of the electrodes in the cathode ray tube of an oscilloscope. Show 'brightness' and 'focus' potentiometers as part of a chain of resistances connecting EHT- to EHT+ and show the connections from this chain of resistances to the appropriate electrodes in the tube.

11 Explain the meaning of 'chop' and 'alternate' as alternative methods of obtaining two traces on a cathode-ray tube screen. State which method is appropriate for very low frequency signals and which for very high frequency, giving a reason in each case.

12 Explain the following, in connection with cathode-ray oscilloscopes:
a) triggering
b) the 'DC-GND-AC' switches
c) the XY mode
d) the 'CAL' settings.

13 Draw the circuit diagram of a 10-to-1 attenuation probe connected to the input circuit of a cathode-ray oscilloscope. Explain how the probe trimmer should be adjusted. Illustrate your answer with sketches of the three waveform shapes which may appear on the screen as the adjustment is carried out.

14 The following measurements were taken from two sinusoidal traces on a cathode-ray oscilloscope screen:

- Trace 1 – peak-to-peak – 7.1 divisions, length of one cycle: 6.3 divisions
  Trace 2 – peak-to-peak – 5.9 divisions, length of one cycle: 6.3 divisions
- With both traces symmetrical about the horizontal centre-line of the screen, it was found that trace 2 was displaced 1.4 divisions to the right of trace 1, measured along the horizontal centre-line.
- The Y-input switch settings were: Y1: 2 V/DIV; Y2: 0.5 V/DIV.
  The timebase switch setting was 0.1 ms/DIV.
- All variable controls were on the CAL setting.

Determine:
a) the amplitude of each signal in volts
b) the RMS voltage of each signal
c) the frequency of the signals
d) the phase angle between the two signals.

15 A cathode-ray oscilloscope is operated in the XY mode with equal sensitivity settings on both inputs, and sinusoidal signals of equal amplitude and frequency are applied to the two inputs. What is the phase angle between the two signals if the display is:
a) a straight line at 45° to the horizontal?
b) a circle?

16 a) Sketch a graph showing the output of the replay head of a magnetic tape recorder, in dB, plotted against recorded frequency.
b) The recorder has a replay head gap of 2.5 μm and its maximum tape speed is 240 inches per second.
  i) Calculate the extinction frequency.
  ii) Estimate an upper limit to the usable frequency range.

17 Explain how a signal is
   a) recorded on
   b) played back from
   c) erased from
   a magnetic tape. Illustrate your answers with a sketch showing the main features of a tape head.

18 For recording on to magnetic tape, compare the relative advantages and disadvantages of
   a) direct recording
   b) FM recording
   c) digital recording.

19 a) In a data acquisition system, explain the purpose of
      i) a signal conditioning module
      ii) a signal processing module.
   b) A data logger has its input channels multiplexed to a single, central amplifier with its gain set for each channel by a microprocessor. What is the main disadvantage of this arrangement?

# Appendix
# A library of control system elements

## Linear hydraulic actuators (hydraulic jacks)

These may be single-acting (i.e. fluid on one side of the piston only), single-acting with return spring, or double-acting. Figure A1 shows the principle of a double-acting hydraulic jack.

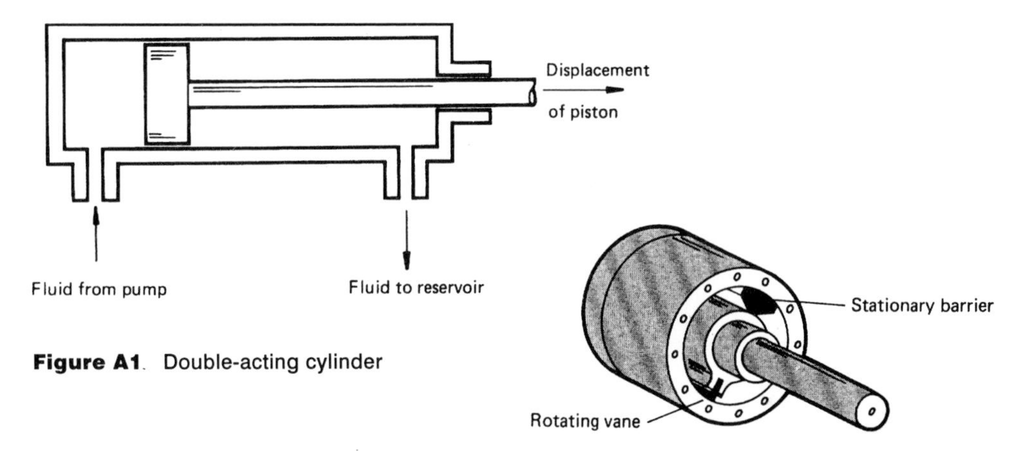


**Figure A1**. Double-acting cylinder

**Figure A2** Vane-type rotary actuator

## Rotary hydraulic actuators

The rotary vane type of actuator is shown in Figure A2. A port on one side of the barrier admits hydraulic fluid to the space at one side of the vane, while a port on the other side receives fluid from the space at the other side for return to reservoir. The rotary vane permits a total rotation of rather less than one revolution. Greater rotation is obtainable from a rack-and-pinion type of actuator, shown in principle in Figure A3, in which rack teeth formed in the plunger of a linear actuator rotate a pinion.

## Spool valve

For motion in alternate directions, each of the two ports of a linear or rotary actuator must be connected alternately to the high-pressure supply and to the fluid reservoir. The connection is usually made through a *spool valve*. This is a rod with cylindrical surfaces which cover the ports in the valve body; in between them the rod is reduced in diameter to allow the passage of the hydraulic fluid. The cylindrical surfaces are ground and lapped to a fluid-tight sliding fit

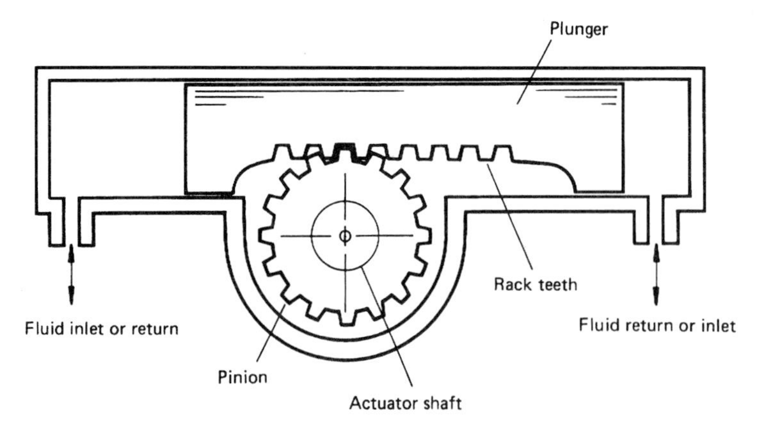


**Figure A3** Rack-and-pinion type rotary actuator

in the bore of the valve body, so that when the valve is in its mid position it seals the ports leading to the actuator, thus locking the actuator piston. Displacement of the valve from this position connects the appropriate ports of the actuator to fluid supply and reservoir. Figure A4 shows an example of a spool-valve and actuator assembly as used in a copying lathe.

Figure A4 also shows a *differential lever* operating the spool valve. An input displacement at the stylus displaces the spool valve from its mid position, admitting fluid to one side of the actuator piston and connecting the other side to the reservoir. This causes the actuator to move in the opposite direction until it has restored the spool valve to its central position.

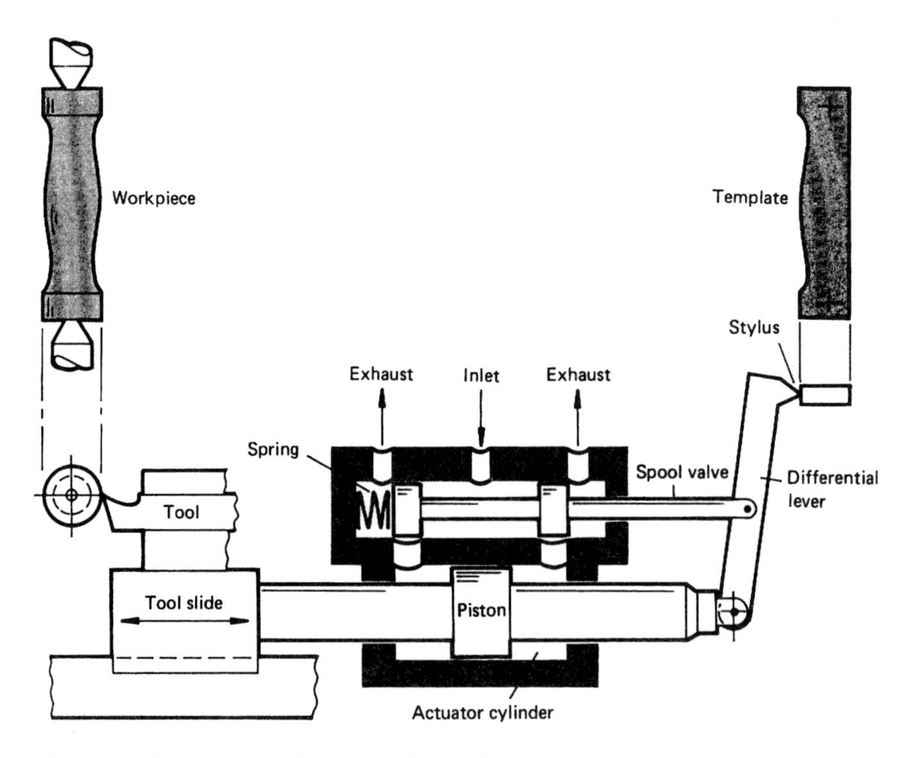


**Figure A4** Hydraulic servo copying system for a lathe

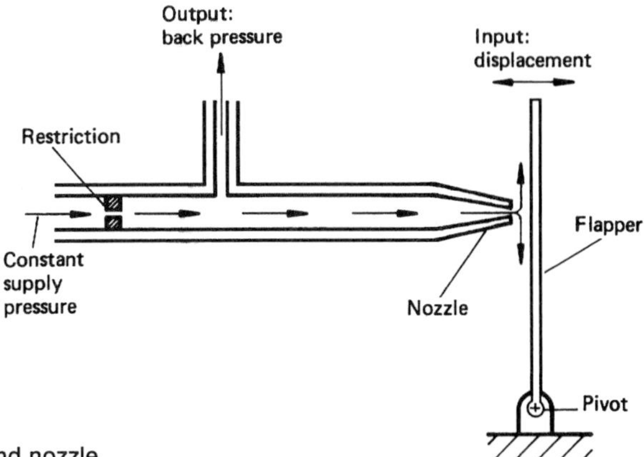


**Figure A5** Flapper and nozzle

## Flapper and nozzle

The principle is shown in Figure A5. Fluid, supplied at constant pressure, passes through a restriction and then escapes through a narrow gap between the outlet nozzle and a perpendicular surface (the flapper). This gives a back pressure which is proportional to the gap, over a limited range as shown in Figure A6. As well as being used for control systems, the principle is the basis of a sensitive displacement-measuring instrument, the *pneumatic comparator*.

In a control system, nozzles are usually arranged in pairs with the jets of fluid striking opposite faces of the flapper. The fluid supplied to the nozzles may be either air or oil. Figure A7 shows an electro-hydraulic valve in which back pressure from a flapper-and-nozzles assembly positions a spool valve.

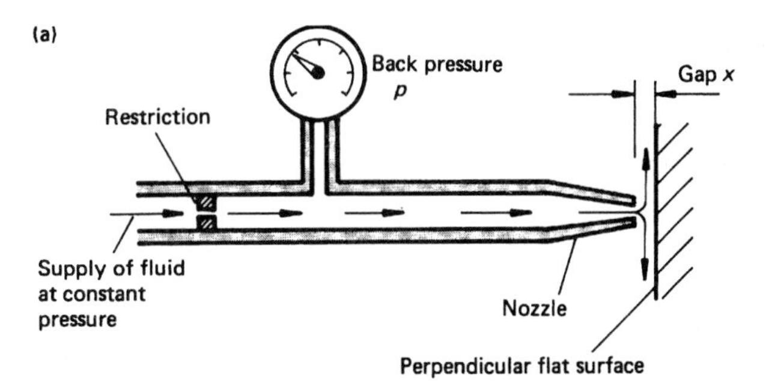


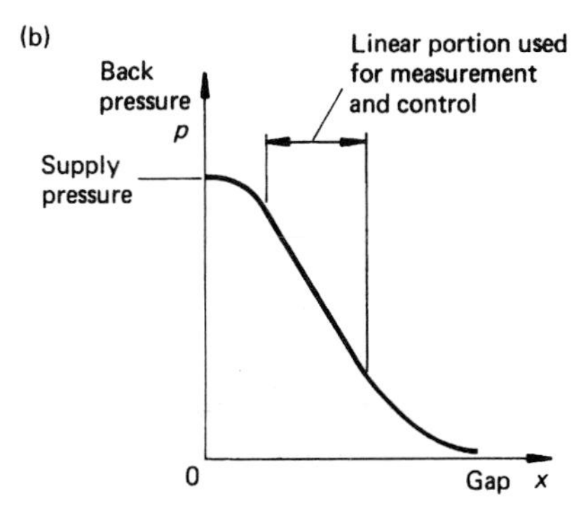


**Figure A6** Relationship between back pressure and flapper-nozzle gap

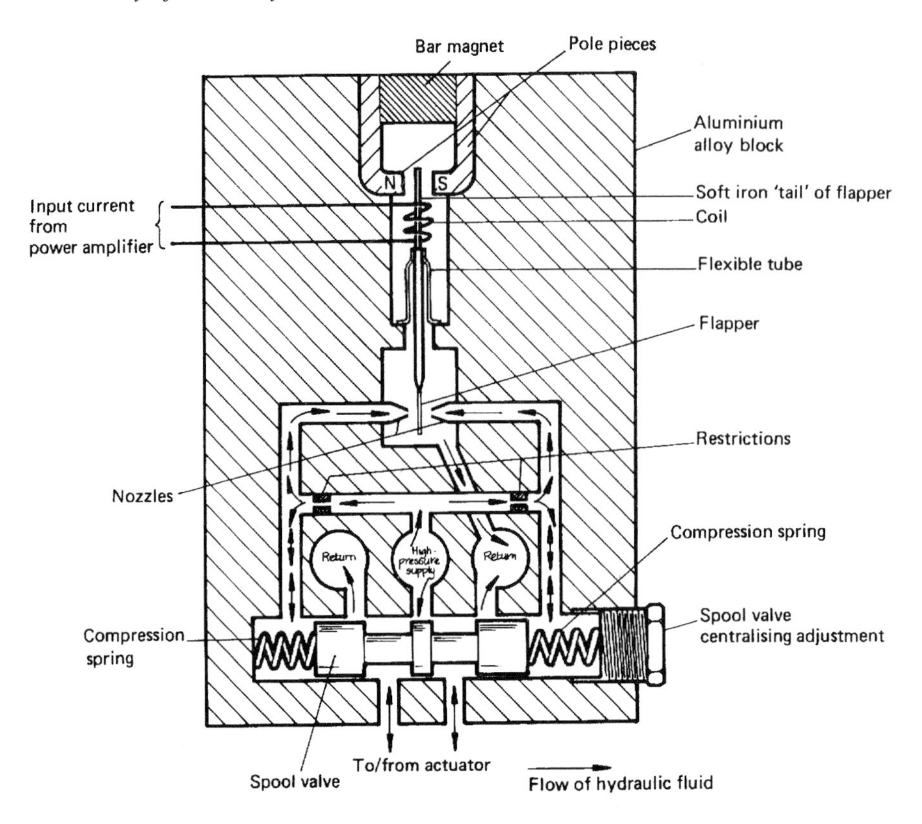


**Figure A7** Section through a servo valve with electrical input and two stages of hydraulic amplification

## Solenoid

This consists of a current-carrying coil and a soft iron rod. The coil is closely wound with several layers of wire on an insulating bobbin. The rod, or *armature*, is an easy sliding fit in the axial hole of the bobbin, and is pulled to the centre of the coil by the magnetic field set up when current is passed through the coil. For a given position of the rod, the pull exerted is proportional to the current, but, for a given current, the relationship between pull and displacement is non-linear, so this is essentially a 'bang-bang' device.

If the rod is fixed in the coil, the device becomes an electromagnet and as such can attract a hinged soft iron plate which can operate switch contacts. This is the principle of the *relay*.

Solenoids and electromagnets can use either DC or low-frequency (e.g. mains frequency) AC, because soft iron obligingly reverses its magnetism when the current reverses, and thus is still attracted.

## DC electric motors

Figure A8 shows a typical small DC motor in section. It has the minimum number of field coils (two); larger motors may have four or more. Figure A9 is a simplified diagram showing how the driving torque is produced in the armature. Current through the field coils magnetises the pole pieces. Current through the armature coils magnetises the armature in a direction at right

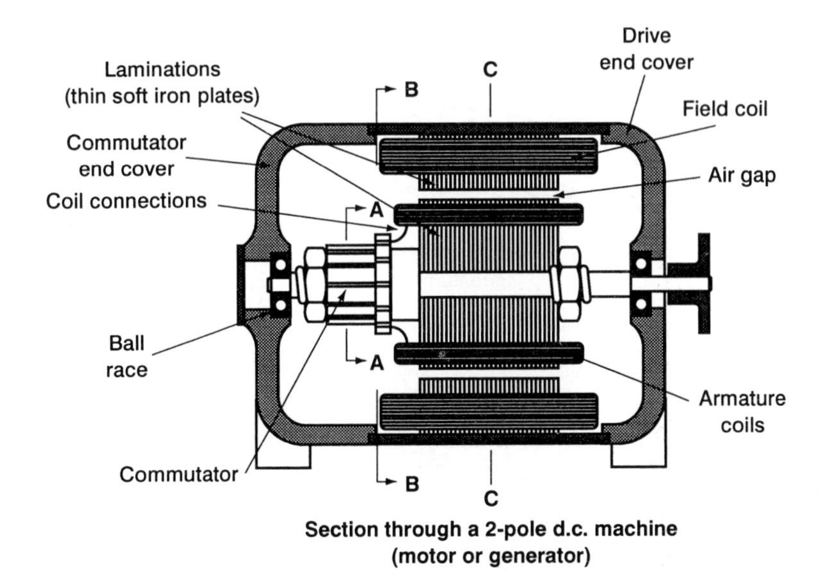


**Section through a 2-pole d.c. machine (motor or generator)**

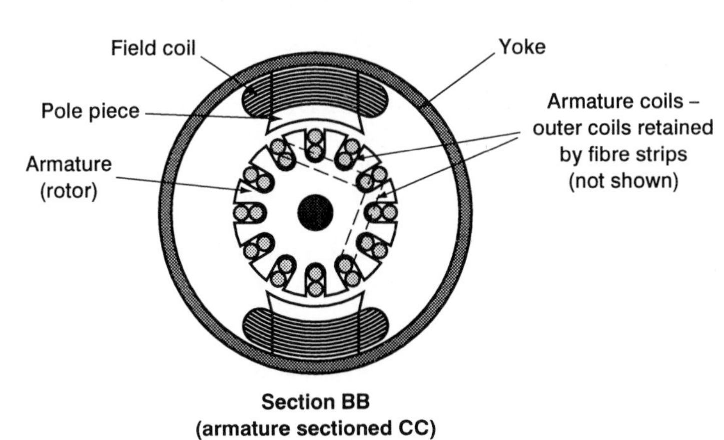


**Section BB (armature sectioned CC)**

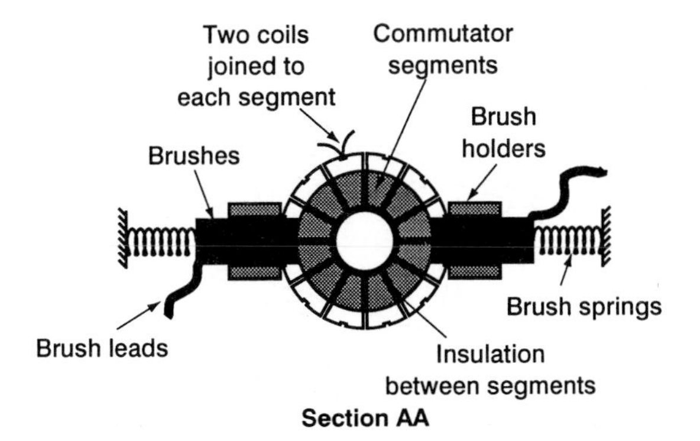


**Section AA**

**Figure A8** The main features of a DC motor

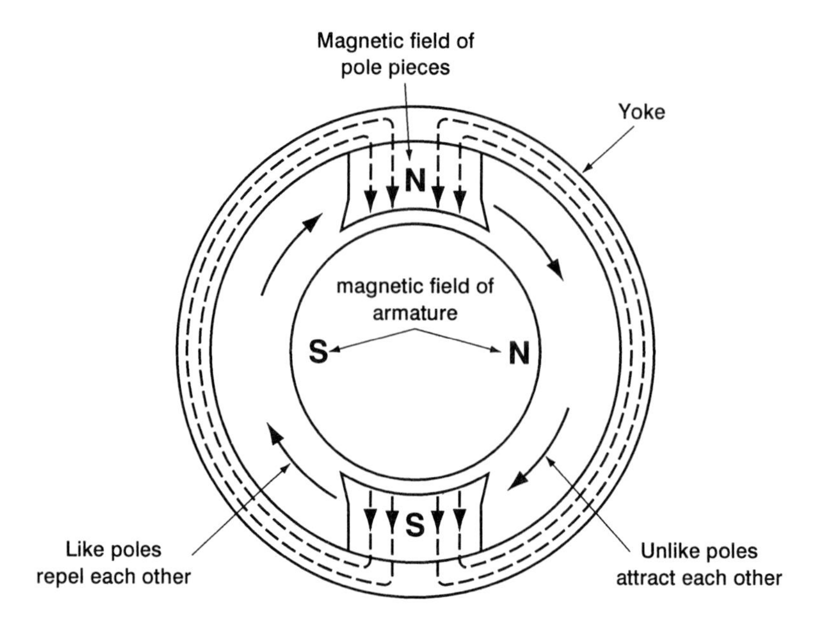


**Figure A9** How a motor produces torque

angles to the field of the pole pieces. Thus there is a magnetic torque on the armature, causing it to rotate. As it rotates, the commutator rotates with it, bringing different segments in contact with the brushes supplying the armature current. This alters the current path through the armature coils in such a way that although the armature is rotating, its magnetic field remains in the position shown.

DC electric motors can be divided into various types, depending on the way current is supplied to the field coils. The main types are described below.

**Permanent magnet motor**
This is the simplest type as it has no field coils at all! The magnetic field in which the armature rotates is provided by two shaped bar magnets, as shown in Figure A10. Permanent magnet motors are used where the power requirement is very small and the motor must be cheap and compact – in cassette recorders, for instance.

The motor's speed is controlled by varying the supply voltage. Any DC motor is also a dynamo; while it is running the armature coils generate a voltage (called *back EMF*), opposing

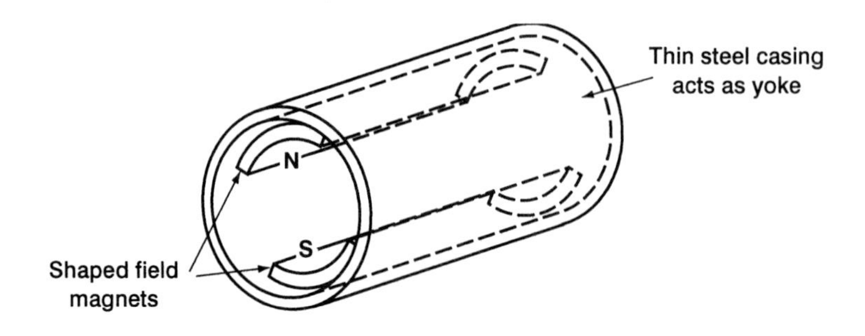


**Figure A10** Arrangement of field magnets in a permanent-magnet motor

the voltage supplied to them and proportional to speed. When switched on, the motor speeds up until its back EMF is about 90% of the supply voltage – it is the remaining 10% which drives the load.

**Shunt-wound motor**
In this type the field coils are connected in parallel with the armature coils (are *shunted* across them), as shown in Figure A11(a). The field coils are of finer wire and have many more turns than the armature coils, so most of the motor current goes through the armature coils.

Like the permanent magnet motor, the shunt-wound motor runs at almost constant speed regardless of load variations if the field current is constant. Its speed can be controlled by a variable resistor inserted in series with the field coils at X in Figure A11(a), to adjust the field current. The power loss in the resistor is small compared with the power consumption of the motor as a whole. Strange as it may seem, reducing the field current makes the motor run faster, because in a weaker magnetic field it takes more RPM to generate the same back EMF, though the motor torque is correspondingly weakened.

**Series-wound motor**
As shown in Figure A11(b), its field coils are *in series* with the armature coils, but this makes it a completely different type of motor. In order not to limit the armature current, the field coils have only a few (5–20) turns and the wire is of large cross-sectional area – it may be in the form of rectangular section strip. Because field and armature coils are in series, at start-up, when the back EMF is zero, this type of motor takes a very large current – it can be several hundred amps – but produces a correspondingly large torque. So it is used for starter motors, to kick petrol and diesel engines into life.

Unlike a shunt-wound motor it is not amenable to speed control; if switched on with no load, it accelerates to a very high speed.

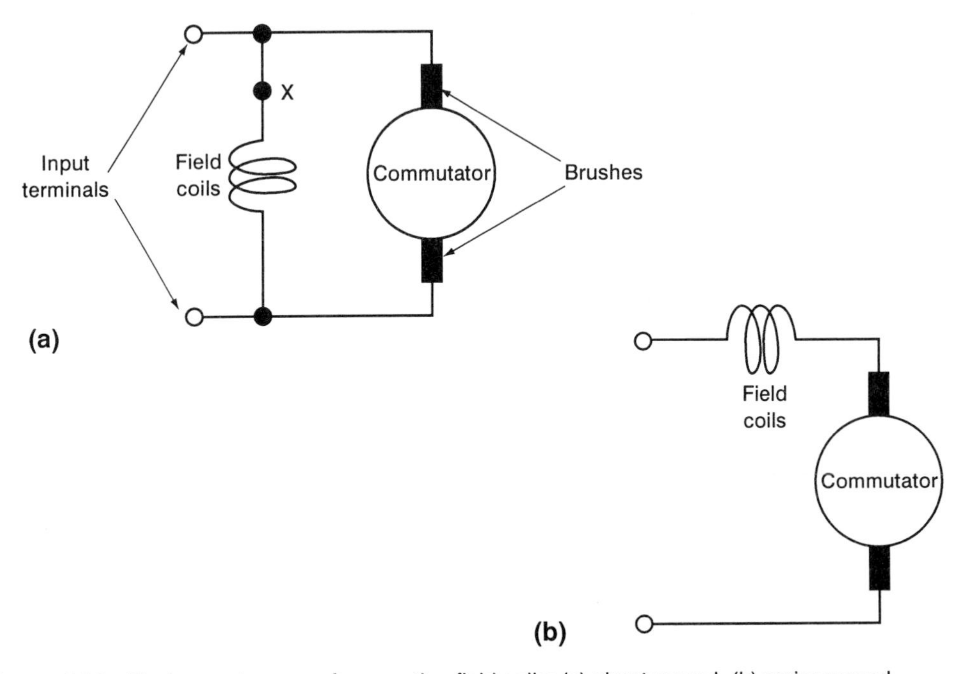


**Figure A11** The two main ways of connecting field coils: (a) shunt wound, (b) series wound

**Compound-wound motor**

This is a combination of shunt and series winding to obtain a compromise between the characteristics of those two types of motor. The main part of the field coils is shunted across the armature, the remainder, from one to ten turns, is wired in series with the armature.

**Separately-excited motor**

In this case the field coils are of the 'shunt' type; a large number of turns carrying a relatively small current, but supplied completely independently of the armature circuit. By varying the field current the motor speed can be precisely controlled, and by reversing the current the rotation of the motor can be reversed.

**DC servo motor**

This is a development of the separately excited motor. The principle is shown in Figure A12. The field coils are divided into two halves by a centre connection. The armature is fed from a constant-current source, its speed and direction of rotation being controlled by the resultant magnetic flux set up by the two halves of the field winding. This is fed from a 'push-pull' type of amplifier; in effect, two identical power amplifiers 'back-to-back', with outputs C and D and a common return connection at E. When equal currents flow as shown in Figure A12, the magnetic effects of the two halves of the field winding cancel out and the armature is stationary. An error signal unbalances this arrangement, causing a greater current to flow through one half of the field winding and a smaller current through the other half. The motor then runs in the direction determined by the greater current flow.

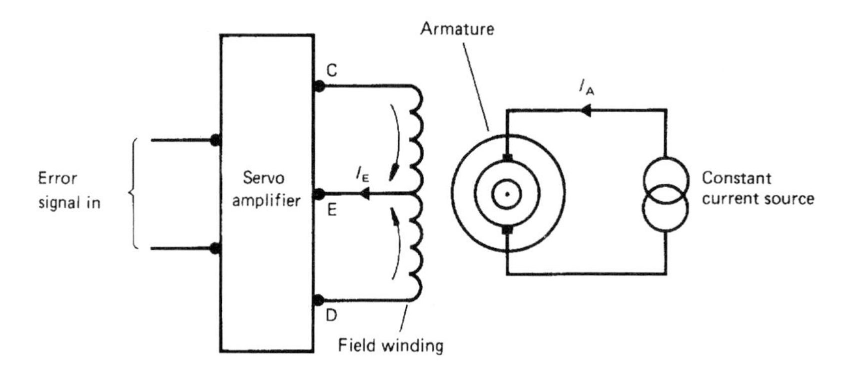


**Figure A12** The essential features of a split-field DC servo-motor

The field coils usually have many turns because the available current is small. Their inductance is therefore considerable, so sudden changes in field current produce a high voltage surge. A well designed split-field servo-motor should therefore have surge protection shunts in parallel with the field windings.

A DC servo-motor usually has a long thin armature, to give the best torque compatible with low moment of inertia.

**Stepper motor**

This type of motor can be made to rotate through any required number of angular steps in either direction and stop instantly. Applications include tape drives, numerically controlled machine tools, computerised engine control systems, robotics. Stepper motors eliminate the need for feedback in many positional control systems because a microprocessor driving a stepper motor can set an output accurately on open loop with no need for the complication of a closed loop system.

A stepper motor has no commutator because the armature (in this case called the *rotor*) has no windings. The only windings are field windings in the outer casing (which is called the *stator*). A series of pulses is applied by a computer or microprocessor to various combinations of these windings, each pulse causing the rotor to rotate through one step of angular displacement. The magnitude of the step depends on the design of the motor; common step sizes are 1.8° or 7.5° but motors giving 2.5°, 3.75°, 15° or 30° are also obtainable. Continuous pulsing gives steady rotation, the rate of pulsing giving exact control of the speed.

The motor output can be taken through a reduction gear, to obtain much smaller angular steps and correspondingly increased torque. Stator windings may be of either the *unipolar drive* or the *bipolar drive* type. Each winding of a **unipolar drive** motor has a centre tapping, and one connection is made to this. The other connection can be switched to either end of the winding, so only half of the winding is in use at a time. This gives a low-cost high-reliability method of control; the output torque at low stepping rates is less than that of a similar bipolar drive motor, but at high stepping rates (high motor speed) the unipolar drive has the greater torque output.

A **bipolar drive** motor makes use of the whole of each stator winding, the direction of current through the winding being reversed as necessary by a suitable drive circuit. The torque advantage of the bipolar drive motor at low speeds is somewhat offset by the fact that it needs a more complicated circuit to drive it.

Stepper motors can be divided into three types: *permanent magnet, variable reluctance* and *hybrid.*

In the **permanent magnet stepper motor** the stator has a number of poles equally spaced around the internal circumference. The rotor is permanently magnetised with alternate north (N) and south (S) poles to match the number of poles or pairs of poles in the stator. In the type of permanent magnet motor described here, there are two stator windings (denoted 1 and 2 below), each of which produces alternate north (N) and south (S) poles around the inside of the stator, one winding's poles being equidistant between those of the other winding. Taking a four-pole motor as a simple illustration (actual motors have more poles than this) and assuming that both windings are energised simultaneously, the arrangement of poles would be as shown in Figure A13(a), each rotor pole being attracted to a pair with its opposite polarity. Figures A13(b) and (c) show the effect of reversing the current in each of the two windings in turn. At each such current reversal the rotor moves through an angle (the *step angle*) equal to the angular spacing of the poles. Figure A13 shows clockwise rotation with a step angle of 45°.

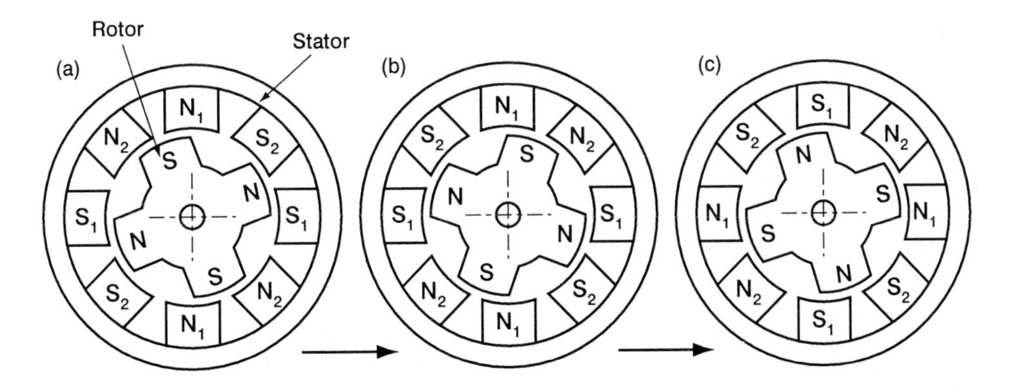


**Figure A13** The operation of a permanent-magnet stepper motor

The direction of rotation of the rotor is reversed by reversing the order in which current reversals take place in the two windings. By switching off one of the two windings before reversing its current the rotor can be advanced by half steps. Thus between (a) and (b) in Figure A13, winding 2 could be switched off and the rotor would advance half a step to the intermediate position where its S poles line up with the N poles on the vertical centre-line.

Small motors can be driven by an integrated circuit which outputs the correct sequence of currents to the two windings. The motor is stepped by applying an input pulse to one pin of the integrated circuit, the direction of rotation being set by a 'high' or 'low' voltage on another pin.

Even when the stator windings are not energised, the poles of the rotor are attracted to those of the stator, so the rotor resists being displaced from its position it reached at switch-off; this resisting torque is called the *detent torque*.

In the **variable reluctance stepper motor**, the rotor is made of soft iron and there are unequal numbers of poles in rotor and stator, to produce small steps by means of a vernier effect. The stator windings are of multiphase construction; Figure A14 shows a three phase example: the windings on poles marked 1 are connected in series to form phase 1, those on poles marked 2 in series form phase 2, and those on poles marked 3 phase 3. The term 'three phase' does not mean that the motor is supplied with three-phase AC. Only DC is used and only one phase at a time is energised. This causes the rotor poles nearest to the energised stator poles to line up with them, rotating to the position of least magnetic reluctance (easiest path for the magnetic flux). Switching off phase 1 and switching on phase 2 causes the rotor to rotate through one step to line up with the phase 2 poles; similarly switching off phase 2 and switching on phase 3 causes a further step – and so on. Switching the phases in this order gives clockwise rotation in the diagram; switching them in the reverse order gives anticlockwise.

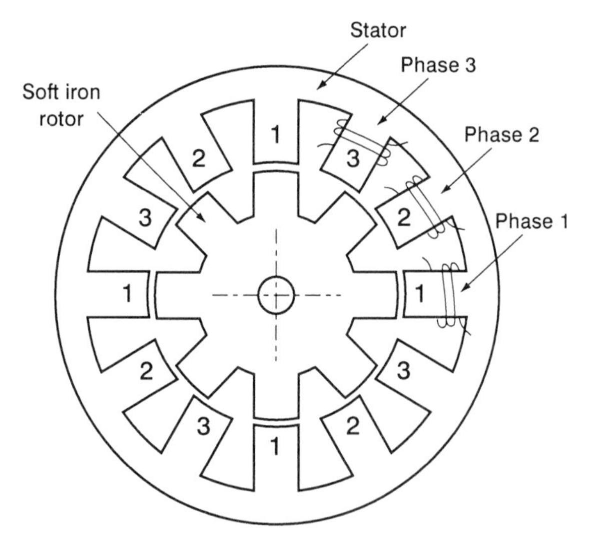


**Figure A14** A three-phase variable reluctance stepper motor

The angle turned through at each step is:

(angle between rotor poles) – (angle between stator poles)

Thus in Figure A14 the step angle is:

$$\frac{360°}{8} - \frac{360°}{12} = 15°$$

Since the soft iron of the rotor is equally attracted to a N as to a S pole, the current in a phase only needs to be switched on and off – it does not need to be reversed. This greatly simplifies the circuit necessary to drive this type of motor.

The detent torque of a variable reluctance motor is negligible because it has no permanent magnetism in rotor or stator.

The *hybrid stepper motor* has a permanently magnetised instead of a soft iron rotor, to augment the variable reluctance effect; it therefore gives the highest torque per unit volume. The construction is shown in Figure A15. The rotor consists of an axial magnet between two iron discs which are grooved longitudinally to form poles. The magnet magnetises the discs so that the poles on one disc are all N while those on the other disc are all S. The discs are staggered by half a pitch so that the poles of one disc are aligned with the grooves of the other. The stator pole-pieces are long enough to bridge both discs of the rotor and are grooved longitudinally to divide the concave surface of the pole piece into separate poles all having the same polarity, which depends on the direction of current through the pole winding.

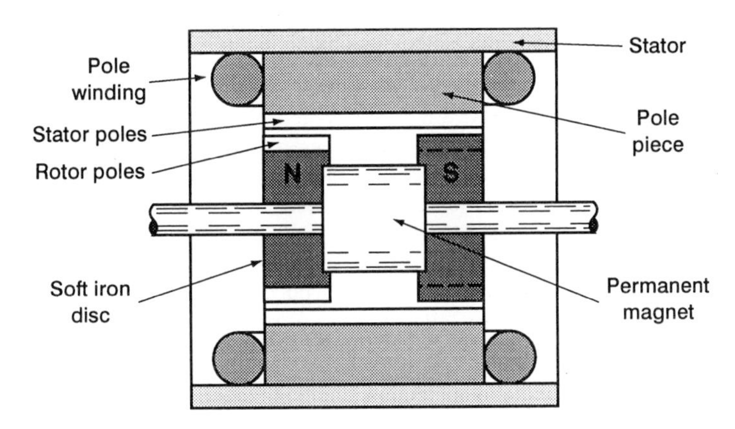


**Figure A15** Section through a hybrid stepper motor

This arrangement results in a high-torque motor with a small step angle. A typical design has a 50-pole rotor and a stator with eight pole pieces, each of which is divided into five poles. The step angle is:

(angle between stator poles) – (angle between rotor poles)

$$\frac{360°}{8 \times 5} - \frac{360°}{50} = 1.8°$$

Besides step angle and detent torque, a stepper motor's specification includes the following:

- maximum working torque
- holding torque (usually about 50% greater than maximum working torque)
- maximum pull-in rate – the maximum step rate at which an unloaded motor can start from rest without losing step. For a small motor this is some hundreds of steps per second
- maximum pull-out rate – the maximum step rate which can be applied to a *running* motor without its losing step. This is higher than the maximum pull-in rate; the difference between the two rates is the slew range
- step angle tolerance – after a step, the rotor's actual position may be at a slight angle to its theoretical position because of friction
- moment of inertia of rotor – combined with the moment of inertia of the load this may cause *ambiguity*.

**Ambiguity** occurs because the rotor tends to oscillate slightly about its final position with one or two cycles of decaying oscillation as it comes to rest. If the next pulse happens to be applied when the rotor is at one of the maxima or minima of its oscillation it may overstep or understep. This can be overcome by damping the oscillation; the rotor is inherently damped by eddy currents and magnetic hysteresis, but if this is insufficient, frictional damping must be applied.

## AC electric motors

Almost all AC electric motors are *induction motors*. They operate on a very simple principle:

1 AC flowing through stator windings creates a rotating magnetic field
2 this generates eddy currents in the rotor
3 the magnetic field of the eddy currents is attracted to the stator's rotating field, thus pulling the rotor round.

The speed of rotation of the stator field is called the *synchronous speed*. To generate eddy currents the rotor must *slip* relative to the rotating field but the amount of slip is quite small; unloaded the motor runs at nearly 100% of synchronous speed whilst at full load it still runs at about 96% of synchronous speed. Thus the induction motor is virtually a constant speed machine the speed of which can only be altered by altering the supply frequency. Figure A16 shows a section through an induction motor. The stator windings lie in slots in the stator ring, which is made up of soft iron laminations. Theoretically, the rotor could be a solid iron cylinder; eddy currents would be generated in it and it would rotate, but it would be a very inefficient motor. The rotor actually consists of soft iron laminations with a 'squirrel cage' of eddy current conductors as shown in Figure A17 embedded in it. The conductors may be of copper or aluminium and insulation from the iron of the rotor is unnecessary. The low melting point of aluminium allows the squirrel cage to be pressure-cast in the passages through the stack of rotor laminations, complete with the end rings and a ring of blades for a cooling fan.

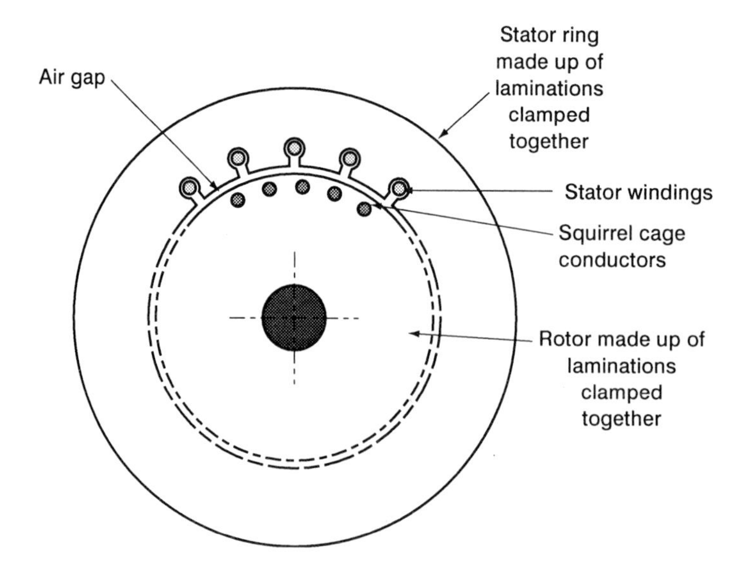


**Figure A16** Section through an induction motor

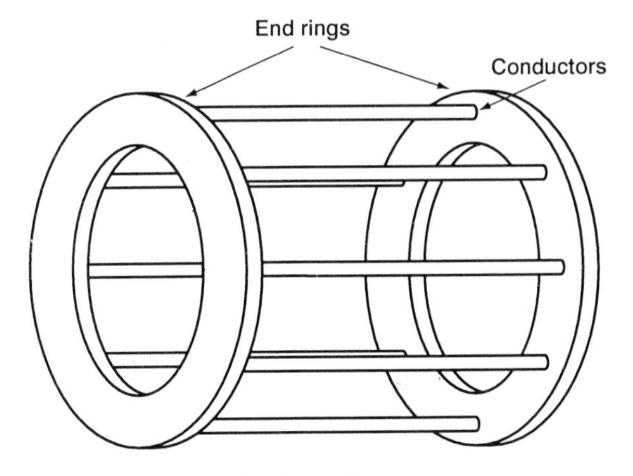


**Figure A17** The 'squirrel cage' of an induction motor

### 3-Phase induction motors

These use three-phase AC as shown in Figure A18. The supply voltage is normally 415 V. Figures A.19 and A.20 show the arrangement of stator coils in a 2-pole and in a 4-pole motor, respectively. The coils are connected to the supply as shown in Figure A21; in this arrangement they are said to be in *delta*. The motor can be reversed by transposing any two of the three phases.

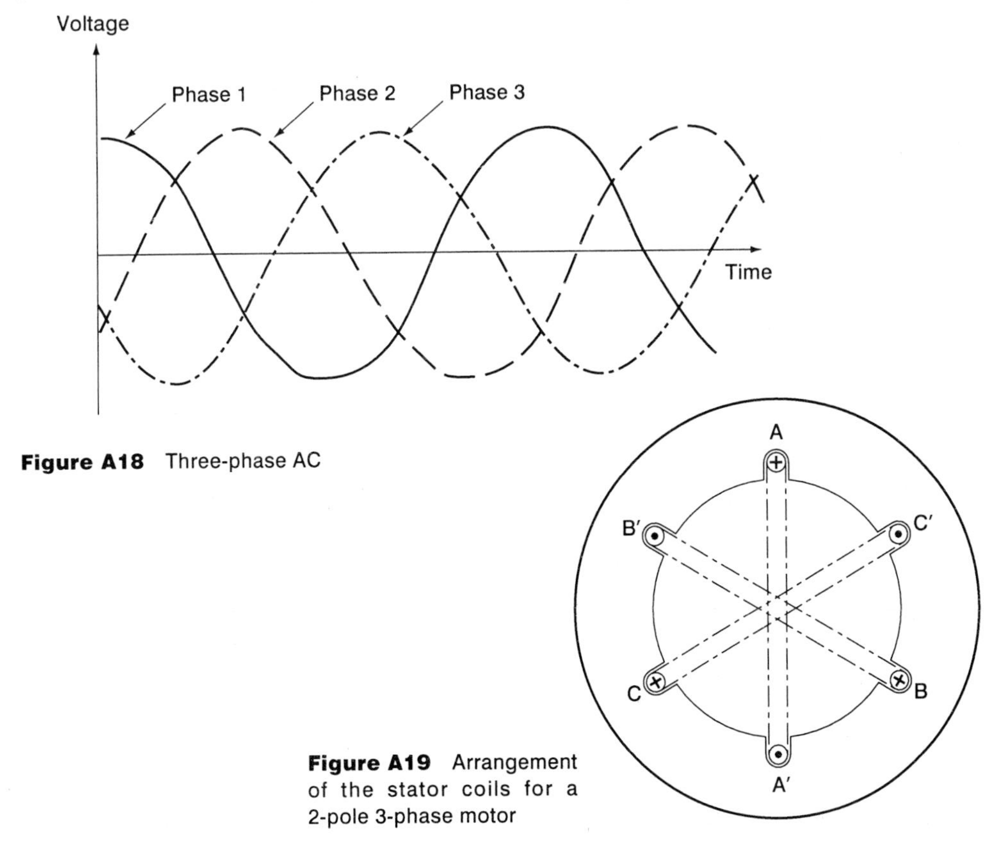


**Figure A18** Three-phase AC

**Figure A19** Arrangement of the stator coils for a 2-pole 3-phase motor

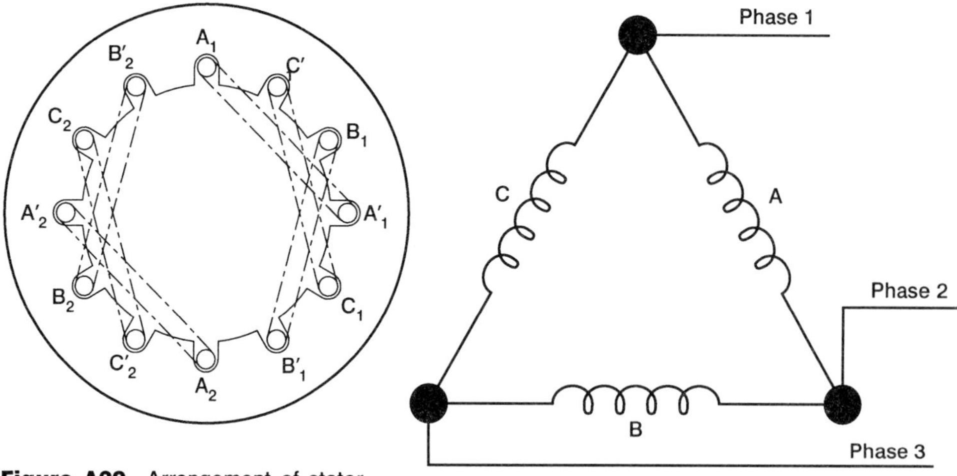


**Figure A20** Arrangement of stator coils for a 4-pole 3-phase motor

**Figure A21** Delta connection

The synchronous speed of an induction motor is the supply frequency (in cycles per minute) divided by the number of pairs of poles. Thus a 2-pole motor supplied with 50 Hz AC has a synchronous speed of $50 \times 60 = 3000$ rev/min. A 4-pole motor has a synchronous speed of:

$$\frac{50 \times 60}{2} = 1500 \text{ rev/min}$$

The starting characteristics of 4-pole squirrel-cage motors are shown in Figure A22. When started direct-on to the supply, their starting current is six times the full-load current, so only sizes up to 5 kW are started in this way. Larger sizes (up to 100 kW) are started by switching

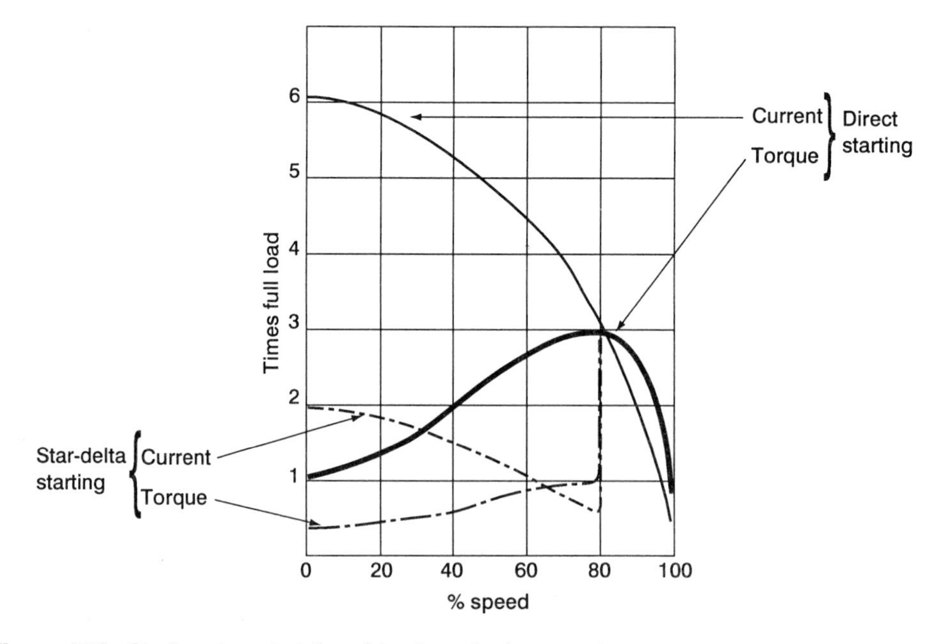


**Figure A22** Starting characteristics of 4-pole squirrel-cage motor

the stator windings so that they are connected in 'star' as shown in Figure A23. This reduces the voltage per phase to $1/\sqrt{3}$ of line voltage (i.e. $415/\sqrt{3} = 240$ V). The phase current is reduced in the same ratio, so the starting torque is only $(1/\sqrt{3})^2 = 1/3$ of what it would be if starting direct-on. The stator windings are switched over to 'delta' when the motor is up to about 80% of full speed.

Motors in excess of 100 kW have rotors with windings brought out to slip rings enabling external resistances to be introduced for starting and 'shorted out' afterwards.

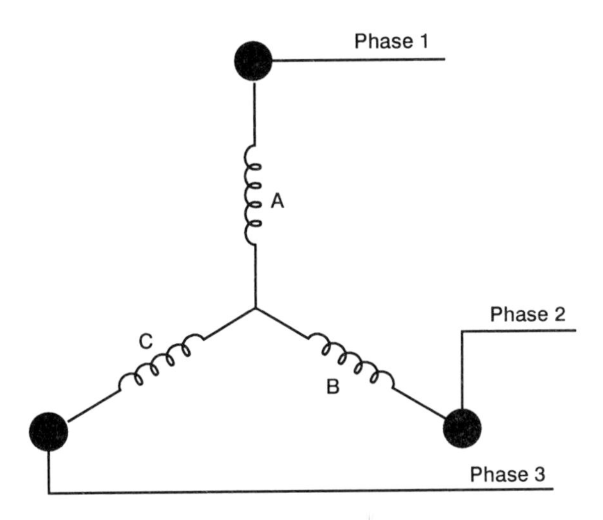


**Figure A23** Star connection

**Single-phase induction motors**

These greatly outnumber 3-phase motors although they are relatively inefficient. They are used in sizes up to 5 kW, in applications where the motor must run indefinitely without attention. House wiring in the UK is single-phase 50 Hz, 240 V, so single phase induction motors are the usual source of power for washing machines and similar domestic equipment.

A single phase can produce only a pulsating, not a rotating magnetic field. If the rotor is already turning, however, the interaction of its own magnetic field increases the stator field in the forward direction and reduces it in the backward direction, thus giving it a rotational tendency which keeps the rotor running once it is started. Methods of starting it include the following.

**Shaded-pole (see Figure A24)**. Used on small motors in which the stator consists of two poles energised by a single coil or by two coils in series. The inductive effect of a copper or brass ring on one section of a stator pole delays the build-up of magnetic flux in that section, giving the effect of a rotating field. This type of motor is simple and cheap but relatively inefficient so it is only found in sizes below about 250 W.

**Phase splitting**. The field winding is split into two sections: a main winding and an auxiliary winding at right angles, connected in parallel. The current through the auxiliary winding can be up to 90° out of phase to that in the main winding, giving a rotating field effect. Motors working on this principle include:

a) *Capacitor start* – the current for the auxiliary winding is taken through a capacitor, which gives it a phase lead of up to 90°. The auxiliary winding is disconnected at about 75% of running speed by a centrifugal switch or by a relay or electronic circuit which senses the corresponding fall in the supply current.

b) *Capacitor start and run* – the auxiliary winding is permanently connected through a *heavy-duty* capacitor.

c) *Resistance start split phase* – the auxiliary winding differs from the main winding in having a different proportion of resistance to reactance; this causes a phase difference between the two currents. As with the capacitor start motor, the auxiliary winding is disconnected as the motor approaches normal running speed to prevent its overheating.

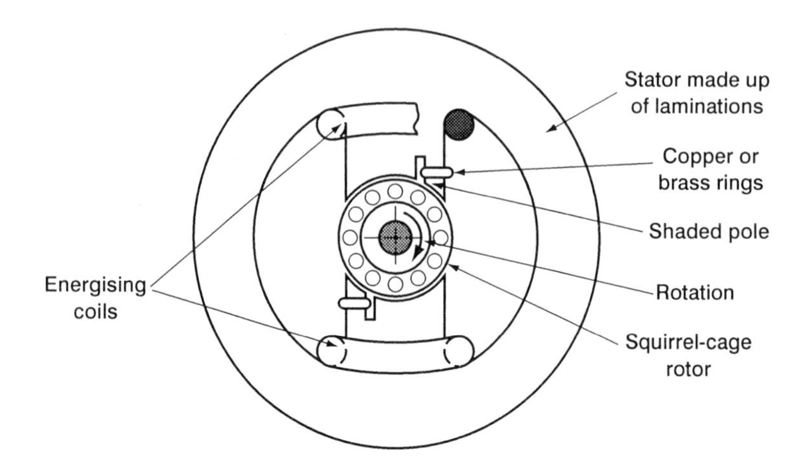


**Figure A24** Shaded-pole induction motor

## AC servo-motors

Most AC servo-motors are 2-phase induction motors in which the two stator windings are in quadrature (i.e. at 90° to each other) but supplied from two independent sources at the same frequency, as shown in Figure A25. One winding, the reference winding, is supplied with a constant AC voltage, $V_R$. The other winding, the control winding, is supplied from an AC amplifier which gives a current $I_C$ either leading or lagging behind the reference voltage by 90°. The lead or lag determines the direction of rotation of the rotor; the speed of rotation is controlled by the amplitude of $I_C$. The conductors of the squirrel cage are close to the surface of the rotor and have high resistance. This gives a high starting torque for rapid response to a step input.

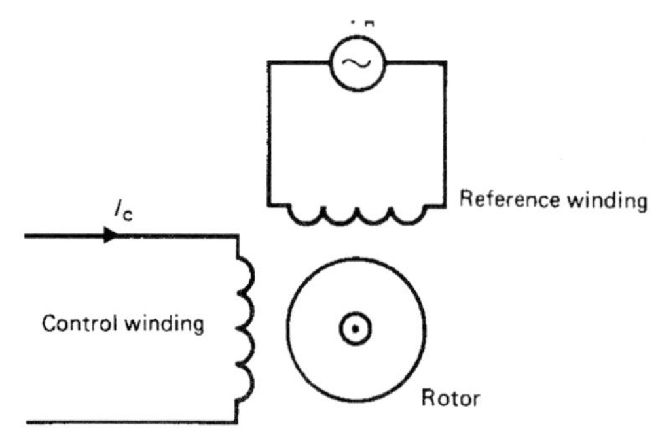


**Figure A25** The essential features of a 2-phase induction-type AC servo-motor

## Synchronous motors

These are constant speed machines, similar to induction motors but their rotors are magnets, so they rotate at the synchronous speed without any slip. They are therefore more efficient than induction motors. Small synchronous motors have permanent magnet rotors; mains electric clocks are driven by motors of this type. In larger motors the rotor is magnetised by windings supplied with DC through sliprings and brushes. Very large motors have brushless excitation, the DC for the rotor windings being rectified from AC produced by an alternator on the same shaft.

## Universal motors

These are series-wound DC motors with their poles and rotors made of laminations, to reduce eddy currents, so that they can work from DC or AC. On AC, the current reverses with each half-cycle, but this reverses the magnetic fields of both the stator and the rotor, so the torque and rotation continue in the same direction. Universal motors are used in portable power tools and domestic appliances such as food mixers, where the series motor characteristics of high speed and high starting torque make them much more suitable than induction motors.

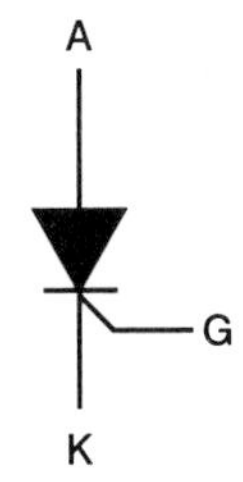


**Figure A26** Thyristor: circuit diagram symbol

## Thyristors (SCRs) and triacs

A *thyristor* (also known as a *silicon controlled rectifier* or *SCR*) is a semiconductor with anode (A), cathode (K) and gate (G) connections (see Figure A26). It acts like a latched relay, a single small pulse of current from gate to cathode switching on a much bigger current from anode to cathode. This main current flows continuously; it can only be cut off by decreasing it below a minimum value, the *holding current*, for the duration of the *turn-off time*, which may be from 5 to 200 μs. The main current is then blocked until another pulse is applied to the gate. The range of values covered by thyristor specifications is:

- maximum voltage which can be blocked: 400 to 1200 V
- maximum current thyristor can pass: 1 to 160 A
- gate current required for switch-on: 0.2 to 150 mA at 1.2 to 3.0 V.

High current, high voltage thyristors must be mounted on heat sinks and may require fan cooling. Increasing the gate current has the effect of turning the thyristor on more rapidly. When the main current is flowing there is a voltage drop of 0.75 to 1.25 V between anode and cathode.

Since the main current cannot be turned off by the gate, the chief use of thyristors is in AC circuits, in which each negative-going half-cycle breaks the main current, which is then turned on again at the required point in the next (positive-going) half-cycle. There is, however, a type of thyristor, the *gate turn-off thyristor* or *GTO thyristor*, which can be can be switched off by the gate, without the necessity for breaking the main current. Its symbol is shown in

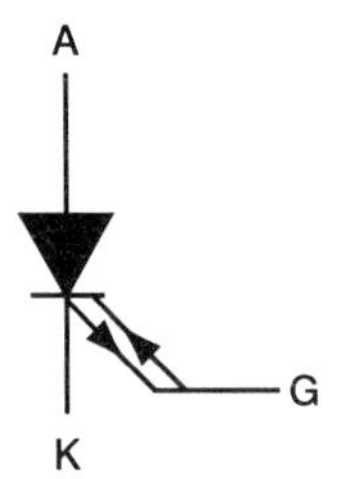


**Figure A27** GTO thyristor

Figure A27. It is switched on by a positive pulse of typically 1.5 V, 100 mA, 10 μs applied to the gate, and switched off by a negative pulse of, typically, 15 V for 1 μs. Figure A28 shows how a thyristor is used to control the AC power supplied to a load. The triggering circuit sets the point in the AC cycle at which a pulse is applied to the gate to turn on the thyristor. The delay angle, $\alpha$, may be varied virtually from 0° (full power) to 180° (zero power) but because the thyristor does not conduct during the negative-going half-cycles, 'full power' uses only half of the available power.

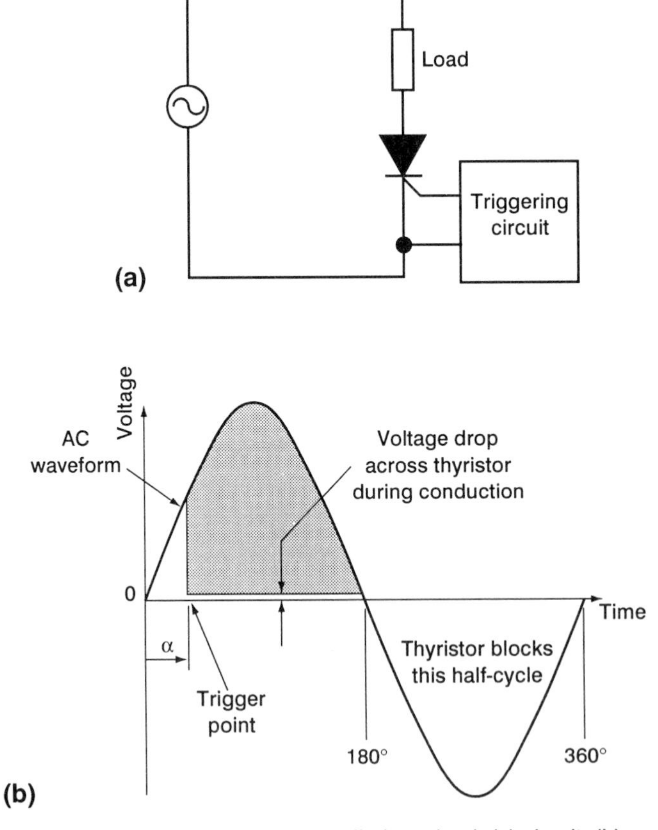


**Figure A28** Thyristor control of the power supplied to a load: (a) circuit, (b) waveforms

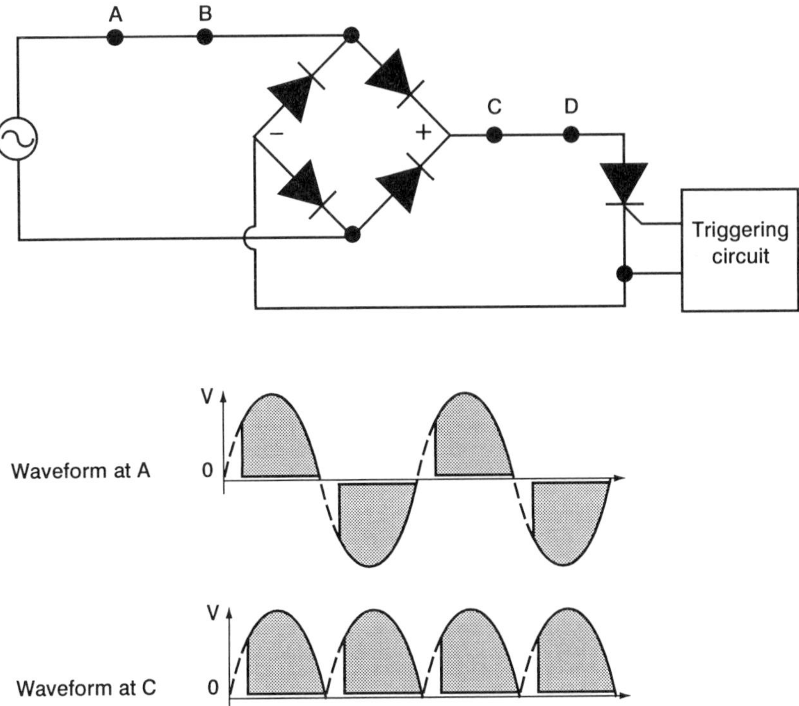


**Figure A29** Full-wave AC power obtained by rectifying the AC before it passes through the thyristor

To utilise both halves of the AC cycle the thyristor current may be taken through a bridge rectifier to convert it to unsmoothed DC. Figure A29 shows a suitable circuit, the load being inserted so that it replaces either link AB or link CD, depending on whether it requires an AC or a DC supply.

Alternatively, a *triac* may be used in place of a thyristor. A triac is a bi-directional thyristor; its circuit diagram symbol is shown in Figure A30. It conducts current in either direction through its main terminals, MT1 and MT2 when triggered by a small pulse of current through the gate, G. The triggering pulse can be of either polarity relative to MT1, regardless of the polarity of MT2 relative to MT1.

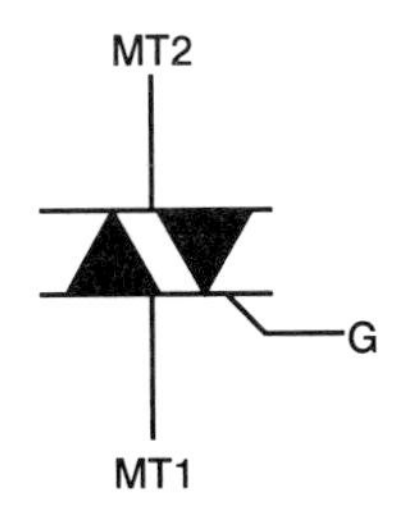


**Figure A30** Triac symbol

A **triggering** voltage can be obtained from the AC power supply voltage by scaling it down by means of a potentiometer but this only enables the delay angle, α, to be varied between 0° and 90°. To vary it over the full 180° range, the timing of the 'switch-on' voltage applied to the gate can be set by the charging of a capacitor through a variable resistance. Figure A31 shows a simple circuit for the control of a triac. $R_2$ and $C_2$ act as a low-pass filter, to eliminate harmonics of the supply frequency. These harmonics are caused by the abrupt switching on of the current as the thyristor or triac is triggered during the half-cycle. The main effect is at the fundamental frequency, but third, fifth, seventh, etc. harmonics are generated up to infinite frequency. The magnitude of the harmonics decreases rapidly with the increase in their numerical order but the higher order harmonics can be of sufficient magnitude to cause radio interference over a wide area, by being radiated from the power supply cables. To prevent this, a suppressor as in Figure 8.15 can be fitted to the supply cables. This can be supplemented by taking the current through an inductance (a *choke*) to further smooth out the waveform of the current taken from the mains.

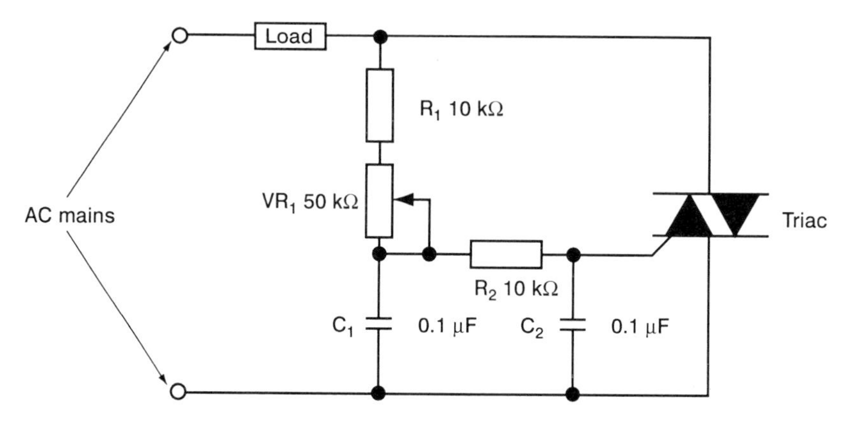


**Figure A31** Circuit using a triac to control the power applied to a load

## Thermostats

A thermostat is a device which maintains a system at a set temperature. In a mechanical type of thermostat the temperature is usually controlled by a bimetallic strip, which operates a valve controlling the flow of the heat source: be it hot water or hot air. Figure A32 is an illustration of this kind of thermostat; it shows the internal construction of a shower control which mixes two incoming flows of water, one hot and one cold, to produce a shower at the set temperature.

Another type of mechanical thermostat, shown in Figure A33, is used to control the cooling water temperature of motor vehicle engines. In this type a container of wax, suspended in the cooling water, carries a valve which allows the cooling water to flow through the radiator when the operating temperature is reached. The wax does this by expanding when it melts. This forces a plunger out of the container, but because the plunger is fixed to the mounting flange, the result is that the container and valve are forced away from the seating.

An electrical thermostat is normally just a switch which controls temperature by switching a heating current on or off. The switch is usually operated by the bending of bimetallic material. To minimise arcing, which would burn the contacts away if they opened slowly or if they 'dithered', the switch must be designed so that the contacts flick open when the operating temperature is reached, and stay open until the temperature has fallen a little. In the simplest form of electrical thermostat, this 'flick' action comes from a bimetallic disc which is slightly dished, so that it suddenly clicks from convex to concave when the operating temperature is

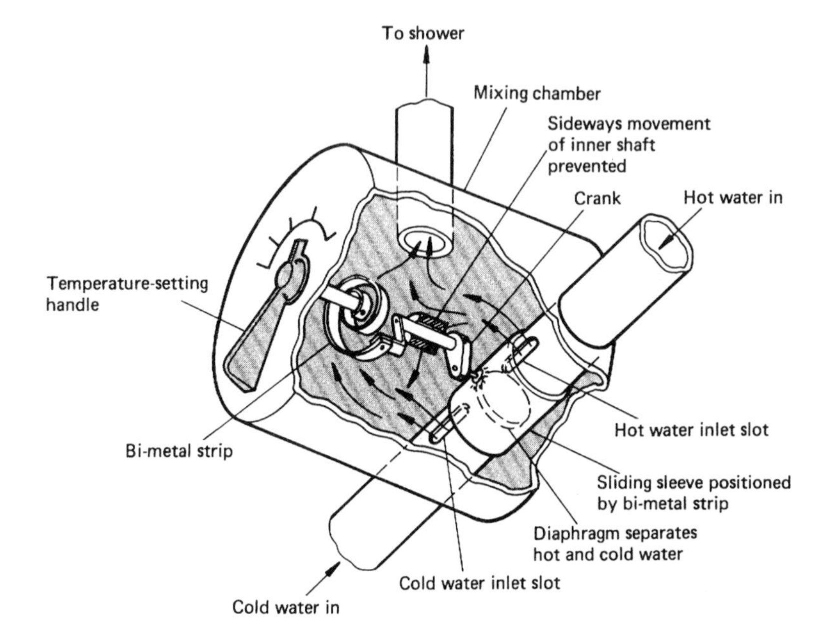


**Figure A32** A mechanical thermostat. A bimetallic strip controlling a mixing valve

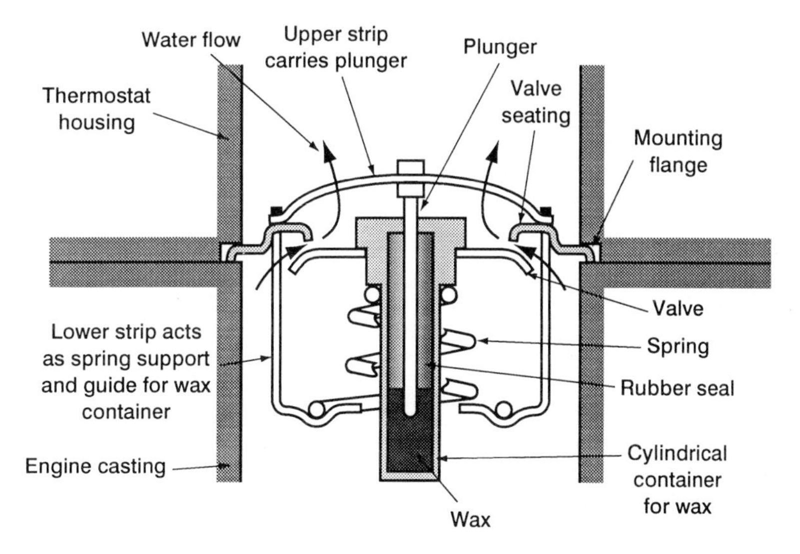


**Figure A33** Section through a thermostat operated by wax

reached and clicks back at a lower temperature. The disc operates the moving contact of the switch through an insulated push-rod, as shown in Figure A34.

When the switch of a thermostat is operated by a bimetallic strip as shown in Figure A35, the same effect is obtained by a small magnet which holds the contacts closed until the bimetallic strip can apply sufficient force to open them. When the magnet has 'unstuck' from its soft iron pad, its attractive force is much less, so the contacts flick open and stay open until a lower temperature is reached.

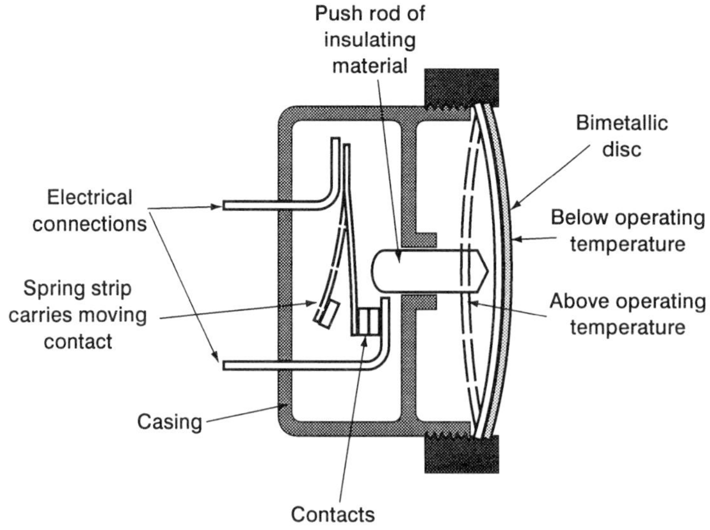


**Figure A34** Section through a bimetallic disc thermostat

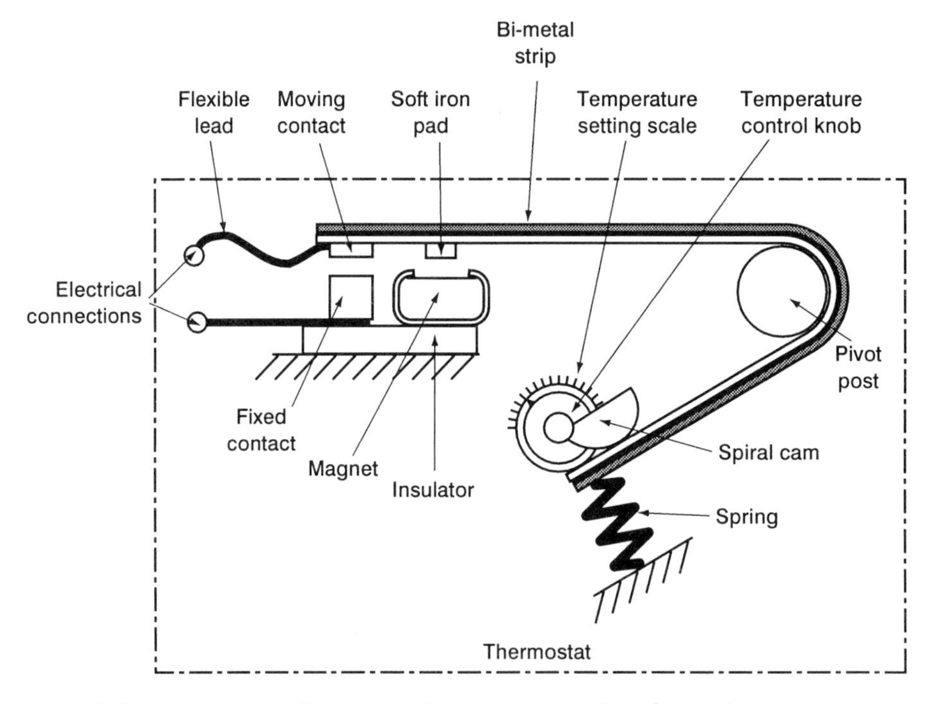


**Figure A35** Bimetallic strip thermostat with magnetic retention of contact

## Process control valves

Process control valves are used for the continuous regulation of the flow of gas or liquid in industry. Various kinds of actuator can be used to position the valve. Figure A36 shows the most common type: the *diaphragm actuator*. The valve position is controlled by compressed air, at a pressure which can be varied between 0.2 and 2 bar, acting on a rubber diaphragm

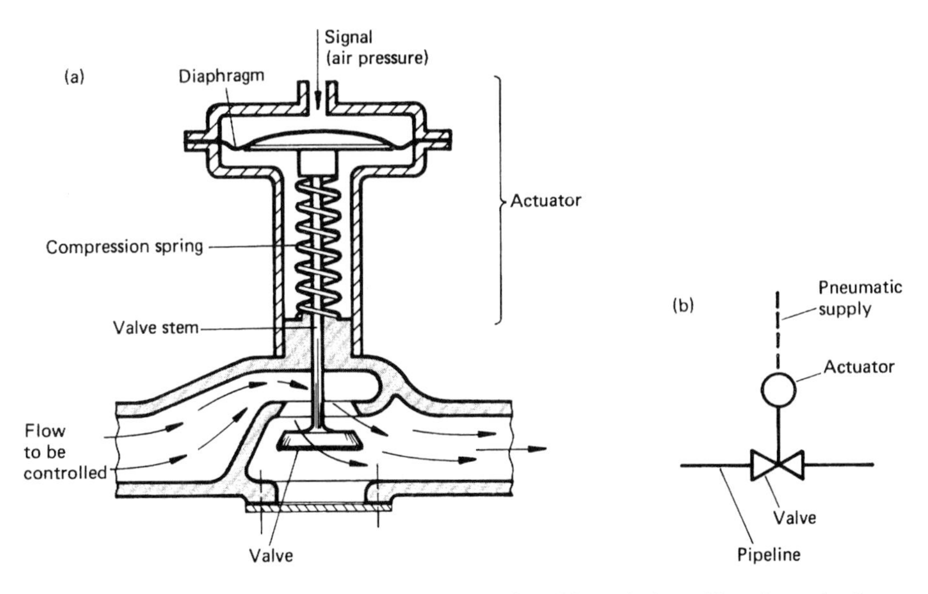


**Figure A36** Process control valve: (a) cross section, (b) symbol used in schematic diagrams

against the force of a spring. This type of actuator is easy to service, the operating pressure can be obtained from pneumatic logic circuits, and the absence of electrical wiring eliminates the risk of explosion where flammable liquids or gases have to be controlled. Its only drawback is that the drag and 'flutter' caused by the flow of fluid past the valve can only be resisted by the stiffness of the spring.

The valve shown in Figure A36 is an 'air-to-open' valve – that is, the controlling air pressure pushes the valve 'off' its seating. Figure A37 shows an 'air-to-close' valve, in which the controlling air pressure pushes the valve *towards* its seating. In deciding which type to use for a particular application, the designer usually works on the principle of 'fail-safe', and chooses the type which would leave the plant in a safe condition if the air supply failed. If for example the valve controls the fuel supply to a furnace, an air-to-open valve would be used, so that in the event of the loss of air pressure, the fuel would be cut off to prevent overheating. Similarly if the valve regulates the flow of cooling water to, say, a heat exchanger, an air-to-close valve would be used so that in the event of air failure the valve would open fully, again preventing damage from overheating.

If the flow is in the direction tending to push the valve open, as in Figures A36 and A37, flow-induced oscillations of the valve tend to die out quickly, but if the flow is in the opposite direction, tending to close the valve, the valve is much more likely to oscillate with increasing

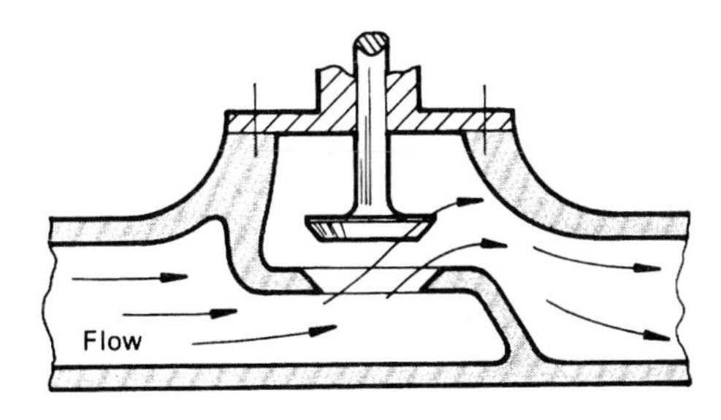


**Figure A37** Air-to-close valve

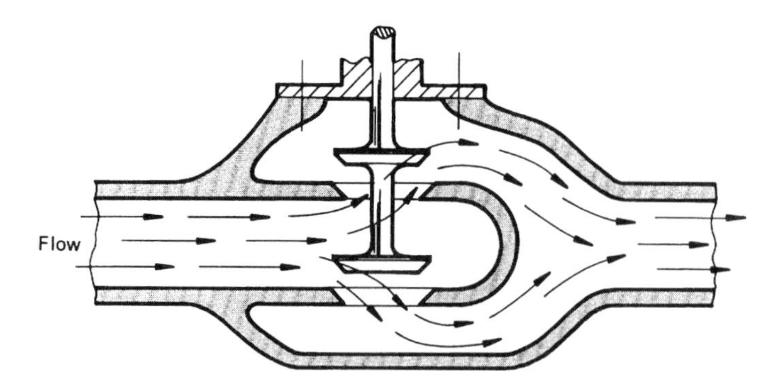


**Figure A38** Section through a double-seated valve

amplitude so that the system becomes unstable. One way of (almost) eliminating the effects of the drag of the fluid on the valve is to use a double-seated valve, as in Figure A38, so that one of the valve plugs tends to be opened by the fluid while the other tends to be closed. Thus the drag forces on the valve plugs almost entirely cancel out and the valve is much more stable. The disadvantage of this type of valve is that it is difficult to ensure that both plugs seat simultaneously, so it cannot be relied on to completely seal off the flow. In place of the mushroom valves shown, an actuator can be arranged to operate a rotary valve by means of a small connecting rod and crank.

Similar to the diaphragm actuator is the *piston actuator*, in which the diaphragm is replaced by a double-acting piston, sliding in a cylinder which takes the place of the diaphragm casing. This type of actuator uses higher air pressures (between 3 and 10 bar). The valve is moved to a new position by a difference in pressure above and below the piston; the forces on the piston are then equalised to maintain the valve in its new position. Fail-safe operation can still be ensured by a compression spring, as in a diaphragm actuator.

Because of the higher air pressure used, the piston actuator is better at resisting the drag forces on the valve, but the position of the valve can not be determined by air pressure alone, so a *positioner* has to be employed. This is operated by a push-rod co-axial with the valve stem and gives a pneumatic or electronic feedback signal of the valve's position for comparison with a demand signal. For this reason the valve has to be positioned by a closed-loop system.

Other types of actuator include *electric* and *electro-hydraulic*. In an **electric actuator**, an electric motor rotates a threaded collar on the valve stem to raise or lower the valve. In an **electro-hydraulic actuator**, the motor operates a pump which applies oil pressure to a piston actuator. These two types provide the ultimate in thrust, speed, frequency of response and stiffness, but they require positioners and limit switches, and they cannot easily be made fail-safe.

# Suggested further reading

Ernest O. Doebelin, *Measurement Systems*. McGraw-Hill
Noel M. Morris, *Advanced Industrial Electronics*. McGraw-Hill
Noel M. Morris, *Control Engineering*. McGraw-Hill
A.J. Bouwens, *Digital Instrumentation*. McGraw-Hill
J.D. Edwards, *Electrical Machines and Drives*. Macmillan
Curtis D. Johnson, *Process Control Instrumentation Technology*. Wiley
Michael Julian, *Circuits, Signals and Devices*. Longman
F.G. Rayer, *50 Projects using Relays, SCR's and Triacs*, Bernard Babani (publishing) Ltd.

# Answers to exercises

## Chapter 1

2 0.2583 Ω

3 a) 0.001 084
b) 75.9 $N/mm^2$
c) 46.7 kN

4 0.497%

5 b) 90.2 $GN/m^2$

7 187.9 $N/mm^2$ tension to 118.5 $N/mm^2$ compression

8 a) 0.566 Ω
b) 120 $N/mm^2$
c) 0.001 667
d) 2.12

9 7.96 mV/kN; hysteresis

## Chapter 2

1 Expansion, thermoelectricity, change of electrical resistance, radiation

3 a) 20 seconds by either method
b) 120 seconds

4 25.1 minutes, 92.6 minutes

5 a) 435.3°C
b) 445.7°C
c) 474.1°C

6 a) 15373 μV
b) 15659 μV
c) 14227 μV

7 a) 45.95 kΩ
b) 1.167 kΩ

8 b) 0.048 Ω, 3380 K. (Slight variations in plotting may cause the value of A to vary by up to 20%.)

10 a) See Figure 2.12

## Chapter 3

2 Classifications: rotary, multiturn, slide; wirewound, carbon track, cermet, conductive plastic; linear, logarithmic

4 a) 2.25°
b) 27.4 V
c) 0.0889 V/degree
d) i) 13.78 V
ii) 12.98 V

8 a) 38.2 pF
b) 4.16 MΩ

9 169.0 kΩ increase

10 b) 7 tracks

## Chapter 4

2 11.4 s

3 a) 18.47 V
b) 10.22 s

4 a) 5.11 V
b) 6.41 s

5 a) Strain gauged load cell or proving ring and LVDT
b) Electromagnetic force balance

6 b) By torque reaction, using a load cell, or by measuring the angle of twist of the shaft (Figures 4.09 or 4.10)

8 a) 127.7 kPa
b) 120.0 kPa

## Chapter 5

1 a) 7.5 m/s$^2$, 1.35 m

2 3.54 m/s

3 a) 26.0 kPa
b) 244 knots

5 23690 rev/min by formula
23750 rev/min by comparison of fractions

9

| | i | ii | iii | iv | v | vi | vii | viii | ix |
|---|---|---|---|---|---|---|---|---|---|
| **a** | Y | N | N | N | N | N | Y | N | N |
| **b** | N | Y | N | N | Y | Y | N | N | N |
| **c** | N | N | Y | N | N | N | N | N | Y |
| **d** | N | Y | N | Y | N | Y | N | N | N |
| **e** | N | Y | N | Y | N | Y | Y | Y | N |

## Chapter 6

1 See Figure 6.1. Displacement, velocity, acceleration

2 See Figure 6.2(a)

3 Using a velocity pick-up, integrating from an accelerometer, differentiating from a displacement pick-up

4 See Figure 6.3. Above

5 c) 0.01

6 See Figure 6.4

7 Fluid or eddy-current. 0.7

10 a) iii
b) ii
c) iv
d) i

## Chapter 7

1 a) 0.0717 m$^3$/s
b) 606000
c) 0.0891 m$^3$/s
d) 55.9 kN/m$^2$

2 c) i) $0.282 \times 10^{-3}$ kg/m s
ii) $0.294 \times 10^{-6}$ m$^2$/s
d) 959 kg/m$^3$
e) 832000

3 a) i) See Figure 7.1
ii) see Figure 7.3(a)

4 See Figure 7.4

6 a) See Figure 7.6

7 a) See Figure 7.7

9 See Figure 7.8.

10 a) See Figure 7.9
b) See Figure 7.10

12 i) b, c, f, i
ii) e, f, g, h, i
iii) h
iv) c, d, e, g, h
v) f
vi) e
vii) c, e, f, h
viii) a, b, d, f, i

## Chapter 8

1 a) 12.87 dB
b) –34.0 dB

2 17.0 dB

3 a) i) 1 V
ii) 17.8 V
iii) 0.25 V
b) i) 0.5 mW
ii) 4 mW
iii) 1 W
c) i) 0.548 V
ii) 1.55 V
iii) 24.5 V

4 b) 2100 Hz

5 a) i) See Figure 8.7
ii) See Figure 8.10
iii) See Figure 8.11
b) 33.4 dB

7 b) See Figure 8.16

8 a) Low-pass: see Figure 8.17 a)
High-pass: see Figure 8.18 a)
b) Low-pass: if $C = 10$ µF, $R = 53$ Ω
High-pass: if $C = 1$ µF, $R = 47$ Ω

9 See Figures 8.21 and 8.22

10 See Figure 8.23

## Chapter 9

1 a) CMOS
  b) CMOS
  c) TTL

2 a)

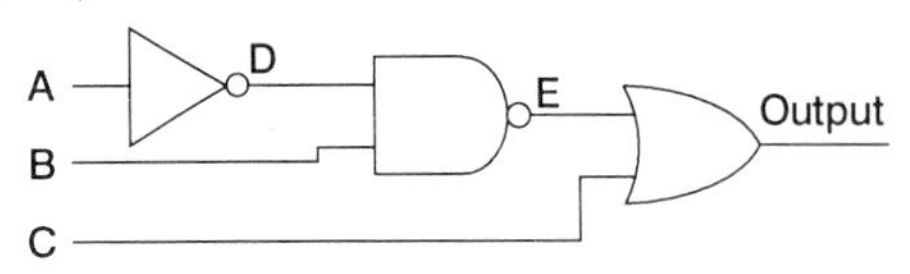


b)

| A | B | C | D | E | Output |
|---|---|---|---|---|---|
| 0 | 0 | 0 | 1 | 1 | 1 |
| 0 | 0 | 1 | 1 | 1 | 1 |
| 0 | 1 | 0 | 1 | 0 | 0 |
| 0 | 1 | 1 | 1 | 0 | 1 |
| 1 | 0 | 0 | 0 | 1 | 1 |
| 1 | 0 | 1 | 0 | 1 | 1 |
| 1 | 1 | 0 | 0 | 1 | 1 |
| 1 | 1 | 1 | 0 | 1 | 1 |

3 a) False
  b) True
  c) False

4 $X_3$ because gate is EXCLUSIVE OR: so output is 'high' only when only one input is 'high'

5 a) or b), but b) is preferable because a) puts three times as much load on the output of the preceding device

6 Binary 1 for 161.6°; binary 0 for 198.4°

7 a) 0010111101
  b) 0001 1000 1001
  c) See Figure 9.6
  d) Taking segments a to g, in order
    1 is 0110000, 8 is 111111
    9 is 1111011 or 1110011

9 a) i) 00005000 (Hz)
    ii) 0.0002000 (s) or 0000200.0 (μs)
  b) i) 0.02%
    ii) 0.05%
  c) 3162 Hz

10 See Figure 9.13

11 a) Twice maximum frequency
  b) Aliasing

12 a) See Figure 9.19
  b) See Figure 9.20

13 a) See Figure 9.16(b)
  b) Slow speed
  c) Digital voltmeters

14 b) 1000, 0100, 0110, 0111, 0110

15 See Figure 9.18. Fastest speed

16 See Figure 9.21

17 b) i) See Figure 9.22
    ii) See Figure 9.23

18 a) 6000 times per second
  b) 0.039 V
  c) 60 kHz

19 b) 00110001 (odd parity)
    00110000 (even parity)

## Chapter 10

1 i) See Figure 10.3. 0.1300 Ω
  ii) See Figure 10.4. 998.7 kΩ

2 See Figure 10.6

4 a) See Figure 10.7
  b) See Figure 10.8. Rise time 20 ms
    Fall time 100 ms

5 a) See Figure 10.9

7 See Figure 10.15

8 See Figure 10.16

9 See Figure 10.17

10 See Figure 10.18

13 See Figures 10.21 and 10.20

14 a) Trace 1: 7.1V. Trace 2: 1.48 V
  b) Trace 1: 5.02 V. Trace 2: 1.04 V
  c) 1.59 kHz
  d) 80.0°

15 a) 0°
  b) 90°

16 a) See Figure 10.29
  b) i) 2.44 MHz
    ii) 1.2 MHz.

17 See Figure 10.28.

# Symbols and SI Units

## Units

SI units are used throughout this book. Symbols are printed in *italic* type, units are printed in normal (roman) type.

**Basic SI units include**

| Concept | Symbol | Name of unit | Abbreviation | Additional information |
|---|---|---|---|---|
| Length | $l$ | metre | m | Distance is usually denoted $s$ |
| Mass | $m$ | kilogramme | kg | The unit may also be written: kilogram |
| Time | $t$ | second | s | |
| Electric current* | $I$ | ampere | A | |
| Temperature | $T$ | kelvin | K | For engineering calculations a temperature in kelvins can be taken as °C + 273 |

**Derived SI units include**

| Concept | Symbol | Name of unit | Abbreviation | Dimensions | Additional information |
|---|---|---|---|---|---|
| Velocity (linear) | $v$, $u$ | metre/second | $\frac{\text{m}}{\text{s}}$ | $\frac{\text{m}}{\text{s}}$ | Also denoted as $\frac{ds}{dt}$ or $\dot{s}$ or $Ds$ |
| Acceleration (linear) | $a$ | metre/second$^2$ | $\frac{\text{m}}{\text{s}^2}$ | $\frac{\text{m}}{\text{s}^2}$ | Also denoted as $\frac{d^2s}{dt^2}$ or $\ddot{s}$ or $\text{D}^2\text{s}$ |
| Angle turned through | $\theta$ | radian | rad | | Radians = revolutions × $2\pi$. Radians are non-dimensional numbers, therefore 'rad' does not appear in the dimensions of angular concepts |

* See footnote on p.209

**Derived SI units (continued)**

| Concept | Symbol | Name of unit | Abbreviation | Dimensions | Additional information |
|---|---|---|---|---|---|
| Velocity (angular) | ω | radian/second | $\frac{\text{rad}}{\text{s}}$ | $\frac{1}{\text{s}}$ | Also denoted as $\frac{d\theta}{dt}$ or $\dot{\theta}$ or $D\theta$ |
| Acceleration (angular) | α | radian/second$^2$ | $\frac{\text{rad}}{\text{s}^2}$ | $\frac{1}{\text{s}^2}$ | Also denoted as $\frac{d^2\theta}{dt^2}$ or $\ddot{\theta}$ or $D^2\theta$ |
| Moment of inertia | $I$ | kilogramme metre$^2$ | kg m$^2$ | $\text{kg} \times \text{m}^2$ | |
| Force | $F$ | newton | N | $\frac{\text{kg} \times \text{m}}{\text{s}^2}$ | Dimensions given by : Force = mass × acceleration |
| Torque | $T$ | newton metre | Nm | $\frac{\text{kg} \times \text{m}^2}{\text{s}^2}$ | Dimensions given by : Torque = force × radius |
| Mechanical work | $W, E$ | joule | J | $\frac{\text{kg} \times \text{m}^2}{\text{s}^2}$ | Dimensions given by : Work = force × distance moved |
| Energy, heat | $W, E$ | joule | J | $\frac{\text{kg} \times \text{m}^2}{\text{s}^2}$ | Dimensions given by : Work = force × distance moved |
| Mechanical power | $P$ | watt | W | $\frac{\text{kg} \times \text{m}^2}{\text{s}^3}$ | Dimensions given by : Power = $\frac{\text{work done}}{\text{time}}$ |
| Pressure, stress | $p$ | pascal | Pa | $\frac{\text{kg}}{\text{m} \times \text{s}^2}$ | Pressure or stress = $\frac{\text{force}}{\text{area}}$. Thus pascals are $\frac{\text{N}}{\text{m}^2}$ and $\frac{\text{N}}{\text{m}^2} = \frac{\text{kg} \times \text{m}}{\text{s}^2} \times \frac{1}{\text{m}^2} = \frac{\text{kg}}{\text{m} \times \text{s}^2}$ by unit cancellation. |

**Derived SI units (continued)**

| Concept | Symbol | Name of unit | Abbreviation | Dimensions | Additional information |
|---|---|---|---|---|---|
| Voltage, EMF* | $V$ | volt | V | $\frac{\mathrm{W}}{\mathrm{A}}$ | |
| Electrical resistance | $R$ | ohm | Ω | $\frac{\mathrm{V}}{\mathrm{A}}$ | Dimensions given by Ohm's law : $R = \frac{V}{I}$ |
| Electrical power* | $P$ | watt | W | V×A | Same unit as for mechanical power, but usually calculated as current × voltage |
| Electrical energy | $E$ | joule | J | W× s | Same unit as for mechanical work, but usually calculated as power × time |
| Electrical charge* | $Q$ | coulomb | C | A × s | Current × time. Also $I = \frac{dQ}{dt}$ |
| Capacitance | $C$ | farad | F | $\frac{\mathrm{C}}{\mathrm{V}}$ | Capacitance $= \frac{\text{charge}}{\text{voltage}}$. Also $I = C \times \frac{dV}{dt}$ |
| Inductance | $L$ | henry | H | | $V = -L \times \frac{dI}{dt}$ |

* In general, in electrical engineering, lower case symbols are used for the instantaneous value of time-dependent quantities, and the corresponding capital letter symbols for values which are not a function of time.

## Multiples and subdivisions

To express multiples and subdivisions of SI units concisely, the following prefixes are used with them:

| *Multiplication factor* | | | *Prefix* | *Symbol* |
|---|---|---|---|---|
| 1 000 000 000 000 | = | $10^{12}$ | tera | T |
| 1 000 000 000 | = | $10^{9}$ | giga | G |
| 1 000 000 | = | $10^{6}$ | mega | M |
| 1 000 | = | $10^{3}$ | kilo | k |
| 100 | = | $10^{2}$ | hecto | h |
| 10 | = | $10^{1}$ | deca | da |
| 0.1 | = | $10^{-1}$ | deci | d |
| 0.01 | = | $10^{-2}$ | centi | c |
| 0.001 | = | $10^{-3}$ | milli | m |
| 0.000 001 | = | $10^{-6}$ | micro | μ |
| 0.000 000 001 | = | $10^{-9}$ | nano | n |
| 0.000 000 000 001 | = | $10^{-12}$ | pico | p |
| 0.000 000 000 000 001 | = | $10^{-15}$ | femto | f |
| 0.000 000 000 000 000 001 | = | $10^{-18}$ | atto | a |

If in doubt, convert all quantities which are in multiples or subdivisions of SI units into the standard SI units before using them in calculations. Then the result when obtained will be in the standard SI unit.

When converting, note that for areas the conversion factor is squared, and for volumes, cubed. Thus:

$1\ \text{mm}^2 = (10^{-3})^2\ \text{m}^2 = 10^{-6}\ \text{m}^2$
$1\ \text{mm}^3 = (10^{-3})^3\ \text{m}^3 = 10^{-9}\ \text{m}^3$

## Algebraic and arithmetic manipulation

If $\frac{A}{B} = \frac{C}{D} = x$ then $A = \frac{B \times C}{D} = B \times x$

If $\frac{A}{Bx + C} = D$ then $A = D(Bx + C)$

If $\frac{Ax + B}{C} = D$ then $Ax + B = C \times D$

and so $x = \frac{C \times D - B}{A}$

## Two resistances in parallel

$$\text{Equivalent resistance} = \frac{R_1 \times R_2}{R_1 + R_2}$$

## The Greek alphabet

The lower-case letters of the Greek alphabet, frequently used in formulae, are given here for purposes of convenient reference. Also included are their pronunciations.

α alpha
β bēta
γ gamma
δ delta
ε epsīlon
ς zēta
η ēta
θ thēta
ι iōta
κ kappa
λ lambda
μ mu

ν nu
ξ xi
ο omīcron
π pi
ρ rho
σ sigma
τ tau
υ upsīlon
φ phi
χ chi
φ psi
ω ōmega

# Index

absolute zero 16
accelerometer 81
accelerometer calibration 84
accelerometer, low-frequency 83
accelerometer, piezoelectric 83
accelerometer, servo 83
acquisition time 136
active gauge 3
aliasing 137
all-active-gauge bridge 10
'alternate' mode 166
ambiguity 190
ammeter 152
amplifiers 106
amplitude modulation 121
analogue 124
analogue-to-digital conversion 137
aperture time 135

balanced-input amplifier 110
band-pass filter 119
band-stop filter 119
bandwidth 105
baud 147
BCD codes 129
Bernoulli's law 86
bimetallic strip 35
binary coded disc 48
binary counter 128
bipolar drive 187
black body 18
bonding 13
Bourdon tube 65
bridge circuit 3

capacitive transducers 46–48
capacitor start motor 193
capstan 174
carbon-track potentiometer 38
carrier wave 120
cathode-ray oscilloscope 164
cermet 38
characteristic temperature 26
charge amplifier 55
chart recorder 163
choke 116, 198
'chop' mode 165
chopper 33, 113
CMOS 126
coaxial cable 114
coefficient of discharge 87
coefficient of viscosity 88
companding 145
compound-wound motor 186
conductive-plastic potentiometer 38
convection 17
corner frequency 117
counter-timer 131
creep 13, 53
cross-sensitivity 12
cut-off frequency 117

damping ratio 79
d'Arsonval movement 160
data acquisition system 174
data logger 175
decade counter 130
decibel 102
demodulation 120
detent torque 188
diaphragm actuator 200
differential lever 180
digital-to-analogue conversion 141
digit display 131

direct reading bridge 7
displacement pickup 79
Doppler effect 69
Doppler flowmeters 97
drag-cup tachometer 73
drift 113
dual-slope A/D converter 138
dummy gauge 3
dynamic viscosity 88
dynamometers 61–65

earth loop 114
eddy-current dynamometer 64
electromagnetic flowmeter 93
emissivity 17
elastic sensors 52
electric dynamometer 62
electromagnetic force balance 57
errors in strain gauge signals 12–14
etched foil gauge 1
extinction frequency 172

filters 116
flapper and nozzle 181
'flash' A/D converter 141
flat grid gauge 1
flowmeter calibration 99
flyback 166
frequency modulation 121

gasometer 99
Gray cyclic code 49
gauge factor 2

Hall-effect transducer 72
heat conduction 17
high-pass filter 117
hot-wire anemometer 94
hybrid stepper motor 189
hydraulic actuators 179
hydraulic bellows 66
hydraulic dynamometer 62
hydraulic load cell 52
hysteresis 13

inductance bridge 45
induction motor, 3-phase 191
induction motor, single phase 193
inductive displacement transducers 44–46
integrated circuit 108
integrator 112
integrated circuit temperature sensor 30
International Temperature Scale (IPTS) 16, 33
invar 35
inverting amplifier 109

kelvin 16
kinematic viscosity 88

laminar flow 88
laser Doppler flowmeter 98
light-emitting diode (LED) 156
linear variable differential transformer 42
liquid crystal display (LCD) 156
liquid-in-metal thermometer 34
logarithmic potentiometer 38
logic gates 126
low-pass filter 117
LVDT 42

magnetic tape recording 171
mechanical displacement transducers 37
mechanical flow meters 91
modulation 120
moving coil meter 151
multiplexing 134
multiturn potentiometer 38

noise 41, 43, 112
non-inverting amplifier 110
notch filter 119
n.t.c. thermistor 26
null-balance accelerometer 83
null method 7

operational amplifier (op amp) 109
orifice plate 90
oscillator 128

parity bit 147
pascal 67
permanent magnet motor 184
permanent magnet stepper motor 187
phase-locked loop 121
phase-sensitive detector 43
phase splitting 193
phototransistor 50
piezoelectric force transducer 54
piezoelectricity 53
piston actuator 202
Pitot-static tube 70

platinum film sensor 29
polarity indication 145
positioner 202
potentiometer (abbreviation: 'pot') 37–39
potentiometer loading 39
pressure head 71
pressure measurement 65–67
primary calibration 33
probe 166
process control valve 200
proving ring 58
PTC thermistor 28
pulse code modulation 146

quantisation noise 144

R-2R ladder 141
radiation 17
radiation pyrometer 30
resistance thermometer 29
Reynold's Number 88
rosette 5
rotameter 92
rotary transformer 59

sample-and-hold amplifier 135
sampling 135
scaler 131
Schmitt trigger 127
SCR 195
screening 113
sealed-fluid thermometer 34
secondary calibration 33
Seebeck effect 22
seismic pickups 78
semiconductor strain gauges 11
separately-excited motor 186
series/parallel data transfer 124
series-wound motor 185
servo motor, AC 194
servo motor, DC 186
servo recorder 161
settling time 143
shaded-pole 193
shot noise 113
shunt 152
shunt-wound motor 185
signal conditioning module 174
single-slope A/D converter 137
sinusoidally varying temperatures 20
slew range 189
slide potentiometer 38
solenoid 182
spool valve 179
squirrel cage 190
stagnation pressure 70
stepper motor 186
strain gauges 14
stroboscopic lamp 71
successive-approximation A/D 140
summing amplifier 111
suppressor 115
synchronous motor 195
synchronous speed 192

tachogenerators 75–76
tachometer 69
telemetry 60
temperature-compensated gauges 6
temperature transducers 16–35
thermal noise 113
thermistor 26
thermocouple 22
thermocouple compensating cable 26
thermocouple materials 25
thermocouple sheathing 26
thermoelectric effect 22
thermopile 32
thermostat 198
thyristor 195
thyristor, GT0 195
time constant 18
toothed wheel and proximity pickup 72
torque cell 60
torque measurement 59–61
torr 67
transformer 108
triac 197
triggering 166, 198
TTL 126
turbine flowmeter 92
turbulent flow 88

ultrasonic Doppler flowmeter 97
ultraviolet recorder 160
unbonded strain gauges 2
unipolar drive 187
universal motor 195
U-tube manometer 67

vacuum measurement 67
vapour-pressure thermometer 34

variable reluctance stepper motor 188
varistor 116
velocity pickup 80
venturi meter 86
voltmeter 152
vortex flowmeter 96

Wheatstone bridge 3
white noise 113
wire-wound potentiometer 38
wrap-around gauges 1

XY plotter 162

zero detector 109